OXFORD MEDICAL PUBLICATIONS

An introduction to
human biochemistry

An introduction to

human biochemistry

C. A. Pasternak

Professor of Biochemistry, University of London
at St. George's Hospital Medical School

OXFORD
OXFORD UNIVERSITY PRESS
NEW YORK TORONTO
1979

Oxford University Press, Walton Street,
Oxford OX2 6DP

Oxford London Glasgow New York Toronto Melbourne
Wellington Ibadan Nairobi Dar es Salaam Lusaka
Cape Town Kuala Lumpur Singapore Jakarta Hong Kong
Tokyo Delhi Bombay Calcutta Madras Karachi

British Library Cataloguing in Publication Data

Pasternark, Charles Alexander
  An introduction to human biochemistry. – (Oxford
medical publication).
  1. Physiological chemistry
  I. Title   II. Series
  612′.015   QPS14.2   78-40649

  ISBN 0-19-261188-7
  ISBN 0-19-261127-5 PbK

Dedicated to the students of the new St George's
Hospital Medical School, London.

Typeset at the Universities Press, Belfast
Printed in Great Britain
by William Clowes Ltd., Beccles

Acknowledgements

I am grateful to many colleagues, especially H. Baum, N. D. Carter, M. J. Clemens, J. A. Edwardson, D. A. Hems, M. R. Holloway, L. Hudson, J. S. Jenkins, P. Knox, T. Northfield, D. J. Perkins, H. P. Rang, and D. H. Williamson for helpful criticism of parts of this book. The careful typing of the manuscript by Helen Withers Green, and the production of the finished work by the staff of the Oxford University Press deserve especial mention. The facilities provided by Dr. S. Grisolia, Director of the Instituto de Investigaciones Citologicas, Valencia, made undisturbed work possible on a number of occasions. Chiefly I am indebted to my family for putting up with so many 'lost' week-ends, during which the weeds in our garden grew considerably faster than the pages of the book that follows.

London and Combe,   C.A.P.
August, 1978.

Contents

Abbreviations

A	adenine
ACTH	adrenocorticotrophic hormone
ADP	adenosine diphosphate
Ala	alanine
AMP	adenosine monophosphate
Ara	arabinose
Arg	arginine
Asn	asparagine
Asp	aspartate
ATP	adenosine triphosphate
C	cytosine
cAMP	adenosine 3',5'-cyclic monophosphate; cyclic AMP
CCK	cholecystokinin
CDP	cytidine diphosphate
CMP	cytidine monophosphate
CoA	coenzyme A
CoQ	coenzyme Q (ubiquinone)
CTP	cytidine triphosphate
Cys	cysteine
DNA	deoxyribonucleic acid
dRib	deoxyribose
ΔG	free energy change
ΔH	enthalpy change
ΔS	entropy change
FAD	oxidised flavine adenine dinucleotide
$FADH_2$	reduced flavine adenine dinucleotide
FFA	free fatty acid
FH_4	tetrahydrofolate
FMN	oxidised flavine mononucleotide
$FMNH_2$	reduced flavine mononucleotide
Fru	fructose
FSH	follicle stimulating hormone
Fuc	fucose
G	guanine
Gal	galactose
GalN	galactosamine
GDP	guanosine diphosphate
GH	growth hormone
Glc	glucose
GlcA	glucuronic acid
GlcN	glucosamine
Gln	glutamine
Glu	glutamate
Gly	glycine
GMP	guanosine monophosphate
GSH	reduced glutathione
GSSG	oxidized glutathione
GTP	guanosine triphosphate
Hb	haemoglobin
HbO_2	oxygenated haemoglobin
His	histidine
Hyp	hydroxyproline

IduA	iduronic acid
IgA	immunoglobulin A
IgD	immunoglobulin D
IgE	immunoglobulin E
IgG	immunoglobulin G
IgM	immunoglobulin M
Ile	isoleucine
kJ	kilojoule (= ~4.2 kilocalories)
Leu	leucine
LH	luteinizing hormone
Lys	lysine
Man	mannose
Met	methionine
mRNA	messenger RNA
MSH	melanophore stimulating hormone
NAD^+	oxidized nicotinamide adenine dinucleotide (DPN^+ in older literature)
NADH	reduced nicotinamide adenine dinucleotide (DPNH in older literature)
NANA	*N*-acetyl neuraminic acid
P_i	inorganic phosphate
Phe	phenylalanine
PP_i	inorganic pyrophosphate
Pro	proline
Rib	ribose
RNA	ribonucleic acid
rRNA	ribosomal RNA
s	sedimentation coefficient
Ser	serine
T	thymine
T_3	triiodothyronine
T_4	thyroxine
TDP	thymidine diphosphate
TG	triglyceride
Thr	threonine
TMP	thymidine monophosphate
tRNA	transfer RNA
Try	tryptophan
TSH	thyroid stimulating hormone
TTP	thymidine triphosphate
Tyr	tyrosine
U	uracil
UDP	uridine diphosphate
UMP	uridine monophosphate
UTP	uridine triphosphate
Val	valine
Xyl	xylose

Introduction

The aim of this book is to present in one volume the biochemistry and cell biology necessary for an understanding of the molecular basis of medicine.

The human body, like all other living organisms, is made up of cells. Cells, most of which are approximately the same size, contain within them distinctive structures such as nucleus, mitochondria, endoplasmic reticulum. The function of these structures is the same in all cells of the body and in the cells of all animals and plants. Part One of this book, which describes the workings of a living cell in terms of such organelles, is therefore as relevant to a mouse as to a human being. Indeed much of the information has been gained by studying mice rather than human beings. The realization that the chemical reactions that occur within cells are the same not only in animals and plants, but in microbial cells also, has led to most of the biochemical details of cellular activity being derived from simple and rapidly dividing bacteria such as *Escherichia coli*. Hence Part One is really an introduction to the function of cells in general. Defects in function that lead to disease are indicated in the appropriate places.

Practically every cell in the human body has the potentiality to synthesize all the proteins and other molecules that make up a human being. But only a *part* of that potentiality is ever expressed in any one cell. Clusters of cells, grouped together in tissues and organs, each express a discrete potentiality, superimposed on the basic biochemical pathways necessary to remain alive: muscle cells, for example, synthesize creatine and myosin, pancreatic cells synthesize digestive enzymes and insulin, lymphocytes synthesize immunoglobulins, and so forth. In Part Two this cellular specialization is described.

In so far as human biochemistry is studied in order to relate it to human disease, then, this book is an attempt to present in molecular terms the basis of that relationship. In order to create a book which is written within the economic as well as within the physical grasp of the student, only the main essentials of each topic have been presented. A short bibliography is appended for those who wish to pursue in greater depth aspects of the subject in which they are particularly interested.

Part One

Cellular
potentiality

1

Cellular constituents and their function

1.1 Composition of cells

Cells, like all living matter, are made up largely of water. The remainder consists of proteins, fats, carbohydrates, nucleic acids, dissolved molecules, and inorganic ions. The approximate amounts that are present in the cells of liver are shown in Fig. 1.1. The distribution is approximately the same in other cells, with minor variations. Fat cells, for example, contain more fat, calcifying tissue, more inorganic ions, and so forth.

The liver is often taken as typical of animal cells, and much of the basic biochemistry of cells is derived from rat liver. However, it should be appreciated (i) that liver tissue contains a mixture of cells, such as hepatocytes, Kupffer cells, and cells of connective tissue, and (ii) that each of these is a specialized cell, secreting specific proteins and other substances. The liver cell is therefore no more typical of animal cells than a cell taken from the kidney or heart.

In fact, there is no such thing as a typical cell. But liver tissue is soft and easily homogenized, and this is an advantage for biochemical studies, which are carried out on solutions or suspensions of cell constituents rather than on the complicated structures of intact cells. Moreover liver does perform all the basic biochemical pathways described in Part One of this book, and hence is as suitable as any other cell. It is because the biochemistry of nucleus, mitochondria, or endoplasmic reticulum is the same in all cells, that it does not matter which type of cell one studies, *provided* that its specialization is not so intense as to mask the basic reactions. The red cell, which contains no nucleus, mitochondria, or endoplasmic reticulum, is an extreme example of the latter situation.

1.1.1 Proteins

1.1.1.1 Primary structure

Proteins are polymers of **amino acids**, linked by **peptide bonds**. The twenty amino acids commonly found in proteins, whether of animal, plant, or microbial origin, are shown in Fig. 1.2, together with a diagram of a peptide bond. Note that all amino acids except glycine have one or more asymmetric carbon atoms (that is, an atom that has each of its four valencies linked to a different substituent), and hence are optically active. All amino acids have the L(+) configuration about the α, or C-2, carbon atom (Fig. 1.2).

The breaking of a peptide bond by hydrolysis is a reaction in which energy is released (**exergonic reaction**); the formation of a peptide bond is the opposite (Fig. 1.3). That is to say, it is a reaction that 'requires' energy (**endergonic reaction**). Endergonic reactions do not take place unless (a) they are somehow coupled to an exergonic reaction, or (b) the concentration of the

3

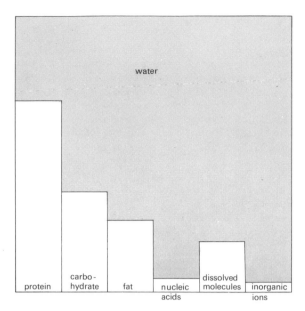

Fig. 1.1 Composition of cells

The approximate composition of liver cells is shown.

reactants is greatly increased. There is no way of 'supplying' energy as such; heat, for example (which is used to 'trigger' reactions – that is to lower the activation energy, see section 1.2.1) cannot be directly used to turn an endergonic reaction into an exergonic one. The use of the phrase 'energy-requiring' (or 'energy-supplying') is therefore not strictly accurate; it does however illustrate the nature of endergonic reactions so well that it is used throughout this book. The way in which endergonic reactions are made to occur in living cells, by (a) coupling to exergonic reactions or (b) increasing the concentration of reactants, is described in the chapters that follow.

Proteins are made up of one or more linear, un-branched chains of amino acids. This is known as the **primary structure** of a protein (Fig. 1.4). The chain is generally some hundreds of amino acids long, yielding proteins of molecular weight between 10 000 and 100 000.

1.1.1.2 Secondary structure

The nitrogen and oxygen atoms of amino acids are in the normal valency states of 3 and 2 respectively, and therefore have one or two pairs of 'spare' unbonded electrons in their first octet. As a result they tend to attract any hydrogen atoms (themselves bonded to nitrogen or oxygen) that are available in the vicinity, to form a **hydrogen bond** (Fig. 1.5). In this bond, the 1S shell of the hydrogen atom becomes distorted in such a way that four (not two) electrons are able to associate with the hydrogen nucleus. The hydrogen

atoms attracted are generally those of a water molecule, since water molecules predominate in biological fluids, but may also be part of another amino acid. Hydrogen bonds between amino acids and water contribute to the solubility of proteins in aqueous media. Hydrogen bonds between one amino acid and another amino acid four residues along the primary sequence give rise to a coiling of the primary structure (Fig. 1.6). The resulting **α-helix** is an example of what is termed the **secondary structure** of a protein. The hydrogen bond is much weaker than a covalent bond. Hence secondary structure is disrupted more easily than primary structure. For example, heating a protein in neutral solution at 60 °C, or adding a strong solution of urea or other compound that competes for hydrogen bonds, disrupts the α-helix. On the other hand, heating a protein in boiling 6N HCl for 24 hours is necessary to disrupt (that is, to hydrolyse, Fig. 1.3) a peptide bond.

The importance of the hydrogen bond to all forms of life cannot be overstressed. It contributes not only to the three-dimensional structure of proteins, as a result of which unique enzyme–substrate (section 1.2), or antibody–antigen (Section 11.2.1.1), interactions take place, but also to the pairing between complementary bases on two strands of nucleic acid, as a result of which the genotype (DNA) is faithfully reproduced at each cell division (Chapter 4) and as faithfully translated into protein via RNA in order to express its phenotypic potential (Chapter 5). In short, the hydrogen bond is crucial to the basic processes of living cells.

When the conformation of the amino acids in part of a protein chain is such as to make coiling into an α-helix difficult, that stretch of protein tends to exist in its straight, uncoiled form. Another manner of coiling through hydrogen bonds results in β-pleated sheets, as in silk fibroin. In proteins such as collagen that have a high content of proline and hydroxyproline (which lack a hydrogen atom attached to nitrogen, Fig. 1.2), hydrogen bond formation cannot occur. Instead a triple helix (called type 2 trans-helix), stabilized by proline–proline interactions, is formed (see section 10.3.2.1).

1.1.1.3 Tertiary structure

A strand of amino acids (whether in α-helix or not) as long as a globin chain (Fig. 1.4) – and globin is one of the smaller proteins known – tends to fold back on itself, producing **tertiary structure** (Fig. 1.7). The bonds involved in tertiary structure are (i) covalent bonds, (ii) hydrogen bonds, (iii) electrostatic bonds, and (iv) hydrophobic bonds.

Covalent bonds: The only covalent bonds that contribute to tertiary structure are those between the sulphur of one cysteine residue and the sulphur of

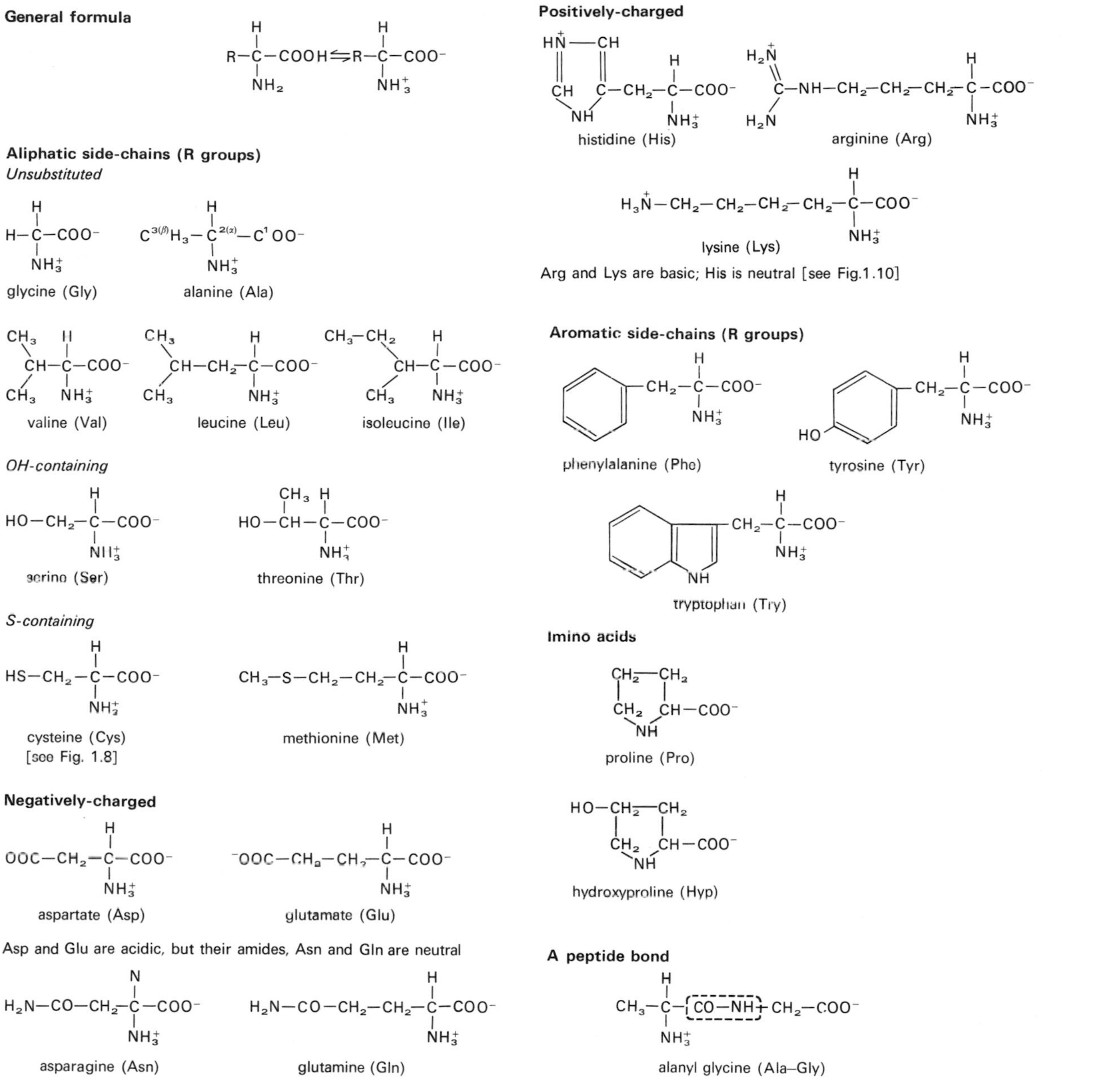

General formula

$$R-\underset{NH_2}{\overset{H}{C}}-COOH \rightleftharpoons R-\underset{NH_3^+}{\overset{H}{C}}-COO^-$$

Positively-charged

histidine (His) arginine (Arg)

lysine (Lys)

Arg and Lys are basic; His is neutral [see Fig.1.10]

Aliphatic side-chains (R groups)
Unsubstituted

glycine (Gly)   alanine (Ala)

valine (Val)   leucine (Leu)   isoleucine (Ile)

OH-containing

serine (Ser)   threonine (Thr)

S-containing

cysteine (Cys)
[see Fig. 1.8]   methionine (Met)

Negatively-charged

aspartate (Asp)   glutamate (Glu)

Asp and Glu are acidic, but their amides, Asn and Gln are neutral

asparagine (Asn)   glutamine (Gln)

Aromatic side-chains (R groups)

phenylalanine (Phe)   tyrosine (Tyr)

tryptophan (Try)

Imino acids

proline (Pro)

hydroxyproline (Hyp)

A peptide bond

alanyl glycine (Ala–Gly)

Fig. 1.2 Structures of common amino acids
The numbering of the carbon atoms is shown in the case of alanine.

another. The resulting **disulphide bond** is rather weak, in the sense that little energy is involved in the oxidation–reduction reaction (Fig. 1.8). Disulphide bonds are formed by bubbling oxygen through a solution of a protein, and are disrupted by addition of competing thiols such as mercaptoethanol (CH_3CH_2SH).

Hydrogen bonds: These bonds formed through the —OH of tyrosine, the free —COOH of aspartate and glutamate, the N of histidine and arginine, or the free NH_2 of lysine, contribute to tertiary structure.

Electrostatic bonds: These are formed between a positively and a negatively charged amino acid. The extent to which an amino acid is ionized depends on the tendency of its groups to ionize at the pH of the environment. The tendency is determined by the dissociation constant, and is expressed as **pK**; this is the pH at which the group is 50 per cent dissociated. It is

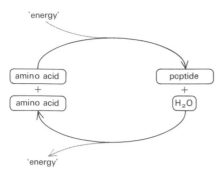

Fig. 1.3 Formation and rupture of peptide bonds

NH_2–Val–His–Leu–Thr–Pro–Glu–Glu–Lys–Ser–Ala–Val–Thr
Gly–Val–Glu–Asp–Val–Asn–Val–Lys–Gly–Trp–Leu–Ala
Gly–Glu–Ala–Leu–Gly–Arg–Leu–Leu–Val–Val–Tyr–Pro
Leu–Asp–Gly–Phe–Ser–Glu–Phe–Phe–Arg–Gln–Thr–Trp
Ser–Thr–Pro–Asp–Ala–Val–Met–Gly–Asn–Pro–Lys–Val
Ser–Phe–Ala–Gly–Leu–Val–Lys–Lys–Gly–His–Ala–Lys
Asp–Gly–Leu–Ala–His–Leu–Asp–Asn–Leu–Lys–Gly–Thr
Leu–Lys–Asp–Cys–His–Leu–Glu–Ser–Leu–Thr–Ala–Phe
His–Val–Asp–Pro–Glu–Asn–Phe–Arg–Leu–Leu–Gly–Asn
Lys–Gly–Phe–His–His–Ala–Leu–Val–Cys–Val–Leu–Val
Glu–Phe–Thr–Pro–Pro–Val–Gln–Ala–Ala–Tyr–Gln–Lys
Lys–His–Ala–Leu–Ala–Asp–Ala–Val–Gly–Ala–Val–Val
Tyr–His–COOH

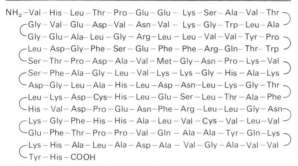

Fig. 1.4 Primary structure of proteins

The sequence of the β-chain of haemoglobin is shown.

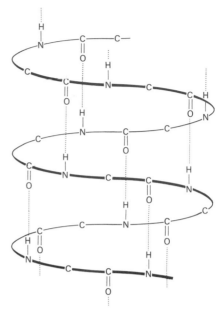

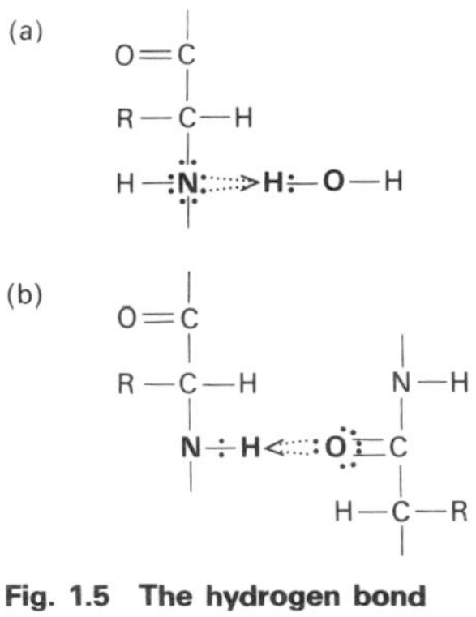

Fig. 1.5 The hydrogen bond

The electrons of the atoms that participate in H bond formation between (a) an amino acid and water and (b) an amino acid and another amino acid, are shown.

also the pH at which the group has maximal buffering capacity (Fig. 1.9). The p*K* values of the ionizable groups of amino acids and other biological molecules are given in Fig. 1.10.

Electrostatic bonds are relatively weak and can be disrupted by solutions of strongly ionized salts such as NaCl, or by high concentrations of H^+ (low pH) or OH^- (high pH) (Fig. 1.11).

Hydrophobic bonds: The term **hydrophobic bond** (literally 'water-loathing') is apt, since the interaction is the result of a mutual avoidance of the aqueous milieu

Fig. 1.6 Secondary structure of proteins: the α-helix

The α-carbon atoms are shown unbonded for clarity; and the H bonds appear longer than they actually are.

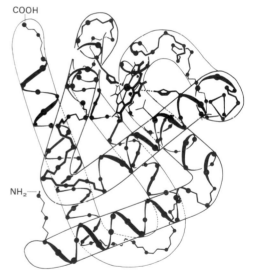

Fig. 1.7 Tertiary structure of proteins

One of the β-chains of haemoglobin is shown. Each amino acid is indicated by a dot. The large dot surrounded by a planar ring in the upper right part of the molecule is the Fe atom and the surrounding porphyrin, known as haem. The 'sausage'-like regions indicate the folding of the chain due to tertiary interactions. Within each region, **sequences of α-helix can be seen. The structure of the α-chains is** similar. Reproduced from a model constructed by Dr. M. F. Perutz, with permission.

by two adjacent groupings. Aromatic groups (of phenylalanine, tyrosine, and tryptophan) and long chains of CH_2 groups (as in leucine, isoleucine, valine, etc.) are the main contributing species. The bonds are

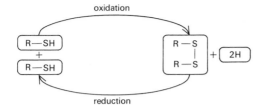

Fig. 1.8   Formation and rupture of disulphide bonds

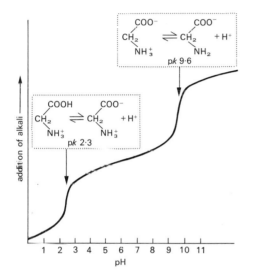

Fig. 1.9   Determination of pK

A break in the titration curve (shown for glycine) gives the pK.

Amino acids

	α-COOH	α-NH₂		R group
glycine	2·3	9·6		
aspartate	2·1	9·8	3·9	(-COOH)
glutamate	2·2	9·7	4·2	(-COOH)
histidine	1·8	9·2	6·0	
cysteine	1·7	10·8	8·3	(-SH)
tyrosine	2·2	9·1	10·1	
lysine	2·2	9·0	10·5	(-NH₂)
arginine	2·2	9·0	12·5	

Phosphorylated compounds

inorganic phosphate	2·1, 7·2, 12·4
AMP	3·7, 6·3
glucose 6-P	1·0, 6·1

Fig. 1.10   pK values of some ionizable groupings in biological molecules

weak and are disrupted by heat or by competition for another hydrophobic compound such as chloroform.

Of these four types of bond, hydrophobic bonds are probably the most important for the determination of tertiary structure. Electrostatic bonds are the least important, since most of the charged amino acids 'face outwards'; that is, they interact with the water molecules of the environment, rather than with each other.

Hydrogen, electrostatic, and hydrophobic bonds are weak in comparison to covalent bonds. What makes the tertiary structure of a protein stable is the *number* of bonds that are involved, rather than the strength of any one bond. Disruption of tertiary structure is caused by the same agents (mild heat, exposure to urea) that disrupt secondary structure, as well as by extremes of pH, by mild oxidation, and so forth. The unfolded protein that results is said to be denatured. Provided that denaturation is carried out in dilute solution and gently enough, so that precipitation, for example, is avoided, the process is reversible, and the protein can be renatured by restoration of the original conditions. In other words, the native (undenatured) conformation of a protein tends to be the most stable. Because no bond involved in secondary or tertiary structure is strong, little free energy is released when a protein is denatured. On the other hand there is a big difference in **entropy** (section 3.3.1) between a highly coiled and folded structure (low entropy) and an unfolded, randomly orientated one (high entropy). In other words, when a protein is denatured the heat (enthalpy) change, ΔH (section 3.3.1), is approximately equal to the entropy change ($T\Delta S$), with only a small **free energy change** (ΔG); when a denatured protein is hydrolysed, there is in addition a big free energy change (ΔG). The concepts of entropy and free energy are more fully discussed in section 3.3.1.

1.1.1.4 Quaternary structure

Several proteins are made up of a number of polypeptide chains or subunits, which may be identical or different. Insulin, for example, consists of different A- and B-chains (Fig. 9.17), chymotrypsin consists of different A, B, and C chains (Fig. 1.29); haemoglobin consists of two α- and two β-chains, each with its own haem residue (Fig. 1.12), lactate dehydrogenase consists of four chains, each of which is either α or β (Fig. 2.10) while immunoglobins consist of two light and two heavy chains (Fig. 11.4). This combination of subunits – by bonds similar to those involved in tertiary structure – is known as **quaternary structure**.

1.1.2 Fats

Fats (also called **lipids**) are either neutral or charged. The neutral fats comprise **triglyceride, cholesterol,**

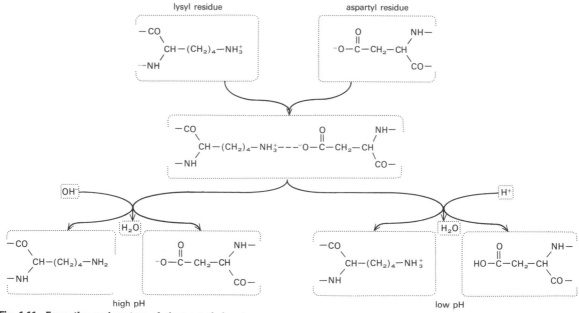

lysyl residue   aspartyl residue

Fig. 1.11   Formation and rupture of electrostatic bonds

high pH   low pH

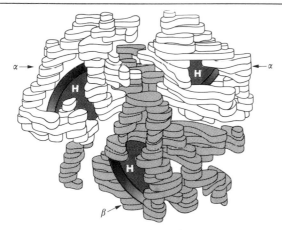

Fig. 1.12   Quaternary structure of proteins

The quaternary structure of haemoglobin consists of two α-chains and two β-chains. In this Figure, two of the α-chains, and one of the β-chains, are seen; the other β-chain is hidden. The three haem groups associated with the chains that are visible, are shown as flat discs labelled H. This illustration, derived from an X-ray crystallographic reconstruction, does not show the individual atoms (as in Fig. 1.7), but rather the 'electron densities'. Reproduced from a model constructed by Dr. M. F. Perutz, with permission.

cholesteryl esters, and certain **glycolipids**. The charged fats comprise **phospholipids** and acidic glycolipids. The structures of these compounds are given in Fig. 1.13. In so far as fats are 'small' molecules (molecular weight less than 1000), they do not have the secondary or tertiary structure of proteins. Their shape is determined, like that of individual amino acid residues, by the nature of the chemical bonds involved. Cholesterol, for example, has the 'puckered' shape typical of

other cyclic, non-aromatic, hydrocarbons; long-chain fatty acids have a flexible shape, with regions of rigidity where double bonds occur.

All organic molecules fall roughly into one of two categories. They are either **polar (hydrophilic)** or they are **apolar (hydrophobic)**. Polar molecules tend to form hydrogen bonds with water, which makes them soluble in aqueous media. Apolar molecules are incapable of forming hydrogen bonds with water, and are therefore insoluble in aqueous media; instead, they tend to form hydrophobic bonds with each other. Neutral fats fall into the latter category. They exist either as fat droplets in intracellular or extracellular fluid (for example chylomicrons), or are embedded within the apolar regions of biological membranes (Fig. 1.14).

Charged fats and neutral glycolipids are molecules having two distinct regions: a polar and an apolar one. As a result they are termed amphipathic molecules. The polar region is in contact with water or other polar molecules; the apolar region is in contact with apolar molecules or with the apolar region of another amphipathic molecule (Fig. 1.14). No covalent bonds are involved in linking lipids to each other or to other molecules. The importance of membranes to living cells is discussed below (section 1.3).

1.1.3 Carbohydrates

The term carbohydrate is used to denote **monosaccharides** such as glucose or fructose, **disaccharides** such as sucrose or lactose, and **polysaccharides** such as

NEUTRAL

triglyceride

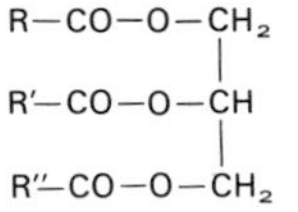

$$R-CO-O-CH_2$$
$$R'-CO-O-CH$$
$$R''-CO-O-CH_2$$

cholesterol

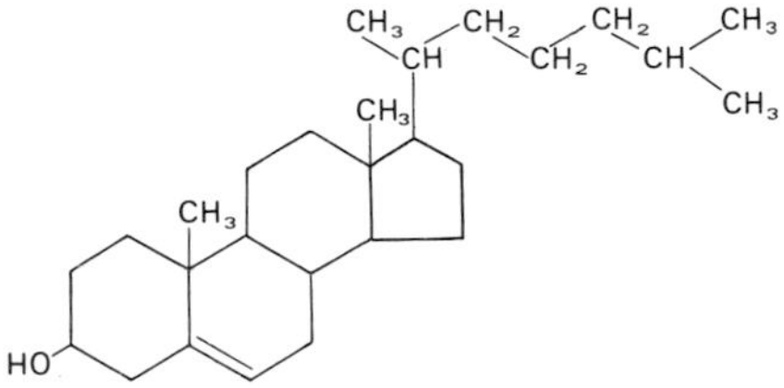

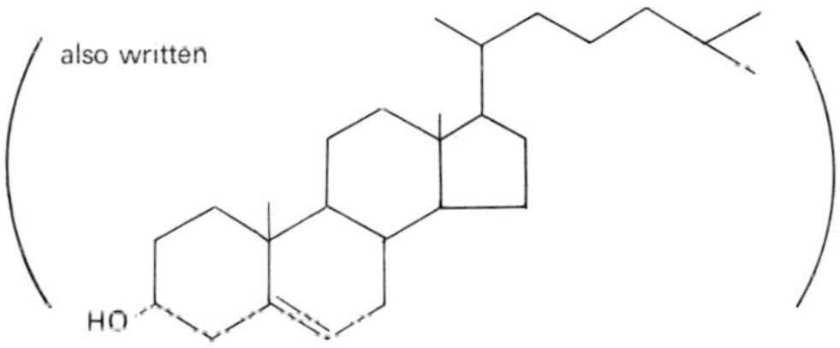

also written

cholesteryl ester

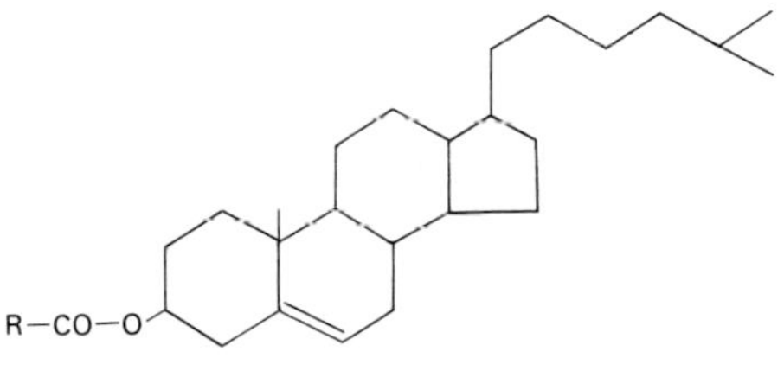

$$R-CO-O-$$

glycolipids

$$CH_3-(CH_2)_{12}-CH=CH-CH(OH)-CH-CH_2-O-(sugar)_n$$
$$R-CO-NH$$

(ceramide)

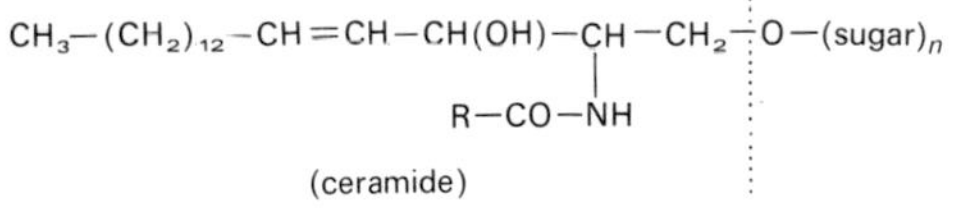

$n = 1, 2, 3,$ or 4 residues; the sugars are generally glucose, galactose, N-acetyl glucosamine, or N-acetyl galactosamine

CHARGED

phospholipids

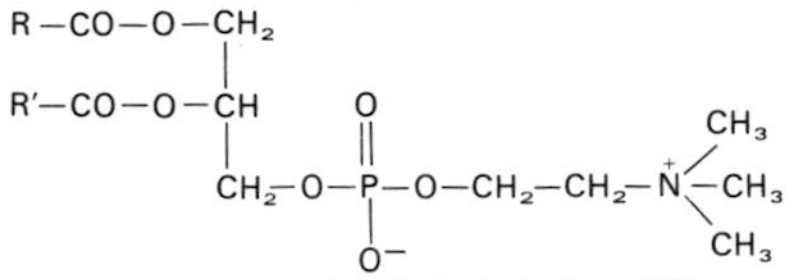

phosphatidyl choline (PC)

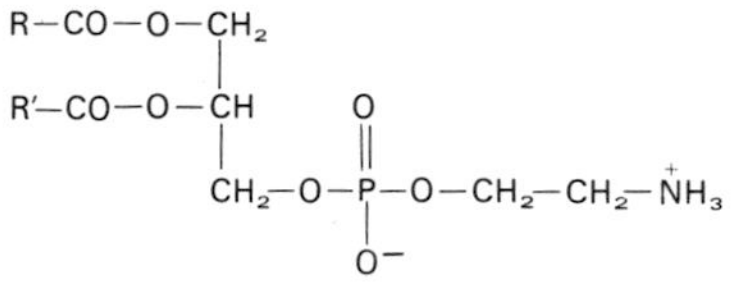

phosphatidyl ethanolamine (PE)

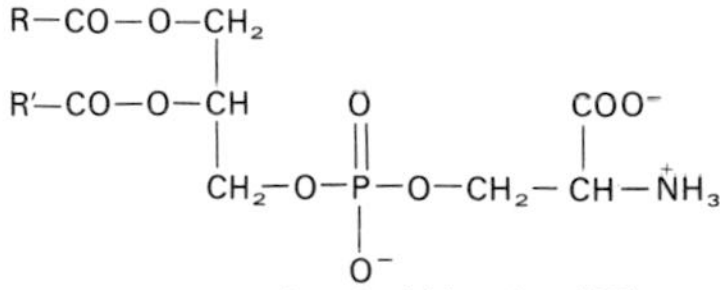

phosphatidyl serine (PS)

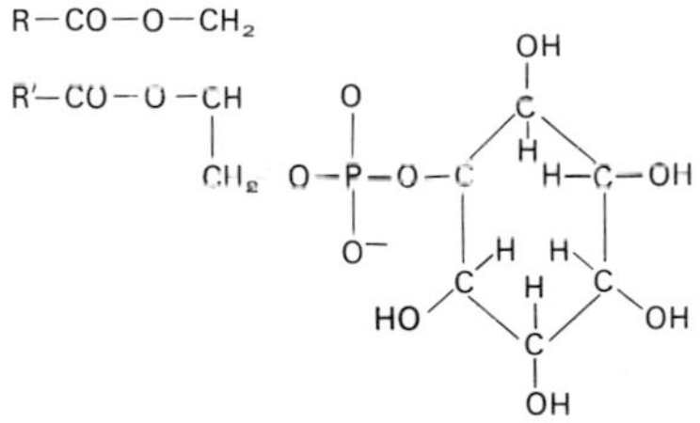

phosphatidyl inositol (PI)

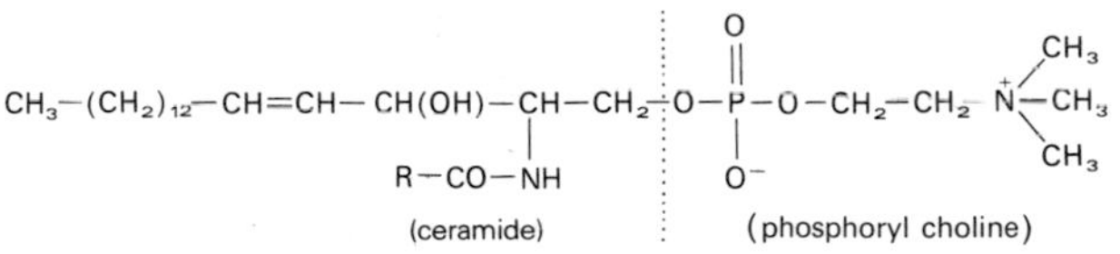

(ceramide)　　(phosphoryl choline)

sphingomyelin (SM)

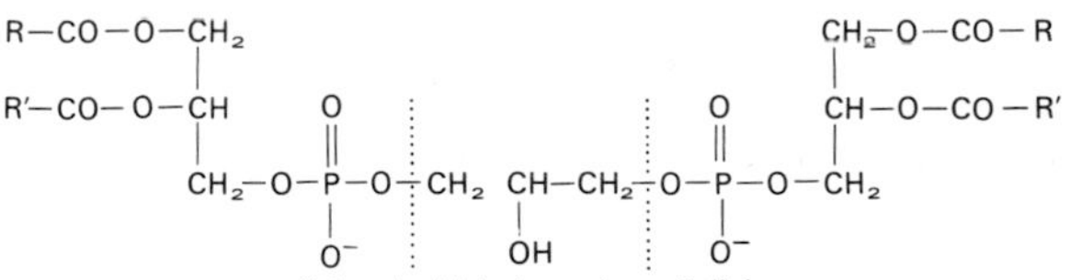

diphosphatidyl glycerol; cardiolipin

acidic glycolipids

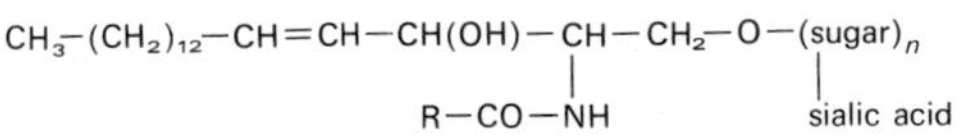

$$CH_3-(CH_2)_{12}-CH=CH-CH(OH)-CH-CH_2-O-(sugar)_n$$
$$R-CO-NH \qquad\qquad\qquad \text{sialic acid}$$

Sialic acid-containing glycolipids are known as gangliosides; generally there are 1–3 sialic acid residues

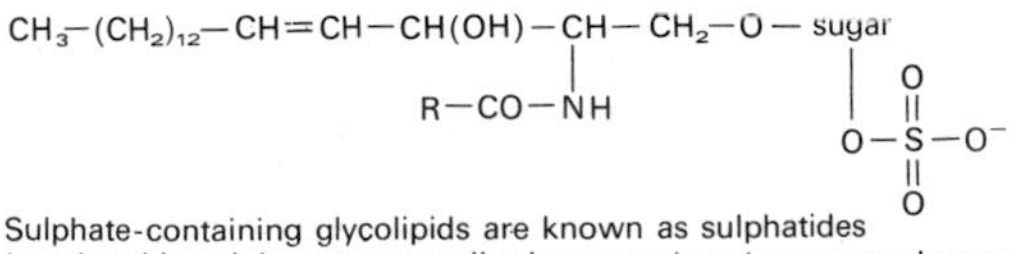

Sulphate-containing glycolipids are known as sulphatides (cerebroside sulphates); generally the sugar is galactose or glucose

The R groups shown in the figure are fatty acyl residues. The commonest fatty acids are:

C_{16}: $CH_3(CH_2)_{14}COOH$ palmitic acid
C_{18}: $CH_3(CH_2)_{16}COOH$ stearic acid
$C_{18:1}$: $CH_3(CH_2)_7CH=CH(CH_2)_7COOH$ oleic acid

Also important are the polyunsaturated (essential) fatty acids:

$C_{18:2}$: $CH_3(CH_2)_4CH=CH-CH_2-CH=CH-(CH_2)_7COOH$ linoleic acid
$C_{18:3}$: $CH_3-CH_2-CH=CH-CH_2-CH=CH-CH_2-CH=$
$\quad CH-(CH_2)_7COOH$ linolenic acid
$C_{20:4}$: $CH_3(CH_2)_4[CH=CH-CH_2]_3CH=CH(CH_2)_3COOH$ arachidonic acid

(see Figs 1.13 and 1.14 for structures of sugars and sialic acid).

Fig. 1.13 Structures of common lipids

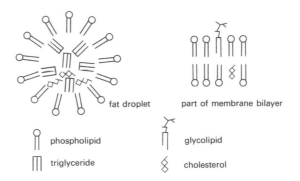

fat droplet   part of membrane bilayer

phospholipid   glycolipid

triglyceride   cholesterol

Fig. 1.14 Arrangement of lipids in aqueous environments

In each case (fat droplet or membrane bilayer), polar groups are in contact with the environment.

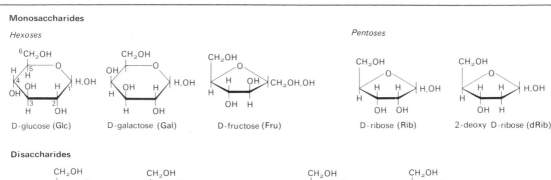

Monosaccharides

Hexoses   *Pentoses*

D-glucose (Glc)   D-galactose (Gal)   D-fructose (Fru)   D-ribose (Rib)   2-deoxy D-ribose (dRib)

Disaccharides

sucrose [D-glucosyl·(1 →2) D-fructoside]   lactose [D-galactosyl·(1→ 4) D-glucose]

Polysaccharides

glycogen (poly D-glucosyl $\left\{ \begin{array}{l} (1\rightarrow4) \\ (1\rightarrow6) \end{array} \right\}$ glucose)

Fig. 1.15 Structures of common carbohydrates

The numbering of the carbon atoms is shown in the case of glucose.

glycogen or starch (Fig. 1.15). In addition, carbohydrate residues linked covalently to lipid or protein are found in most cells of the body. The glycolipids, having the characteristics of fats, have already been described (Fig. 1.13). **Glycoproteins**, which have the characteristics of proteins, contain some carbohydrate. Glycolipids and glycoproteins are important components of the plasma membrane (section 6.2.1). Protein–carbohydrate complexes that are predominantly carbohydrate, known as **proteoglycans**, occur largely in the spaces outside cells (section 10.3). They are heteropolymers containing amino sugars and uronic acids (Fig. 1.16) in addition to simple sugars.

The sugar residues of carbohydrates do not interact with each other in the way that the amino acid residues of proteins do. Instead they have a tendency, because of their many oxygen atoms, to form hydrogen bonds with water. As a result the sugar residues are spread out in random manner and the molecules become rather voluminous. The high viscosity and

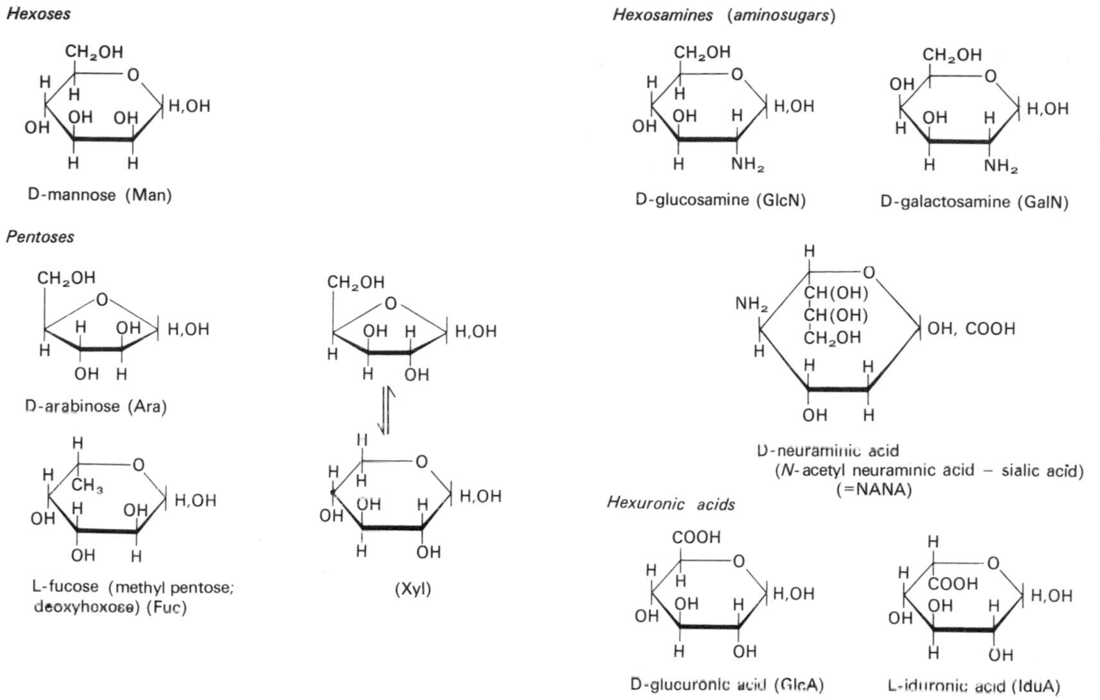

Hexoses

D-mannose (Man)

Pentoses

D-arabinose (Ara)

L-fucose (methyl pentose; deoxyhexose) (Fuc)

(Xyl)

Hexosamines (aminosugars)

D-glucosamine (GlcN)   D-galactosamine (GalN)

D-neuraminic acid
(N-acetyl neuraminic acid – sialic acid)
(=NANA)

Hexuronic acids

D-glucuronic acid (GlcA)   L-iduronic acid (IduA)

Fig. 1.16   Structures of less common carbohydrates

'jelly'-like character of mucous secretions or of synovial fluid is due to a high content of carbohydrate residues.

1.1.4 Nucleic acids

Deoxyribonucleic acid (**DNA**) is the hereditary (genetic) material of all animals, plants, and microbes. In certain viruses it is replaced by ribonucleic acid (RNA). Genes are translated into proteins via the participation of **RNA**. Nucleic acids are heteropolymers of alternating residues of ribose (or deoxyribose) and phosphate. The C-1 position of the sugar is linked to a nitrogenous base (thymine, cytosine, adenine, or guanine in DNA; uracil, cytosine, adenine, or guanine in RNA; Fig. 1.17).

The three-dimensional structure of nucleic acids is determined largely by interactions between the nitrogenous bases. The interactions are mainly hydrophobic. Hydrogen bonding also occurs, and it is this that creates a specificity of interaction between adjacent chains. In the case of DNA, the interaction leads to a double helix of two complementary chains (Fig. 1.18a); in the case of RNA molecules like transfer RNA, it leads to an intramolecular 'clover-leaf' shape (Fig. 1.18b). Because of the extreme length of nucleic acids, especially of DNA which has molecular weights well in excess of a million, the strands are wound around each other to form a super-coiled structure (Fig. 1.19). As in proteins, the ordered three-dimensional structure is lost by heating to approximately 60 °C.

1.1.5 Dissolved molecules

Most cells contain the biosynthetic enzymes necessary to synthesize their constituents, whether macromolecules (protein, polysaccharide, and nucleic acid) or lipid molecules, from low molecular weight precursors. Hence such precursors, which are amino acids in the case of proteins, sugar phosphates in the case of polysaccharides, and nucleotides (compounds containing a nitrogenous base, pentose sugar, and phosphate, with or without additional substituents) in the case of nucleic acids and lipids, are present within a cell. Because these molecules are merely intermediates in a metabolic sequence, they tend to be present in rather small amounts. Even the starting compounds, which are sugar, amino acid, or fat derived from the diet, are at low concentration in most types of cell, because of relatively restricted entry from the blood-stream. In other words the intracellular concentration of **dissolved molecules** is rather low (Fig. 1.1). Although the amount of dissolved gases (oxygen and carbon dioxide) is considerably higher, the contribution in terms of weight is minimal.

11

(a) The bases

pyrimidines

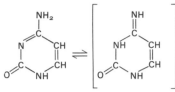

uracil (U)   cytosine (C)   thymine (T)

purines

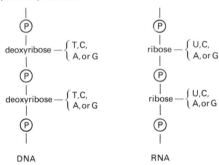

adenine (A)   guanine (G)

(b) Primary structure of nucleic acids

(c) Hydrogen-bonding between bases

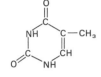

$CH_3(H)$ O····H—N

sugar - P chain

thymine (uracil)   adenine

DNA

deoxyribose — { T,C, A,or G }

deoxyribose — { T,C, A,or G }

RNA

ribose — { U,C, A,or G }

ribose — { U,C, A,or G }

sugar-P chain

cytosine   guanine

Fig. 1.17 Structures of DNA and RNA

Note that T, A, and G exist in isomeric (tautomeric) forms similar to those shown for U and C. Such tautomerism is responsible for the fact that nucleic acids absorb light in the ultraviolet region (~260 nm).

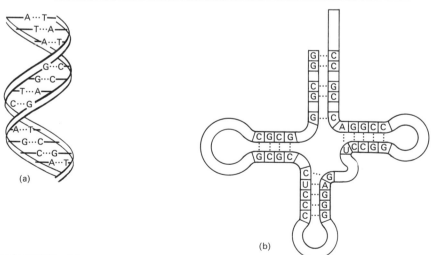

(a)

(b)

Fig. 1.18 H bonding in nucleic acids

(a) Intermolecular bonding (double helix of DNA) (b) Intramolecular bonding (tRNA).

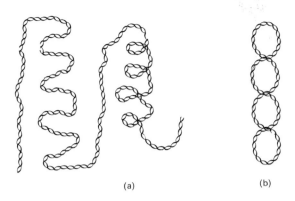

Fig. 1.19 Supercoiling in DNA

The exact way in which the DNA of chromosomes is super-twisted and super-coiled is not known; possible ways are shown in (a). In the case of mitochondrial DNA (section 3.2), the DNA is known to be circular, and coiled as in (b).

1.1.6 Inorganic ions

The predominant **cations** within cells are K^+, Na^+, Mg^{2+}, and Ca^{2+}; the predominant **anions** are Cl^-, HCO_3^-, PO_4^{3-} and phosphorylated intermediates, SO_4^{2-}, and negatively charged proteins. The distribution of ions in extracellular fluid is different from that of intracellular fluid in that Na^+ exceeds K^+, and Ca^{2+} and Cl^- are each considerably higher (Fig. 1.20). K^+, Na^+ and Cl^- are essentially in free solution, whereas

Mg^{2+} and Ca^{2+} are largely bound (to phosphorylated and carboxylated intermediates, and to phospholipids, proteins, and $—COO^-$-containing carbohydrates).

1.1.7 Disease

In disease, the composition of a cell changes. For example in certain hereditary defects called lipidoses, the fat content increases. The cause is not an over-production of lipid, but a failure to degrade lipid normally (Fig. 2.25). In very obese people, the number and fat content of adipose cells is high, due to saturation of the degradative pathway. In another group of hereditary defects called glycogen storage diseases (Fig. 2.57), the content of glycogen, which is the main carbohydrate of cells, is altered. It may be abnormally high, due to a failure of the degradative pathway, or it may be abnormally low, due to failure of the synthetic pathway. In most instances, the changes are not so obvious. Lack of one type of protein or substitution by another, for example, is unlikely to be revealed by gross analysis of cells. The only reason why the lipidoses or glycogen storage diseases, each of which is in fact due to lack of one specific enzyme, are detectable is that they happen to result in abnormal accumulation or depletion of reaction product.

An even greater difficulty for biochemical diagnosis of disease, however, is the fact that one cannot just

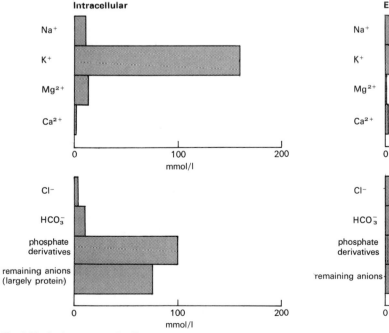

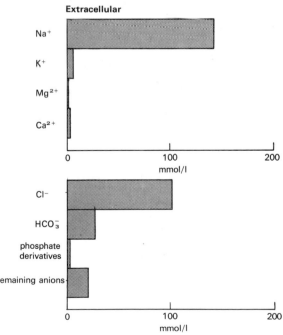

Fig. 1.20 Ionic content of cells

The approximate concentrations in liver cells is shown (intracellular); the concentration in serum (extracellular) is shown for comparison.

remove pieces of liver, muscle, or fat for routine biochemical analysis. The amount obtainable by surgical biopsy is very limited and is attended by discomfort and risk to the patient. Most biochemical investigations are therefore restricted to analysis of blood or urine. The consequent drawbacks are: (i) it is usually difficult to relate an altered level of blood or urinary constituent to a specific cellular defect and (ii) many cellular defects do not result in an altered composition of blood or urine. What is needed are methods for assessing enzymic function within tissues.

1.2 Enzymes

1.2.1 Function

The prime function of most proteins is to act as enzymes. Other proteins have a structural role, such as the collagen fibrils of connective tissue; some proteins have both a structural and an enzymic role, as in the myosin ATPase of muscle fibres. Several proteins have a functional role that is similar to that of enzymes in that they are able to bind certain substances (ligands), but different in that the resulting complex has no catalytic activity. Examples are haemoglobin, which binds oxygen with kinetic characteristics that are exhibited by many enzymes (Fig. 7.19), and immunoglobins (antibodies), which bind antigens (other proteins, lipids, or carbohydrates) with the same degree of specificity shown by enzymes (section 11.2.1.1). A cell may contain a hundred or so different non-enzymic proteins, which on a weight basis constitute the bulk of cellular proteins; the number of different enzymes, on the other hand, is likely to be in the region of ten to twenty thousand. Hence the diversity of protein structure that is created by the many ways of linking the twenty different amino acids in chains of a hundred or more, is most fully expressed in enzymes. And it is the catalytic power of enzymes that distinguishes living organisms from dead or inorganic matter.

The nature of biological **catalysis** is the same as that of non-biological catalysis: to increase the rate of chemical reactions by lowering the **activation energy**. All reactions, whether they are exergonic (release of free energy) or endergonic (requiring free energy) proceed only if the reactants are in a sufficiently 'active' state; the energy which the reactants need to undergo reaction is known as the activation energy (Fig. 1.21). The reactants achieve their activation energy by being heated. In the presence of a catalyst, the activation energy is reduced, and less heat is required; that is, the reaction is able to proceed at a lower temperature.

Enzymes reduce the activation energy to the point that all reactions occur at temperatures below 40 °C

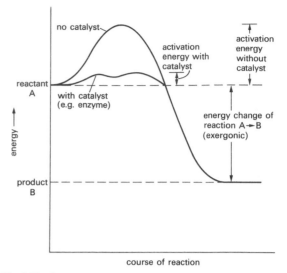

Fig. 1.21 Concept of activation energy

(body temperature). The extent to which enzymes reduce the activation energy may be illustrated by an example. The decomposition of hydrogen peroxide has an activation energy of 76 kJ/mol; in the presence of colloidal platinum the activation energy is 49 kJ/mol; in the presence of the enzyme catalase, it is <8 kJ/mol.

The other difference between enzymic and non-enzymic catalysis lies in the **specificity** of enzymes. Treating a cell extract with concentrated acid at 100 °C hydrolyses all the ester bonds that are present. Treating it with a phosphatase enzyme hydrolyses only certain types of phosphate ester. Some enzymes have high specificity towards the reactants (substrates), such that they can distinguish between D- and L-lactic acid, or between D-glucose 6-phosphate and D-glucose 1-phosphate. Other enzymes have low specificity, such that hydrolysis of any protein or of any nucleic acid is initiated.

The function of enzymes, then, is to catalyse the reactions occurring within cells. What are these reactions? They can be summarized as:

i. the oxidation of nutrients such as sugars, amino acids and fats to release energy and to supply macromolecular precursors,

ii. the coupling of the energy released by (i) to the synthesis of cellular components, and

iii. the function of those components.

In short, the chemical workings of a cell. Without any enzymes the rates of cellular reactions would be so slow as to virtually come to a stop; the result would be cellular death. If only one or other enzyme is inactive, only one or other metabolic pathway is affected. The result may lead to cellular injury. Just as enzymes are the basis of life, so also are they the basis of disease.

To understand medicine in molecular terms is to recognize the enzymic defects of diseased tissue. As yet such recognition has been largely restricted to analysis of tissues removed at surgery or death. The development of techniques for measuring the activity of enzymes within a living cell should be the aim of every clinical biochemist.

1.2.2 Properties

Since enzymes are proteins, their properties are those of proteins. The agents that denature proteins (section 1.1.1) denature enzymes also, and a denatured enzyme is an inactive one. Indeed, because enzymic activity is a more sensitive index of secondary and tertiary structure than is a physical property such as light scattering, most of our knowledge regarding the reversibility of denaturization of proteins comes from enzymic studies. Conversely the use of mild denaturing agents gives much information on what part of the enzyme molecule is involved in the catalytic act.

The effect of altering pH is an example. Several enzymes have maximal activity around neutrality (Fig. 1.22). Exceptions are the proteolytic enzyme pepsin, which functions maximally at the low pH (1·1–2·6) of gastric juice, or the hydrolytic enzymes of lysosomes, that have maximal activity around pH 4–5.

Fig. 1.22 shows that the activity of enzymes falls at pH changes *insufficient* to cause complete denaturation. Presumably this is because charged groups are involved, and these are maximally active in either the acidic or the basic form. But why does enzymic activity

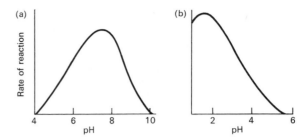

Fig. 1.22   Effect of pH on rate of reaction

(a) pH optimum near neutrality, e.g. trypsin; (b) pH optimum at low pH, e.g. pepsin.

fall if the pH is increased *or* decreased? In other words, why are some activity curves bell-shaped? The most likely explanation is that more than one charged group is involved. In ribonuclease (optimal activity around pH 7), for example, two histidine residues (pK approximately 6) are known to participate in catalysis: one is in the ionized form, the other in the non-ionized form. But in other enzymes having a pH optimum around 7, it is not necessarily histidine residues that are involved. The carboxyl group of glutamate (pK 4·3 in free solution), for example, may have a pK of 7 when it is 'buried' within a protein. In other words, identification of specific amino acids as playing a role in catalysis requires information in addition to pH–activity curves.

The rate of a chemical reaction increases as the concentration of reactants is increased. Enzymically catalysed reactions are no exception. The only difference is that the rate approaches a maximum at high substrate concentration (Fig. 1.23). This is due to the

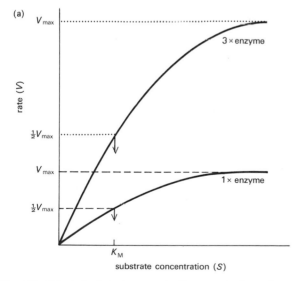

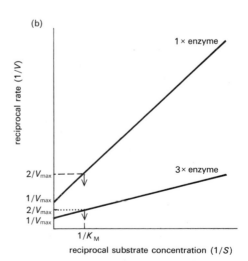

Fig. 1.23   Effect of substrate concentration on rate of reaction
(a) Plot of V against S at two concentrations of enzyme. Note that V always refers to the *initial* rate of reaction; (b) plot of $1/V$ against $1/S$ at the same concentrations of enzyme. See text for details.

fact that the enzyme becomes saturated with substrate; in other words, the amount of enzyme, not of substrate, becomes rate limiting. If more enzyme is added, a new maximal rate is reached. Whenever a rate–concentration curve departs from linearity in this way, one can deduce the participation of some saturable component. This is not necessarily an enzyme. Compare, for example, the rate–concentration curve for facilitated uptake (Fig. 6.8) with that of Fig. 1.23.

The maximum rate of a reaction (V_{max}), which is expressed in units of enzymic activity per milligram of enzyme protein, represents the *catalytic power* of an enzyme. The amount of substrate required to achieve it is a measure of the *affinity* between enzyme and substrate; because it is impossible to place a precise figure on a value reached asymptotically (Fig. 1.23) the affinity is expressed as the amount of substrate required to achieve *half* the maximum velocity. This concentration is referred to as the Michaelis constant, or K_M, for that particular enzyme and that particular substrate. The lower the K_M, the greater the affinity. Note (Fig. 1.23) that the K_M is independent of enzyme concentration.

An alternative method for measuring V_{max} and K_M from experimental data is to plot the reciprocals of rate and substrate concentration against each other. In that way the rectangular hyperbola of Fig. 1.23 becomes a straight line, which makes graphical determination of K_M and V_{max} easier. The reciprocal of V_{max} is the point of intersection on the ordinate. By doubling this value and reading against the abscissa, the recip-

rocal of K_M is determined. The slope of the line is K_M divided by V_{max}. Students who desire an algebraic derivation of this relationship should consult any larger textbook of basic biochemistry.

Changes in the shape of the curves of Fig. 1.23 may be interpreted in terms of changes in K_M and V_{max}. The presence of an inhibitor or drug, for example, may change the slope of the curve without affecting the intercept (Fig. 1.24). In other words V_{max} remains the same but K_M is increased. That is, the drug competes with substrate for its binding site, but any substrate that *is* bound has the same activity. At sufficiently high concentrations of substrate, no drug is bound at all. Hence V_{max} remains the same. Such inhibition, which is known as competitive, is shown by drugs that are analogues of the substrate. The inhibition by malonate of succinate dehydrogenase (Fig. 3.16) is an example:

$$\begin{array}{ccc} \text{CH}_2\text{—COO}^- & & \text{CHCOO}^- \\ | & \xrightarrow{\text{CH}_2 <^{\text{COO}^-}_{\text{COO}^-}} & || \quad +2\text{H} \\ \text{CH}_2\text{—COO}^- & & {}^-\text{OOC—CH} \end{array}$$

substrate   analogue   products

The dissocation constant of the enzyme–inhibitor complex is termed K_i and may be calculated by plotting reciprocal curves such as that of Fig. 1.24 at different inhibitor concentrations.

Another type of inhibitor may affect V_{max} but not K_M (Fig. 1.25). In this case the inhibitor presumably binds to a part of the enzyme *other* than the substrate

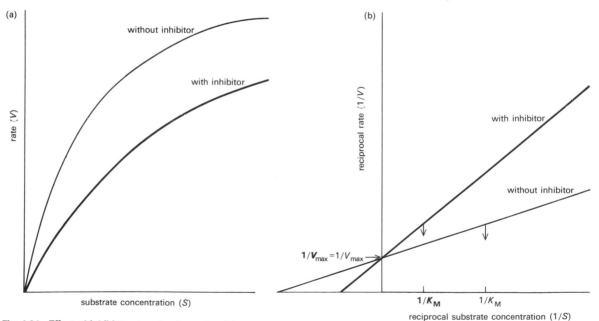

Fig. 1.24   Effect of inhibitors on rate of reaction (V_{max} unchanged)

(a) Plot of V against S: (b) plot of $1/V$ against $1/S$. See text for details.

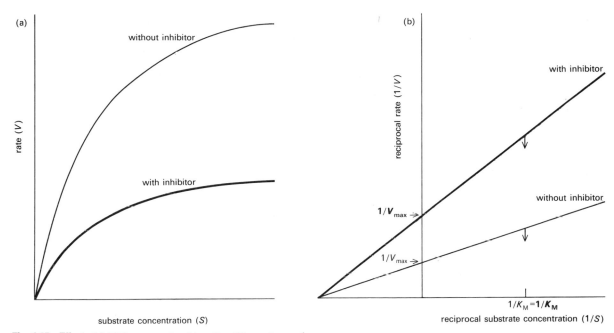

Fig. 1.25   Effect of inhibitors on rate of reaction (K_M unchanged)

(a) Plot of V against S; (b) plot of $1/V$ against $1/S$. See text for details.

binding site, and has a direct effect on catalytic activity. Bound substrate therefore has a lower activity, no matter how much of it is present. It is as though less enzyme is present; compare Fig. 1.25 with Fig. 1.23. An example of such inhibition, which is known as non-competitive, is shown by heavy metals like mercury or its derivatives, that bind to thiols (that is cysteine residues). Enzymes such as glyceraldehyde

3-phosphate dehydrogenase, in which a cysteine residue is part of the active site (Fig. 2.8), are affected.

In some instances the shape of the reciprocal curve departs from linearity (Fig. 1.26). The reason for this may be seen in the rate–concentration plot, which is S-shaped. This has been interpreted in terms of **cooperativity**, which may be positive (activation) or negative (inhibition). It results from the presence of an

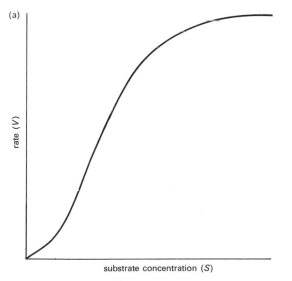

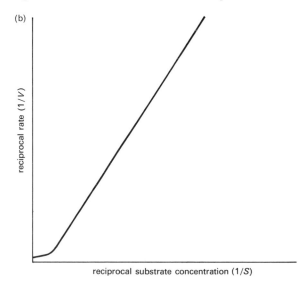

Fig. 1.26   Effect of co-operativity on reaction rate

(a) Plot of V against S; (b) plot of $1/V$ against $1/S$. See text for details.

inhibitor or an activator, or may be due to an effect of the substrate itself. The latter situation arises if the enzyme has quaternary structure, that is, is made up of more than one subunit. In that case, binding of substrate to one subunit may affect (positively or negatively) the binding of substrate to another subunit. In other words one binding site 'is aware' of what is happening at another. An example of positive cooperativity is shown by pyruvate kinase of liver; one of negative cooperativity by glyceraldehyde 3-phosphate dehydrogenase or by aspartate transcarbamylase (Fig. 1.27). Note that the binding of oxygen to haemoglobin (Fig. 7.19) also shows cooperativity.

Where S-shaped kinetics are due to an activator or an inhibitor, the enzyme does not necessarily have to be made up of subunits, though it generally is. Sometimes the subunits are identical, sometimes not. The enzyme aspartate transcarbamylase, for example, has two different types of subunit (Fig. 1.27). One binds substrate, the other binds an endogenous inhibitor (that is, one normally present in cells). This inhibitor is CTP, which is actually an end-product of the sequence of pyrimidine biosynthesis (section 2.3.4.1), which begins with the enzyme aspartate transcarbamylase. Since the two types of subunit are different, they can be separated. The subunit that binds substrate has catalytic activity on its own; the subunit that binds CTP has no catalytic activity. When it is added to the catalytic subunit and substrate, in the presence of CTP, the rate of reaction is decreased. Note that the maximum velocity remains unaltered; in other words the effect is largely on the affinity between substrate and enzyme.

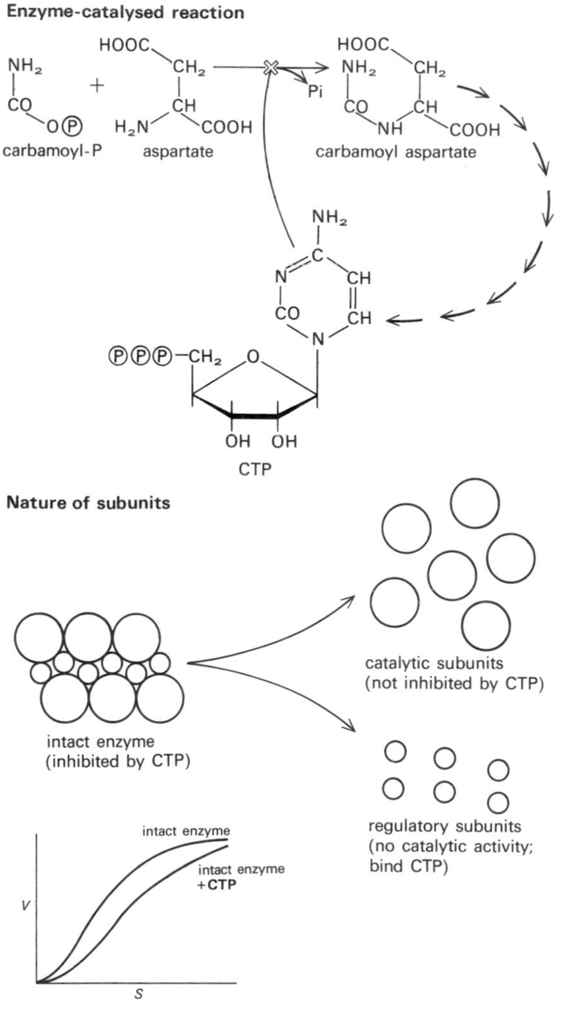

Cooperativity, or S-shaped kinetics, then, results when the binding of substrate to enzyme at one site is affected by the binding of a compound (substrate, inhibitor, or activator) at another site. This situation is referred to as **allosteric** inhibition (or activation), implying that the first substrate binding site has changed to another configuration as a result of the events at the other site.

1.2.3 Mechanism

Enzymes, like inorganic catalysts, become saturated with substrates (reactants) at high substrate concentrations (Fig. 1.23). Saturation implies the formation of some type of complex between enzyme and substrate. It is by bringing the reactants together that enzymes lower the activation energy. The chances of a successful encounter between two molecules is increased greatly if each is bound to a common third molecule (Fig. 1.28). The ability of proteins to bind substrates at specific points, such that their interaction is virtually inevitable, makes them ideal molecules for catalysis. The area at which enzyme and substrate are in contact, and at which catalysis takes place, is known as the **active site**. The active site generally forms a cleft within the tertiary structure of the enzyme; the cleft is so arranged that substrate just fits into it.

The exact mechanism by which catalysis takes place has not been worked out for many enzymes. Knowledge of the tertiary and quaternary structure of the enzyme, and of the amino acids that comprise the active site, is required. This has been largely achieved in the case of chymotrypsin, which will be used as an example.

Chymotrypsin is a protease; that is, it hydrolyses peptide bonds. It has specificity for the amino acid on one side, but not on the other side, of a peptide bond. Thus peptides having a hydrophobic amino acid (phenylalanine, tyrosine, tryptophan, leucine,

Enzyme-catalysed reaction

Nature of subunits

intact enzyme (inhibited by CTP)

catalytic subunits (not inhibited by CTP)

regulatory subunits (no catalytic activity; bind CTP)

Fig. 1.27   Sub-units of aspartate transcarbamylase

(a)

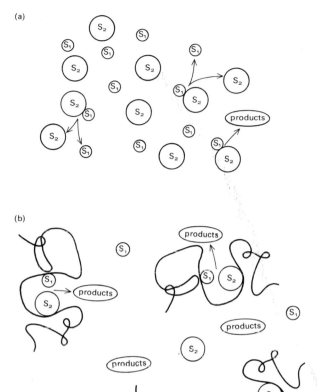

(b)

Fig. 1.28 Basic mechanism of enzyme catalysis

(a) Without enzyme; (b) with enzyme.

In (a), few collision between S_1 and S_2 result in reaction. In (b), in which S_1 and S_2 are each bound to a third molecule (the enzyme), most collisions result in reaction.

isoleucine, valine, or methionine) on the carboxyl side of a peptide bond are hydrolysed (Fig. 1.29).

The enzyme is made up of three polypeptide chains called A, B, and C. These are derived by hydrolysis of a single-chain precursor of the enzyme, known as chymotrypsinogen (Fig. 1.29). Chymotrypsinogen has no enzymic activity. The three chains become arranged in chymotrypsin in such a way that Ser-195 (C-chain), His-57 (B-chain), and Asp-102 (B-chain) lie close to each other on one side of a cleft; opposite, on the other side, lie a series of amino acids that form a hydrophobic pocket, into which fits the hydrophobic amino acid of the peptide bond that is being hydrolysed.

Ser-195 is known to be particularly reactive towards substrate and its analogues; of the 28 serine residues present in chymotrypsin, Ser-195 is the only one that reacts under mild conditions with a reagent known as DFP (diisopropyl fluorophosphate). The reactivity of Ser-195 may be related to the proximity of His-57 and of Asp-102. Together the three amino acids could form a 'charge transfer' system, pulling a proton away from serine towards aspartate, and then pushing it back again; certainly the three amino acids are situated so that hydrogen bonding between them would facilitate charge transfer.

This part of the mechanism is by no means proved, however. An ionic bond between the NH_2-terminal Ile-16 (Fig. 1.29) and Asp-194 (adjacent to Ser-195) may also be involved. Another reason why Ser-195 might become activated is the following. When the substrate (that is, a hydrophobic residue on a polypeptide chain) becomes anchored in its hydrophobic pocket, it tends to distort the enzyme molecule so that Ser-195 moves a little closer towards the substrate and makes reaction with it easier. At the same time the substrate becomes 'strained' towards Ser-195. In other words binding and catalysis are not separate events. On the contrary, as a result of specific binding, substrate becomes 'activated' and already committed towards undergoing reaction. Such a mechanism of '**Induced fit**' may underlie the action of many, if not all, enzymes.

Whatever the reason for the increased reactivity of Ser-195, the result is that the peptide bond becomes broken, with concomitant transfer of the acyl part of the substrate on to the serine residue (Fig. 1.29). A covalently linked enzyme–substrate intermediate is transiently formed. This is extremely susceptible to attack by water, and hydrolysis of the intermediate completes the overall hydrolysis of the peptide bond.

The sequence of steps outlined in Fig. 1.29, namely the activation of a serine residue (perhaps through an adjacent histidine residue), followed by attack on a bond susceptible to hydrolysis, is believed to be common to many hydrolytic enzymes. It is interesting to note that this mechanism of hydrolysis is similar to non-enzymic hydrolysis catalysed by acids or bases. There are several reasons why an enzyme such as chymotrypsin reduces the activation energy of hydrolysis further than does an acid. For one, the substrate is 'anchored' in an optimal position relative to the attacking serine residue. For another, some kind of 'charge transfer system' may make the serine a particularly effective attacking group. Together, these and other considerations explain not only the efficiency of enzymic catalysis, but its specificity also.

In many instances enzymes catalyse an initial reaction between a substrate and a **coenzyme**, which is a small molecule present in catalytic amount that is

(a)
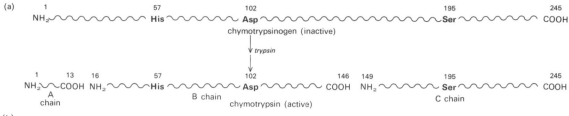

chymotrypsinogen (inactive)

↓ *trypsin*

chymotrypsin (active)

(b)
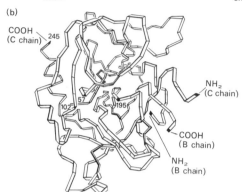

Fig. 1.29 Mechanism of action of chymotrypsin

(a) Activation of chymotrypsinogen. The numbers refer to amino acid residues on the chymotrypsinogen chain; some of the amino acid residues that participate in catalysis are shown in bold.

(b) Quaternary structure of chymotrypsin. The three chains are arranged as shown; the A chain is 'buried' in the molecule, so that its terminals are not clearly discernible. Note the proximity of His 57, Asp 102, and Ser 195. Redrawn from a model constructed by Professor D. Blow, with permission.

(c) Function of chymotrypsin in hydrolysis of peptide bonds. Bonds that are adjacent to a hydrophobic grouping (Tyr, Try, Phe, Met) are hydrolysed. The reaction proceeds in two stages. First the hydrophobic group is 'anchored' in a hydrophobic pocket of the enzyme, such that the substrate is particularly susceptible to attack by the activated Ser-195. Next the enzyme-substrate intermediate is hydrolysed by water. Components of the enzyme are shown in bold.

(c)
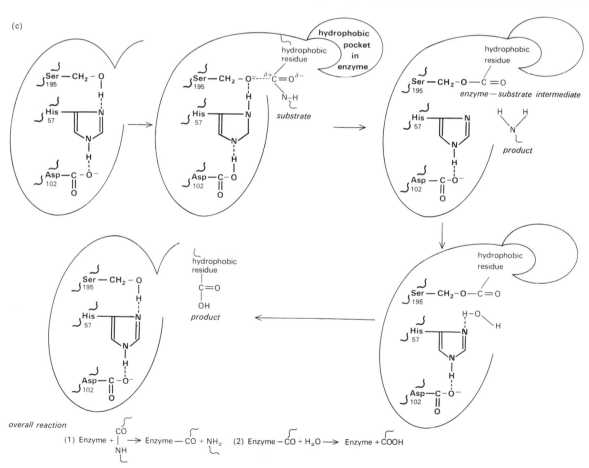

Fig. 1.30 Mechanism of action of coenzymes

substrate 1 + coenzyme → product A + altered coenzyme

altered coenzyme + substrate 2 → product B + coenzyme

net: substrate 1 + substrate 2 → product A + product B

attached to the enzyme or is in free solution. The enzyme (or a different enzyme) then catalyses a second reaction to regenerate the coenzyme (Fig. 1.30). When a coenzyme is bound to an enzyme, the resulting complex is referred to as a holo-enzyme; the enzyme without the co-enzyme is referred to as an apo-enzyme. Coenzymes such as nicotinamide and flavin (vitamin B_2) nucleotides are involved in oxidation–reduction reactions (section 3.4), coenzymes such as thiamin pyrophosphate (vitamin B_1) and biotin are involved in decarboxylation and carboxylation reactions, and a coenzyme such as pyridoxal phosphate (vitamin B_6) is involved in reactions affecting the α-

carbon atom of amino acids (section 2.4). In short, coenzymes participate in most of the major pathways of intermediary metabolism (Fig. 1.31). They are derived from the various B group vitamins, and their metabolic role reflects the dietary need for the vitamins.

1.2.4 Disease

Enzymes are the basis of cellular function. If an enzyme is missing or is incorrectly synthesized, cellular function becomes impaired. The underlying cause may be hereditary, or it may be environmental. Enzymes can be divided into three categories:

i. enzymes that are *crucial* to the functioning of cells: without them, cells die;

ii. enzymes that are *necessary* for the proper functioning of cells; without them metabolism is deranged;

Fig. 1.31 Origin and function of coenzymes

Parent vitamin	Coenzyme	Major metabolic role
thiamin (B_1)	thiamin pyrophosphate	oxidative decarboxylation (e.g. of pyruvate and α-ketoglutarate)
riboflavin (B_2)	flavin nucleotides	H transfer ($\diagup$CH—CH$\diagdown$ → $\diagup$C=C$\diagdown$)
niacin (B_5)	nicotinamide nucleotides	H transfer 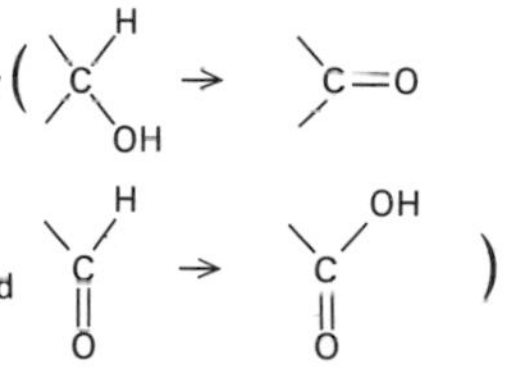 ($\diagdown$C$\diagup$ (H, OH) → $\diagdown$C=O) and $\diagdown$C$\diagup$ (H), ‖O → $\diagdown$C$\diagup$ (OH), ‖O)
pyridoxin (B_6)	pyridoxal phosphate	transamination and other reactions involving α-carbon atom of amino acids
pantothenic acid	coenzyme A	oxidative decarboxylation (e.g. of pyruvate and α-ketoglutarate;) β-oxidation of fatty acids; metabolism of acetyl CoA (to CO_2 + H_2O via tricarboxylic acid cycle; to fatty acids; to cholesterol)
biotin	biotin (enzyme bound)	carboxylation (e.g. of acetyl CoA during fatty acid synthesis)
folate	tetrahydrofolate	1–carbon transfer' (e.g. serine → glycine + '1–carbon' → synthesis of purine nucleotides; synthesis of TMP from dUMP)
B_{12} (cyanide derivative)	B_{12} (deoxyadenosine derivative)	? reduction of nucleotides

iii. enzymes that appear to be *unnecessary* to the functioning of cells: their absence leads to no overt defect.

Defects in the first category, which includes enzymes necessary for the synthesis of DNA, RNA, and protein, and for the synthesis of ATP, are never observed. A cell in which such an enzyme is missing cannot grow and divide, and is therefore eliminated. This may occur at the first cell division during foetal development, or it may occur at any later stage in life.

Defects in the second category result in progression towards disease. In the chapters that follow, many examples will be presented. In the case of hereditary defects, the more serious the outcome, the less common the fault. For if the life expectancy is so low that maturity is rarely reached, the faulty gene will only rarely be transmitted; it will therefore tend to be eliminated from the population. Most of the diseases in which a missing enzyme leads to serious consequences occur at frequencies of less than 1 in every 100 000 of the population. On the other hand such enzyme deficiencies comprise the bulk of the hereditary disorders that have been recognized to date. For if a disease is not serious, it is much more difficult to pin-point the enzymic lesion. Yet it is a fault in just those enzymes, which is likely to underlie many common diseases.

In the third category are enzymes that appear to play no obvious role. Few are known, but as mentioned above, the less important the activity of an enzyme appears to be, the less likely it is to be investigated. One might add to this category nondeleterious changes in enzymes that belong to groups (i) and (ii). That is, structural changes that appear not to affect the catalytic function of enzymes. Such changes, which are known as polymorphic variation, are very common indeed. They are of less relevance to disease than to evolution (section 4.4.2).

1.3 Membranes

Membranes are as fundamental a component of cells as enzymes. Without a membrane there is no cell. A biochemical description of nucleus, endoplasmic reticulum, mitochondrion and cell surface is a description of different types of membrane.

1.3.1 Composition

All membranes contain phospholipid. The reverse is also true: all phospholipids are in membranes. Mostly the membranes are **bilayers**, in the form of double layered sheets, tubes or vesicles, with the polar groups of the phospholipid in contact with the aqueous medium (Fig. 1.14). In the case of fat droplets, a single sheet of phospholipid, its polar groups in contact with the aqueous medium, surrounds a mixture of neutral lipids (Fig. 1.14). These arrangements are not due to a carefully programmed, energy-requiring mechanism. On the contrary: it is an inevitable result of the amphipathic character of phospholipids, and represents the structure of lowest energy, and hence of greatest stability.

Although membranes consisting only of phospholipid are entirely stable, biological membranes invariably also contain protein. Many of the so-called structural proteins of a cell are in membranes. The type of phospholipid that is present varies from membrane to membrane. Plasma membrane, for example, has relatively more sphingomyelin than other membranes; mitochondrial membranes contain cardiolipin. The type of protein present in a membrane also varies. Endoplasmic reticulum, for example, contains NADPH–cytochrome C reductase, plasma membrane contains 5′-nucleotidase, and so forth. While few of the non-enzymic proteins have as yet been characterized, their structure is known to vary from membrane to membrane.

Several membranes contain other components. The plasma membrane is particularly diverse. In addition to protein and phospholipid it contains cholesterol, glycoproteins, and glycolipids. Parts of the endoplasmic reticulum have ribosomes, which are made up of RNA and protein, attached; such portions are known as rough endoplasmic reticulum, in contrast to ribosome-free portions known as smooth endoplasmic reticulum. Mitochondria also have two structurally distinct membranes: an inner and an outer membrane. It is the former that contains cardiolipin and the components of the electron transport chain.

The composition of the membranes of a cell is shown in Fig. 1.32. The distribution is similar in most cells, since the function of the organelles of which the membranes are a part is the same in most cells. Where function differs, composition differs. In cells such as liver that synthesize free glucose, for example, the endoplasmic reticulum contains glucose 6-phosphatase, an enzyme absent from other cells. The endoplasmic reticulum of liver also contains enzymes capable of detoxifying foreign compounds (Fig. 5.4). In muscle the endoplasmic reticulum (sarcoplasmic reticulum) contains a calcium pump for regulating muscular contraction (section 10.2.3). The main functions of the plasma membrane are the same in all cells: to insulate the interior from the environment, to regulate the entry of nutrients, and to control cellular activity by interaction with the environment. In order to achieve selectivity of the latter functions, the composition of the plasma membrane varies. Intestinal and

Fig. 1.32 Composition of cellular membranes. (percentage composition by weight)

	Plasma membrane	Endoplasmic reticulum		Mitochondrial membrane		Nuclear membrane
		'Smooth'	'Rough'	'Inner'	'Outer'	
Protein	68	71	79	79	73	71
Phospholipid	19	25	20	20	26	28
Cholesterol	7	3	<1	<1	1	<1

Data for (rat) liver recalculated from Colbean *et al.* (1971). *Biochim. Biophys. Acta* **249,** 462 with permission. Note that the percentage distribution according to *moles* differs considerably, since the molecular weight of proteins is between 10 000 and 100 000, of phospholipids around 700, and of cholesterol 387; a protein : phospholipid ratio of 80 : 20 by weight, for example, implies a molar ratio of 1 : 4–1 : 40.

renal epithelial cells contain enzymes for accumulating sugars, amino acids, and ions (e.g. Figs 6.9, 6.10, 8.12). Muscle and fat cells contain an insulin receptor for controlling the entry of glucose (Fig. 9.11). Lymphocytes contain an antigen receptor for triggering synthesis of antibody (Fig. 11.9). Much of the specialized function of cells is due to a specialization of their plasma membrane.

1.3.2 Structure

The disposition of phospholipids in biological membranes has been described (Fig. 1.14). How are the proteins arranged? The original suggestion of a layer of protein on either side of the phospholipid bilayer is no longer tenable. Some proteins certainly do lie on either side of the membrane, held by electrostatic and hydrogen bonds. They are known as **peripheral** or **extrinsic** proteins (Fig. 1.33) and are released by treatment with strong solutions of salt or urea (see section 1.1.1). But many proteins are embedded in the phospholipid bilayer and require treatment with detergents such as deoxycholate or Triton X100, or with solvents such as acetone or mixtures of methanol and chloroform, to release them. These proteins are called **integral** or **intrinsic** (Fig. 1.33). The bonds that hold them in position are predominantly hydrophobic. It

might be thought that such proteins are especially rich in the hydrophobic amino acid residues. But this is not so. Moreover once extracted, such proteins are often soluble in aqueous media. How is this possible?

The answer is that the tertiary structure of intrinsic proteins (and of other proteins also) depends on their environment. In aqueous media, where the predominant molecule capable of forming hydrogen bonds is water, the polypeptide chains fold so as to expose as many of the polar amino acid residues as possible; that is, to make the maximum number of hydrogen bonds with water. In non-aqueous media, where the only possibility to form hydrogen bonds is between the amino acid residues themselves, the most stable structure becomes the reverse; that is, one in which the polar residues face inwards, so that the possibility for hydrophobic interaction with the environment is maximal (Fig. 1.34).

The difference between the two conformations is not as clear-cut in membrane proteins as indicated in Figure 1.34. For one thing some polar interaction with phospholipids – especially at the edge of the bilayer – is possible. For another, many intrinsic proteins are

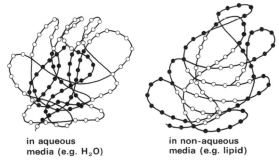

in aqueous
media (e.g. H$_2$O)

in non-aqueous
media (e.g. lipid)

o—o polar amino acids (Arg, Asp, Cys, Glu, Gly, His, Lys, Ser, Thr,)

•—• apolar amino acids (Ala, Ile, Leu, Met, Phe, Pro, Try, Tyr, Val)

Fig. 1.34 Conformation of proteins in aqueous and non-aqueous media

Note that most proteins are not as soluble in lipid as in aqueous media; the structure shown on the right represents the limited amount that *is* soluble. Polar and apolar amino acids are shown grouped together, for clarity; generally the sequence is more mixed.

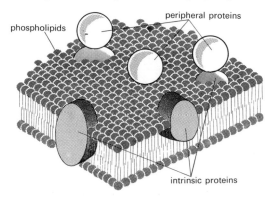

phospholipids

peripheral proteins

intrinsic proteins

Fig. 1.33 Disposition of proteins in membranes

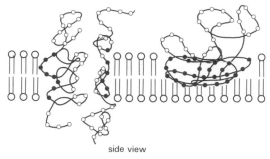

side view

○——○ polar amino acids [see Fig. 1.34]

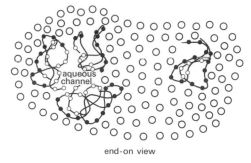

end-on view

●——● apolar amino acids [see Fig. 1.34]

Fig. 1.35 Conformation of intrinsic proteins in membranes
Phospholipids are shown as circles in end-on view.

made up of subunits and it is probable that in the region between them, polar not apolar residues are exposed (Fig. 1.35). Such an arrangement is particularly attractive as it explains (i) how water-soluble compounds such as ions, sugars, and amino acids cross biological membranes and (ii) how the transport of such compounds may be accelerated by binding to specific sites on the proteins (see section 6.3.1).

The arrangement of cholesterol, glycolipids and glycoproteins in membranes is that expected from their properties. Cholesterol is buried within the hydrophobic region of the bilayer, while glycolipids and glycoproteins are inserted such that the carbohydrate portions extend into the aqueous medium (Fig. 1.36); the lack of internal interaction between the rather bulky sugar residues and their preponderance for forming hydrogen bonds with water has already been mentioned.

The arrangements depicted in Figs 1.33, 1.35, and 1.36 are not static. Just as solvent and solute molecules of an aqueous or a non-aqueous solution are in continuous motion relative to each other, so solvent and solute molecules of a lipid bilayer move relative to each other. That is, phopholipids (solvent), and proteins and other components (solute), have rotational and translational freedom within the plane of membrane (Fig. 1.37). But just as solvent and solute molecules remain essentially within the liquid phase (except above the boiling point), so the components of the membrane do not move from polar to non-polar phase or vice versa. That is, proteins and phospholipids do not 'tumble' into the interior, or across the membrane, since to do so they would have to change phase.

A second analogy between the components of a membrane and the components of a solution concerns the effect of temperature. Just as below a certain temperature the movement of solvent and solute molecules virtually ceases (that is, at the freezing-point), so below a certain temperature the movement

of membrane components ceases also. This is known as the transition temperature. Above it the components (mainly cholesterol and phospholipids) are said to be in a fluid ('liquid-gel') state; below it, in a solid ('crystalline-gel') state. Just as the freezing-point of a solution depends on the nature of the dissolved molecules, so the transition temperature of a membrane depends on its components also. The content of

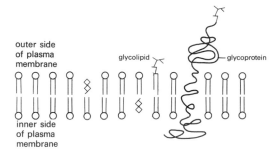

outer side
of plasma
membrane

glycolipid

glycoprotein

inner side
of plasma
membrane

Fig. 1.36 Disposition of minor components in membranes
Some components of the plasma membrane are shown; cholesterol is indicated by the same symbol as in Fig. 1.14.

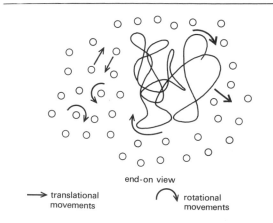

end-on view

→ translational movements

↷ rotational movements

Fig. 1.37 Movement of components within membranes

cholesterol and the length and degree of unsaturation of the phospholipid acyl chains are particularly important determinants; in fact the composition of cellular membranes is adjusted so that at the temperature at which the organism lives, the lipids within its membranes are in the fluid state. Generally, the transition temperature is not sharp, but rather diffuse. This is due to the fact that the components are not really in true solution, but are segregated to a certain extent in patches; different proteins may have different phospholipids in their immediate vicinity. The movement of proteins is less extensive than that of lipids. Moreover their motions are determined by factors other than the state of the surrounding lipids. Many intrinsic proteins of the plasma membrane, for example, appear to be 'anchored' to other proteins at the cytoplasmic side of the membrane.

The important point, then, is that biological membranes are in a rather **fluid** state with respect to their components. Such fluidity is quite distinct from, and superimposed on, gross movement of the membrane such as the locomotion of cells by 'spreading', or the pinching in half of a cell during mitosis, or the engulfing of particles by phagocytosis.

1.4 Summary

Composition of cells

1. Cells consist of **water, protein, fat, carbohydrate, nucleic acid, dissolved molecules**, and **inorganic ions**. The overall composition varies relatively little between different types of cell.

2. Proteins are made up of a mixture of twenty different l-amino acids joined in linear sequence by amide (that is, **peptide**) bonds (primary structure). Stretches of amino acids may be coiled in an α-helical structure by **hydrogen bonds** (secondary structure). Further coiling results from interactions between amino acid residues by covalent **S–S bonds**, by hydrogen bonds, by **hydrophobic bonds**, and by **ionic bonds** (tertiary structure). A protein may consist of two or more chains, each with secondary and tertiary structure, linked together by covalent S–S bonds, by hydrogen bonds, by hydrophobic bonds or by ionic bonds (quaternary structure).

3. Fats (lipids) consist of neutral and charged compounds. **Cholesterol** and **triglyceride** are the main neutral lipids; they are soluble only in organic, apolar solvents and exist in cells as fat droplets or within the apolar region of membranes. **Phospholipids** are the main charged lipids and have two distinct regions: a polar, charged, water-soluble region, and an apolar, neutral region. Hence phospholipids arrange themselves in cells such that the polar region is in contact with water, and the apolar region is in contact with another apolar molecule; this results in the formation of double layered membranes. **Glycolipids**, which have polar carbohydrate residues attached to a lipid chain, are a minor component of cellular membranes.

4. The main carbohydrate of cells is **glycogen**, a polymer of d-glucose residues. Other carbohydrates, containing a mixture of sugars, are linked to protein. Those compounds in which protein predominates, called **glycoproteins**, occur chiefly in membranes. Those compounds in which carbohydrate predominates, called **proteoglycans**, occur chiefly in extracellular fluids.

5. Nucleic acids are made of a chain of alternating **sugar–phosphate** residues; each sugar has a nitrogenous base (**purine** or **pyrimidine**) attached. In ribonucleic acid (**RNA**) the sugar is d-ribose; in deoxyribonucleic acid (**DNA**) the sugar is 2-deoxy d-ribose. In DNA, which is the genetic material of living organisms, two chains are coiled around each other in a double helix, held together by hydrogen bonds between the nitrogenous bases. RNA exists as double or single chains. Both DNA and RNA are further coiled as a result of hydrogen and other bonds.

6. The cytoplasm of cells contains dissolved molecules of **low molecular weight**. These are mainly precursors of proteins, fats, carbohydrates, and nucleic acids and are a mixture of amino acids, phosphorylated sugars, and other derivatives.

7. The main inorganic ions are **sodium, potassium, magnesium, calcium, chloride, bicarbonate, phospate**, and **sulphate**.

Enzymes

1. The main function of many proteins is to act as **enzymes**. Enzymes are biological **catalysts**, capable of increasing the rate of reactions so that they occur at body temperature. As a result of specific binding to reactants (substrates), each enzyme catalyses a specific reaction.

2. The rate of an enzyme-catalysed reaction is sensitive to changes of **pH** and to the presence of **inhibitors** and **activators**. The way in which the rate is affected yields information about the nature of catalysis.

3. The effect of a missing or malfunctioning enzyme is to alter metabolism. This in turn may lead to disease.

Membranes

1. Cells are divided into discrete compartments (nucleus, mitochondria, endoplasmic reticulum) by membranes. The basic structure of membranes is a **phospholipid bilayer**, with proteins embedded in it (**intrinsic proteins**) or more loosely attached to its surface (**extrinsic proteins**). Membranes, in particular the plasma membrane, contain in addition cholesterol, glycolipids, and glycoproteins.

2. The components of membranes are in a fluid state, able to move within the plane of the bilayer.

FURTHER READING

Structure and function of cell constituents,

These two books are illustrated with great clarity:

P.C.Hanawalt and R.H.Haynes (1973). *The chemical basis of life: an introduction to molecular and cell biology.* (Readings from *Scientific American*) W.H.Freeman and Co., San Francisco.

L.Stryer (1975). *Biochemistry.* W.H.Freeman and Co., San Francisco.

Biochemistry of membranes

J.B.Finean, R.Coleman, and R.H.Michell (1978). *Membranes and their cellular functions* (2nd edition). Blackwell Scientific Publications, Oxford.
R.Harrison and G.G.Lunt (1976). *Biological membranes.* Blackie and Son Ltd., Glasgow.

P.J.Quinn (1976). *The molecular biology of cell membranes.* Macmillan, London.

Enzymology

A general account is given by:
K.G.Scrimgeour (1977). *Chemistry and control of enzyme reactions.* Academic Press, New York.

A stimulating article requiring some knowledge of thermodynamics:
W.P.Jencks (1975). Binding energy, specificity and enzymic catalysis: the circe effect. *Adv. Enzymol.* **43,** 219.

2

Cytoplasm: intermediary metabolism

2.1 Introduction

The metabolic reactions within a cell are analogous to the production line in a factory. Raw materials are fed in, processed, and formed into a finished product. In the case of the animal cell, the raw materials are molecules such as sugars, fats, and amino acids. The processing involves degradation to common intermediates, from which cellular constituents such as protein, nucleic acids, carbohydrates, and fats are synthesized. Just as a car factory may use old scrapped cars as a source of raw materials, so an animal cell uses largely dietary proteins, nucleic acids, carbohydrates, and fats (all degraded outside the cell) as a source of the amino acids, sugars, etc. required for synthesizing cell-specific proteins, nucleic acids, carbohydrates, and fats. In order to be able to manufacture cars from raw materials, a factory requires a source of energy. The form in which this is supplied is largely as electricity, itself produced in another sort of factory, that is, a power station, that converts coal into electricity. An animal cell is both car factory and power station at the same time. In fact it uses the same raw materials (sugars, fats, and amino acids) both for the synthesis of cell constituents and for the production of energy necessary to drive that synthesis. It also uses the raw materials for the generation of heat; most of the energy inherent in food is used to maintain body temperature at 37 °C.

Just as a power station generates electricity for consumption by organizations other than car factories, so a cell generates energy for consumption by processes other than biosynthesis. Most of the energy generated by muscle cells, for example, is used to drive muscular contraction (see section 10.2), while much of the energy generated by kidney cells is used to drive ion pumps that underly the basis of tubular reabsorption (see section 7.2). It is another feature of animal cells that the form of energy used to drive biosynthesis, muscle contraction, and ion movements is the same, namely chemical energy. The nature of the energy-yielding reaction is the same also: hydrolysis of the terminal bonds of ATP. The picture that emerges of an animal cell is that of a highly efficient system, using common metabolic pathways for the synthesis of its constituent parts and for the generation of a single type of molecule that drives the diverse processes underlying a living cell (Fig. 2.1). Of the processes of synthesis and energy production, energy production is by far the greater; over 90 per cent of the food that is eaten is normally converted into energy, (approximately 25 per cent as ATP and the rest as heat), and only a small amount is used for the synthesis of cell constituents.

It might be thought that ATP produced by the degradation of food in one organ could be used to

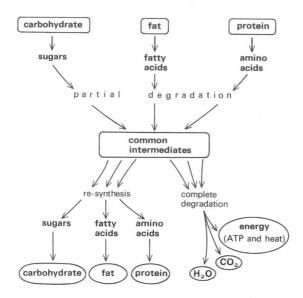

Fig. 2.1 **Role of common intermediates in metabolism**

drive energy-requiring processes in another organ. In other words, that ATP can circulate in the blood-stream as an energy source. This is not so. ATP crosses membranes rather poorly, and the ATP that is required to drive a reaction in a muscle or kidney cell is formed within that cell. Instead, it is compounds such as glucose, fatty acids, amino acids, and ketone bodies that circulate in the blood-stream, and that are taken up and oxidized, to generate ATP, by energy-requiring cells. Some cells, such as muscle, can use any one of these nutrients as a source of 'metabolic fuel'; other cells, such as nerve cells, can use only glucose or, after a period of adaptation, ketone bodies.

The site of ATP formation in cells is the mi-tochondria; the mechanism by which ATP is synth-esized, and its relation to mitochondrial structure, are considered in section 3.4. In this chapter the metabolic transformations that occur predominantly in the cytoplasm of animal cells are discussed. The pathways can be considered under four headings:

i. degradative pathways leading to the formation of common intermediates;

ii. the synthesis of 'building blocks' intermediate in the synthesis of cellular constituents;

iii. the synthesis and function of cytoplasmic cofac-tors, and

iv. the terminal stages in the biosynthesis of cellular constituents.

The synthesis of carbohydrates and fats only will be considered under (iv); the biosynthesis of nucleic acids and proteins is described in Chapter 4 and 5 respectively.

2.2 Degradation of food stuffs and cell constituents to common intermediates

It is an oversimplification to describe all metabolic processes as either degradative (catabolic) or synthetic (anabolic). Several reactions, such as transamination between amino acids and their corresponding keto acids, and reactions that maintain steady-state levels of tricarboxylic acid cycle intermediates, are neither. Some of these reactions have been termed anaplerotic ('filling-up'). Any definition is rather restrictive, and in the discussion that follows no attempt will be made to categorize each reaction sequence precisely. In any case, as has already been indicated, the same reaction sequence may lead both to the formation of a cellular end-product and to the formation of ATP.

The type of cell in which the reaction sequence occurs, as well as the physiological state of the cell at that particular time, influences the end-result. In heart muscle, for example, the formation of acetyl coenzyme A (CoA) from glucose is the first stage in the overall degradation of glucose to CO_2 and H_2O, for the concomitant production of the ATP necessary to maintain the heartbeat. In liver the formation of acetyl CoA from glucose may likewise be the first stage in the overall degradation sequence leading to the pro-duction of ATP — in this case used for biosynthetic reactions such as gluconeogenesis, the urea cycle, or the synthesis of plasma proteins, as well as for the operation of ion pumps; alternatively, however, the formation of acetyl CoA may be the first stage in the conversion of glucose to cholesterol or to fatty acids.

The situation is best described by a simple diagram (Fig. 2.2). The '**common intermediates**' to which refer-ence has been made are compounds such as pyruvate, acetyl CoA, and the intermediates of the tricarboxylic acid cycle. It must be emphasized that Fig. 2.2 is a

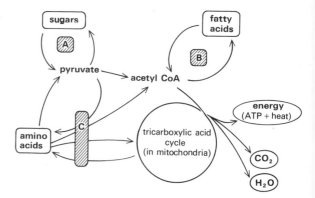

Fig. 2.2 **Degradation of foodstuffs and cell constituents to common intermediates**

A, B, and C refer to the pathways by which sugars, fatty acids and amino acids are degraded (section 2.2) and re-synthesized (section 2.3).

simplified diagram, and that other common intermediates such as glycerophosphate or phosphoenolpyruvate have been omitted for clarity. Moreover all twenty amino acids have been grouped together, although the particular sequence that is followed depends on the structure of the amino acid. As a result, pathway C comprises three broad categories. Note that only the long-chain fatty acid component of triglyceride and phospholipid is denoted; the other components (glycerol and nitrogeneous base) account for relatively minor amounts of organic material (5–10 per cent of total molecular weight).

It can be seen that all foodstuffs — amino acids, sugars, and fats — are ultimately degradable by oxidation to CO_2 and H_2O. Hence the word 'common intermediate'. It can also be seen that **sugars** are convertible to **fats**, via acetyl CoA, but that fats cannot be converted back to sugar, because of the irreversibility of the pyruvate→acetyl CoA reaction. (Although intermediates of the tricarboxylic acid cycle such as α-ketoglutarate and oxaloacetate can be converted to sugar by the reactions summarized in Fig. 2.27, a net conversion of acetyl CoA to sugar is not possible. This is because a molecule of oxaloacetate is required before acetyl CoA can enter the cycle; after one turn of the cycle, a single molecule of oxaloacetate is regenerated; hence no net conversion of acetyl CoA into oxaloacetate has occurred.)

The inability of acetyl CoA to be converted to sugar is of fundamental importance to an understanding of animal metabolism. It means that when **carbohydrate** is eaten, it can be

(a) burnt to CO_2, H_2O, and energy,

(b) laid down as glycogen, or

(c) converted into fat.

When **fat** is eaten, it can only be

(a) burnt to CO_2, H_2O, and energy or

(b) laid down as fat.

Since sugars are essential ingredients of cellular metabolism, it is clear that whereas a fat-free diet is acceptable, a carbohydrate-free one is not. (It may be worth noting that plants and certain microbes are able to by-pass the irreversible pyruvate→acetyl CoA reaction by means of a sequence of reactions known as the glyoxalate cycle, and hence are able to convert fat into sugar.)

With regard to **amino acids**, the situation is a little more complex (see Fig. 2.22). Those that are degraded directly to acetyl CoA cannot, like fats, be converted to sugar; they are known as ketogenic, because the ketone bodies to which they sometimes give rise are derived from acetyl CoA. Those that are degraded to pyruvate can be converted into either sugar or fat; conversion into sugar is reversible. Those that are degraded into intermediates of the tricarboxylic acid cycle are likewise convertible, by reactions not shown in Fig. 2.2, to sugar or fat. Amino acids that are convertible to sugar are known as glucogenic. In general, the reactions leading to acetyl CoA are not reversible whereas those leading to pyruvate or to tricarboxylic acid cycle intermediates are. Hence the ketogenic amino acids are largely those known as 'essential' (that is, required in the diet), whereas the glucogenic amino acids are the non-essential ones (able to be synthesized from sugars and a source of nitrogen).

The reader will note that **nucleic acids** and their low molecular weight components have been omitted from this discussion and from Fig. 2.2. The reasons are as follows. First, the amount of nucleic acids present in most foods is low compared to that of protein, fat, or carbohydrate. Hence their degradation to CO_2, H_2O, and energy is quantitatively of minor importance. Secondly, DNA does not 'turn over' in cells. Once synthesized it is stable for as long as that cell or its progeny remains viable. RNA does turn over, and very rapidly at that, but its degradation stops largely at the level of nucleotide, from which fresh RNA (often of different structure) is resynthesized. In other words, relatively little of the components of RNA and DNA enter the metabolic pool of common intermediates. That amount of nucleic acid that is eaten is degraded as far as the nucleosides, which are then absorbed and used for the synthesis of RNA and DNA. Only in cases of starvation or rapid growth is synthesis of nucleotides (from amino acids and sugars) important.

Before proceeding to a more detailed discussion of pathways A, B, and C (Fig. 2.2), the origin of the amino acids, sugars, and fatty acids that are present in cells has to be considered. In a healthy, well-fed individual these compounds are derived predominantly from dietary protein, carbohydrate, and fat. The mechanism by which foodstuffs are broken down into their low molecular weight components and absorbed into the circulation is described in Chapter 8. In addition, however, there is a significant and continuous breakdown of cellular protein, carbohydrate, and fat: the so called 'wear and tear', or **turnover** (section 2.2.4), of cellular constituents. During starvation the breakdown of cellular constituents becomes the major source of metabolic intermediates required to maintain essential energy supplies. Under these conditions, resynthesis of cellular constituents does not occur and the individual loses weight. Every cell therefore contains enzymes capable of degrading protein, carbohydrate, and fat. Such intracellular enzymes are similar to, but not identical with, the digestive enzymes. The mechanism by which intracellular enzymes act will be described in the relevant parts of this section.

2.2.1 Carbohydrates

2.2.1.1 Degradation of glycogen

The main reserve of cellular carbohydrate is in the form of **glycogen**. Glycogen is found in two types of tissue: liver and muscle. Approximately a third of the body glycogen is in liver (for example 108 g in the 1.8 kg of liver in a 70 kg man), with the other two-thirds distributed among the various types of muscle (for example 245 g in the 35 kg of muscle in a 70 kg man). Glycogen is a highly branched polymer of $\alpha 1$–4 and $\alpha 1$–6 D-glucose units (Fig. 1.15). As a result of its high molecular weight (it varies from 5×10^6 to 5×10^8 or higher), it exists (despite the large number of —OH groups) in the form of cytoplasmic granules, soluble only in boiling alkali. Degradation is partly by phosphorolysis and partly by hydrolysis. The 1–6 branches are degraded by transferase and hydrolase enzymes to yield a 1–4 backbone which is then degraded by phosphorolysis, starting at the nonreducing end (Fig. 2.3). It is obvious that by phosphorolysis the energy inherent in the glycosidic bond is maintained—as glucose 1-phosphate—whereas by hydrolysis it is lost.

The nature of **phosphorylase** enzymes is complex. They have been purified from the two main tissues that contain glycogen, namely liver and skeletal muscle. The enzymes consist of an active and an inactive form (Fig. 2.4). In each case, the trigger for conversion of inactive to active form is an increase in the concentration of an intermediate compound known as **cyclic AMP** (cAMP). The increase is itself the result of

hormonal activation of an enzyme known as adenyl cyclase (Fig. 6.14). The important feature of a 'cascade' sequence of successive enzymes as depicted in Fig. 2.4 is that the original stimulus is amplified many fold. In other words relatively few molecules of hormone cause the conversion of large amounts of glycogen to glucose 1-phosphate.

The enzymes from liver and skeletal muscle differ in some of the details shown in Fig. 2.4; they are also coded for by different genes. That is, they are cell-specific iso-enzymes. The most striking evidence of this is seen in the heriditary defect of phosphorylase deficiency. In one type of disease (glycogen storage disease Type VI, or Hers' disease), patients suffer from an inability to degrade liver glycogen, whereas degradation of muscle glycogen is normal; the symptom is an enlarged liver. In the other type of disease (glycogen storage disease Type V, or McArdle's disease) the disability lies in the degradation of muscle glycogen, the degradation of liver glycogen being normal; the symptoms are muscle cramps during exercise. Deficiency of phosphorylase is but one of a number of heriditary defects of glycogen metabolism. Other enzymes that are affected are the debranching hydrolases and other enzymes (Fig. 2.57). In these cases clear distinctions between liver and muscle enzymes have not been described.

2.2.1.2 Glycolysis

The further degradation of **glucose** units, by which a certain amount of metabolic energy is converted to

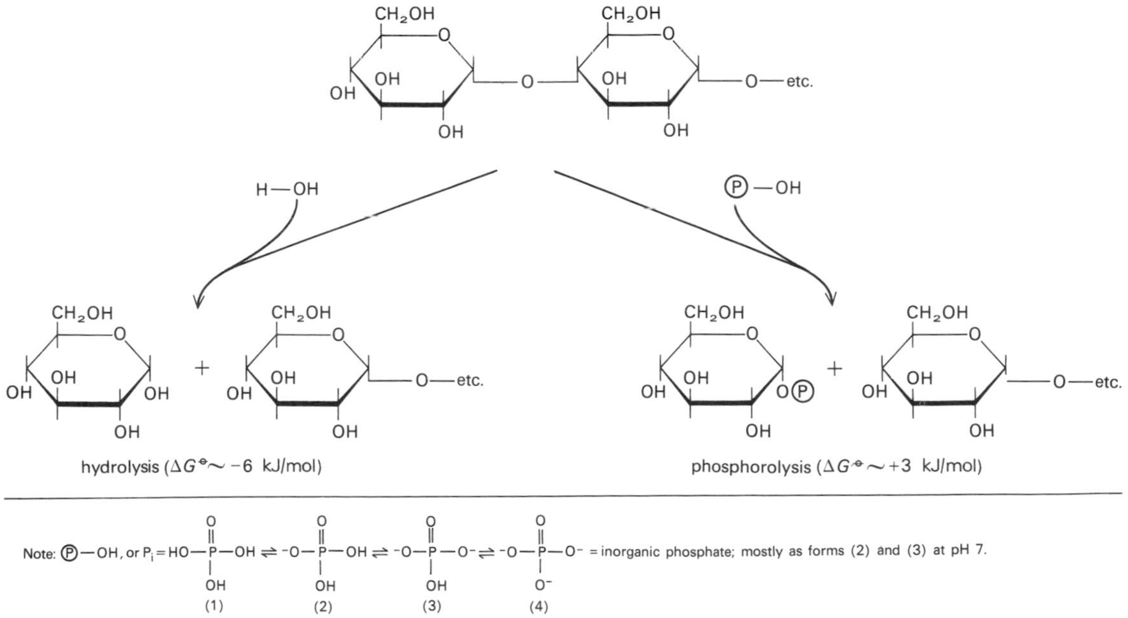

Fig. 2.3 Breakdown of glycogen

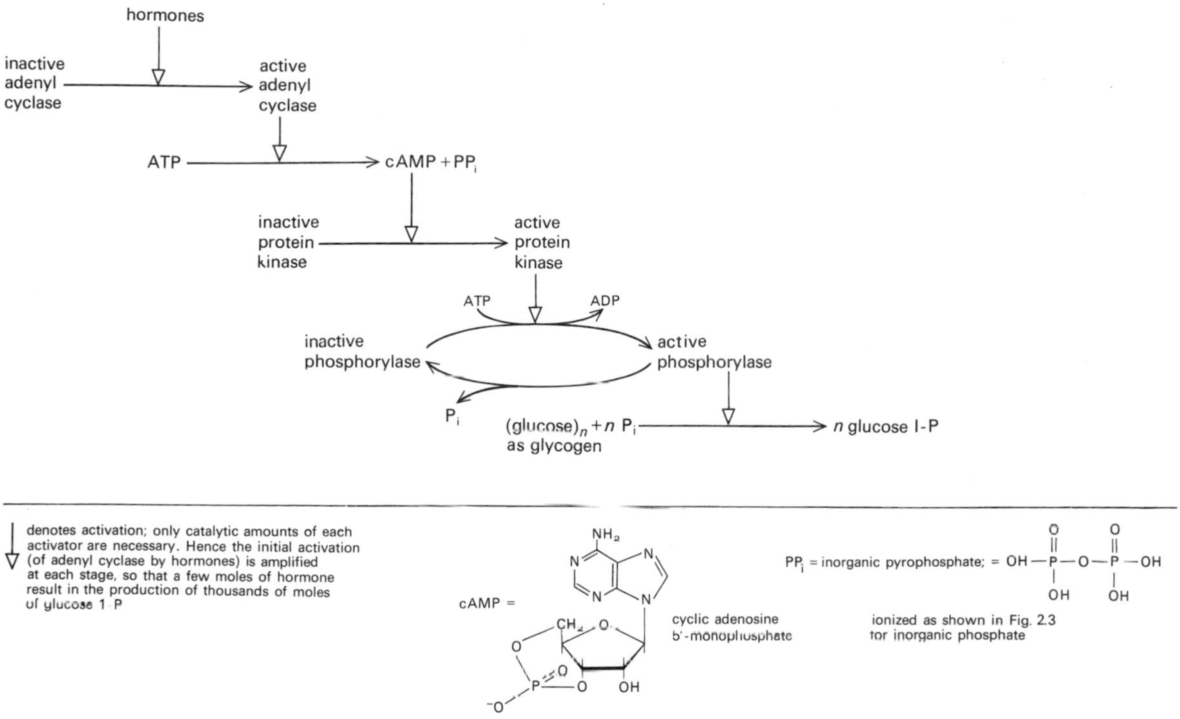

Fig. 2.4 Activation of phosphorylase

ATP, occurs by the pathway known as glycolysis (Fig. 2.5). Glucose units not derived from intracellular glycogen (that is derived from dietary carbohydrate or from glucose synthesized in the liver; section 2.3.1) enter cells as free glucose. This is phosphorylated with ATP by an enzyme called hexokinase to yield glucose 6-phosphate (Fig. 2.6). Several hereditary disorders of hexokinase deficiency are known. The reason for this diversity is due to the fact that hexokinase exists in different isoenzymic forms. In each case the main symptom of enzyme deficiency is anaemia; the destruction of red blood cells results from failure to maintain adequate supplies of ATP, generated by glycolysis, within the red blood cell. The reason why red blood cells are particularly affected is that in other cells alternative supplies of ATP (by degradation of intracellular glycogen, as well as by mitochondrial degradation of amino acids or fats) exist. Only in the red cell is glycolysis the sole means of generating ATP. Another glucose phosphorylating enzyme exists in liver. This is called glucokinase, because of its greater specificity (Fig. 2.6). Glucokinase is further distinguished from hexokinase by its lower affinity for glucose, and by the fact that it is inducible (section 2.6) by hormones such as insulin.

Glucose 6-phosphate is next converted to fructose 6-phosphate, which is further phosphorylated by ATP to form fructose 1,6-diphosphate (Fig. 2.6). Fructose 1,6-diphosphate is cleaved in half by an enzyme termed aldolase, the reaction being the reverse of the familiar aldol condensation of aldehydes and ketones (Fig. 2.7). Three isoenzymes exist: aldolase A is concentrated in muscle, aldolase B in liver and kidney, and aldolase C in brain and heart. The products of aldolase action (dihydroxyacetone phosphate and glyceraldehyde 3-phosphate) are interconvertible, and it is glyceraldehyde 3-phosphate (triose 3-P for short) that is next oxidized to 3-phosphoglyceric acid; at the same time **ATP** is formed. The reaction is complicated (Fig. 2.8). The important points to note are (a) that it is an oxidation (dehydrogenation) and (b) that concomitant with oxidation, inorganic phosphate becomes covalently bound so as to form a 'high-energy' bond (Fig. 3.5; p. 86), which is then used to generate ATP. 3-Phosphoglyceric acid is metabolized to 2-phosphoglyceric acid. Dehydration of this creates a 'high energy' bond in phosphoenolpyruvate, which is used to generate another molecule of ATP (Fig. 2.9). Glycolysis is completed by the reduction of **pyruvate** to **lactate** (Fig. 2.5).

Each stage of glycolysis has been associated with a specific disease, in which the enzyme necessary for that step is missing. As with hexokinase deficiency, the main symptoms are anaemia, for the reasons mentioned above.

The enzyme catalysing the conversion of pyruvate to

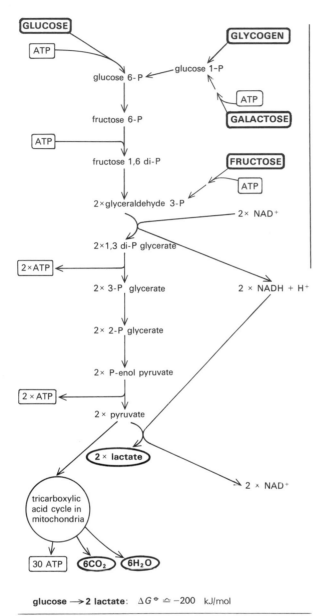

Fig. 2.5 Pathway of glycolysis

glucose $\rightarrow$ 2 lactate: $\Delta G^{\ominus} \simeq -200$ kJ/mol

ADP formed or utilized has been omitted for clarity

$$\text{glucose} + \text{ATP} \xrightarrow[\substack{(K_M \text{ for glucose} \\ \simeq 0.1 \text{ mM})}]{\text{hexokinase}} \text{glucose 6-P} + \text{ADP}; \Delta G^{\ominus} \simeq -17 \text{ kJ/mol}$$
(fructose, mannose, glucosamine) → (fructose 6-P, mannose 6-P, glucosamine 6-P)

$$\text{glucose} + \text{ATP} \xrightarrow[\substack{(K_M \text{ for glucose} \\ \simeq 10 \text{ mM})}]{\text{glucokinase}} \text{glucose 6-P} + \text{ADP}; \Delta G^{\ominus} \simeq -17 \text{ kJ/mol}$$

$$\text{galactose} + \text{ATP} \xrightarrow{\text{galactokinase}} \text{galactose 1-P} + \text{ADP}$$

$$\text{fructose} + \text{ATP} \xrightarrow{\text{fructokinase}} \text{fructose 1-P} + \text{ADP}$$

$$\text{fructose 6-P} + \text{ATP} \xrightarrow{\text{phosphofructokinase}} \text{fructose 1,6 di-P} + \text{ADP}; \Delta G^{\ominus} -14 \text{ kJ/mol}$$

Fig. 2.6 Some sugar kinases

of each form is characteristic of different cell types (Fig. 2.10). Those tissues that are capable of withstanding anaerobiosis (see below) have high amounts of LDH5; those that are not have relatively more LDH1.

During the process of glycolysis there is a net gain of 2 molecules of ATP (2 molecules consumed; 4 molecules synthesized). Since the free energy change (see section 3.3.1) is approximately -200 kJ, and the generation of two molecules of ATP is approximately equivalent to 60 kJ, the overall efficiency of glycolysis in energetic terms is approximately 30 per cent. If glycogen rather than free glucose is used, 3 molecules of ATP are formed per sugar residue, bringing the

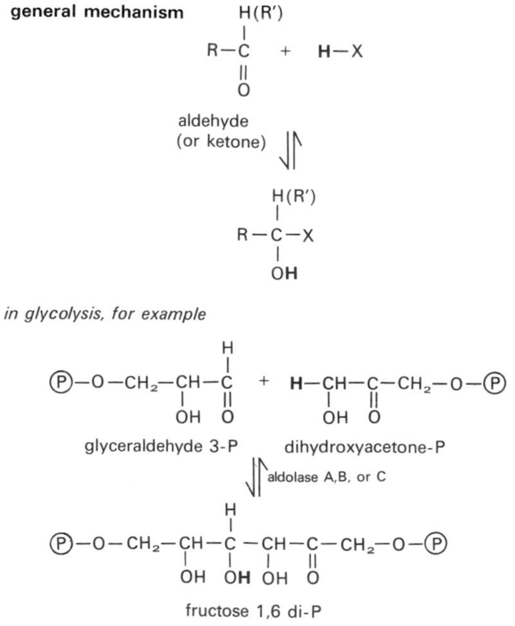

Fig. 2.7 Aldol condensation

lactate, lactate dehydrogenase (LDH), exists in a number of **isoenzymic** forms. As in phosphorylase, separate genes are involved in the synthesis of each of these forms. In this instance, however, the tissue distribution is less clear cut. The quaternary structure of the enzyme is made up of four subunits, each of which is derived from one of two genes, coding for an α- or a β-polypeptide chain (Fig. 2.10). The resulting enzymes are easily separated by electrophoresis and can be revealed by an enzymic 'stain'; the relative amount

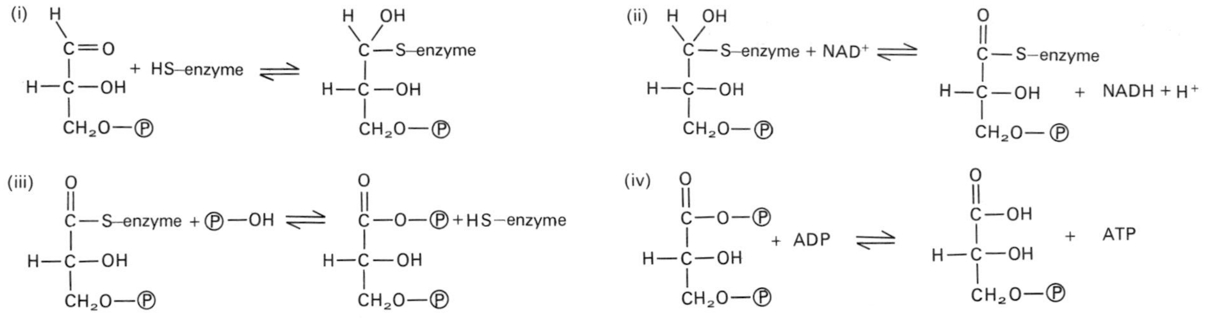

(i)

$$\begin{array}{c} H \\ | \\ C=O \\ | \\ H-C-OH \\ | \\ CH_2O-\textcircled{P} \end{array} \quad + \; HS\text{--enzyme} \; \rightleftharpoons \quad \begin{array}{c} H \quad OH \\ \backslash \; / \\ C-S\text{--enzyme} \\ | \\ H-C-OH \\ | \\ CH_2O-\textcircled{P} \end{array}$$

(ii)

$$\begin{array}{c} H \quad OH \\ \backslash \; / \\ C-S\text{--enzyme} \\ | \\ H-C-OH \\ | \\ CH_2O-\textcircled{P} \end{array} + NAD^+ \rightleftharpoons \begin{array}{c} O \\ || \\ C-S\text{--enzyme} \\ | \\ H-C-OH \\ | \\ CH_2O-\textcircled{P} \end{array} + NADH + H^+$$

(iii)

$$\begin{array}{c} O \\ || \\ C-S\text{--enzyme} \\ | \\ H-C-OH \\ | \\ CH_2O-\textcircled{P} \end{array} + \textcircled{P}-OH \rightleftharpoons \begin{array}{c} O \\ || \\ C-O-\textcircled{P} \\ | \\ H-C-OH \\ | \\ CH_2O-\textcircled{P} \end{array} + HS\text{--enzyme}$$

(iv)

$$\begin{array}{c} O \\ || \\ C-O-\textcircled{P} \\ | \\ H-C-OH \\ | \\ CH_2O-\textcircled{P} \end{array} + ADP \rightleftharpoons \begin{array}{c} O \\ || \\ C-OH \\ | \\ H-C-OH \\ | \\ CH_2O-\textcircled{P} \end{array} + ATP$$

reactions (i)–(iii): $\Delta G^\ominus \simeq +\;6$ kJ/mol

reaction (iv): $\Delta G^\ominus \simeq -19$ kJ/mol

overall (i)–(iv): $\Delta G^\ominus \simeq -13$ kJ/mol

Fig. 2.8 Formation of 3-phosphoglycerate from glyceraldehyde 3-phosphate: generation of ATP

Reactions (i)–(iii) are catalysed by the enzyme glyceraldehyde 3-P dehydrogenase; reaction (iv) is catalysed by the enzyme P-glyceric acid phosphokinase.

HS–enzyme denotes a cysteine residue of glyceraldehyde 3-P dehydrogenase. The dehydrogenation step (ii) creates a thioester bond, which has a high energy of hydrolysis; in step (iii) this bond is converted to a mixed anhydride, which also has a high energy of hydrolysis (Fig. 3.5). In step (iv) the energy inherent in the mixed anhydride bond is utilized to form a pyrophosphate bond, i.e. ATP. Each step is reversible.

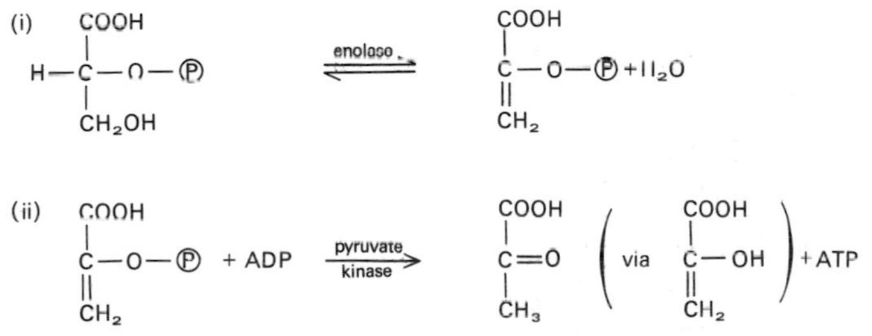

(i)

$$\begin{array}{c} COOH \\ | \\ H-C-O-\textcircled{P} \\ | \\ CH_2OH \end{array} \quad \underset{}{\overset{enolase}{\rightleftharpoons}} \quad \begin{array}{c} COOH \\ | \\ C-O-\textcircled{P} \\ || \\ CH_2 \end{array} + H_2O$$

(ii)

$$\begin{array}{c} COOH \\ | \\ C-O-\textcircled{P} \\ || \\ CH_2 \end{array} + ADP \quad \overset{pyruvate\; kinase}{\longrightarrow} \quad \begin{array}{c} COOH \\ | \\ C=O \\ | \\ CH_3 \end{array} \left(via \begin{array}{c} COOH \\ | \\ C-OH \\ || \\ CH_2 \end{array} \right) + ATP$$

reaction (i) $\Delta G^\ominus \simeq +\;2$ kJ/mol

reaction (ii) $\Delta G^\ominus \simeq -31$ kJ/mol

overall (i)–(ii) $\Delta G^\ominus \simeq -29$ kJ/mol

Fig. 2.9 Formation of pyruvate from 2-phosphoglycerate: generation of ATP

A dehydration (step i) creates an enol phosphate bond, which has a high energy of hydrolysis (Fig. 3.5). In step (ii) the energy inherent in the enol phosphate bond is utilized to form a pyrophosphate bond, i.e. ATP; there is a further release of energy, as the enol form of pyruvic acid is converted into the keto form, making step (ii) irreversible. In fact this is the only reaction known that *synthesizes* ATP, yet is not reversible.

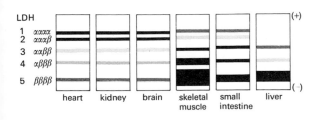

LDH

1 $\alpha\alpha\alpha\alpha$
2 $\alpha\alpha\alpha\beta$
3 $\alpha\alpha\beta\beta$
4 $\alpha\beta\beta\beta$
5 $\beta\beta\beta\beta$

heart   kidney   brain   skeletal muscle   small intestine   liver

(+) ... (−)

Fig. 2.10 Lactate dehydrogenase iso-enzymes

Electrophoresis of lactate dehydrogenase (LDH) from various (mouse) tissues (after C. L. Markert). The quaternary structure of LDH is made up of 4 polypeptide chains (α or β) as shown.

efficiency of the process up to approximately 45 per cent. This is one of the most efficient processes of generating ATP. Note that the overall sequence from glucose to lactate is neither an oxidation nor a reduction. It can be most simply described as the splitting in half of a hexose sugar, with retention of up to 30 per cent of the energy (in the form of ATP) inherent in the original molecule. It is in fact the only means of generating ATP independently of oxygen supply in animals. Hence tissues such as skeletal muscle, intestinal epithelium, liver, erythrocytes, embryonic cells, and many malignant tumours, all of which have high levels of the glycolytic enzymes and the necessary cofactors (phosphate and nicotinamide–adenine dinucleotide, NAD$^+$), are able to withstand considerable periods of oxygen deprivation, (due either to a failure of blood-flow (ischaemia) or to a failure of blood oxygenation (anoxia)); this is in contrast to tissues such as heart-muscle or brain, that are rapidly and often irreversibly damaged by a failure of the oxygen supply.

The similarity between embryonic cells and malignant tumours in having high rates of anaerobic glycolysis underlies the fact that each is able to grow in the absence of a blood-supply. The synthesis of haemoglobin and the establishment of an embryonic cardiovascular system occur many cell divisions after fertilization of the ovum has taken place. Likewise malignant cells are sometimes able to grow in areas, such as the peritoneal cavity, devoid of blood-supply; in general, though, tumours do become vascularized since they require nutrients, if not oxygen, for growth. Rapidly growing tumours resemble embryonic cells in another respect. Each type of cell has a high rate of DNA synthesis, whereas the synthesis of less essential products is depressed. The very structure of malignant cells resembles that of embryonic cells in the disposition of certain cell surface proteins. What is different is the underlying cause. In an embryonic cell, the genes which are responsible for its biochemical characteristics are under specific control, and become modulated as the embryo develops. In a malignant cell, the capacity to exercise that control is lost.

Under normal conditions, little lactic acid accumulates in healthy cells. Instead, most of the pyruvic acid is oxidized by the tricarboxylic acid cycle in mitochondria; NAD$^+$ is regenerated from its reduced form NADH in the cytoplasm, not by reduction of pyruvate to lactate, but by reduction of oxaloacetate to malate coupled to the carboxylic acid 'shuttle' between cytoplasm and mitochondria (Fig. 2.32). Under conditions of intensive muscular activity such as sprinting, on the other hand, insufficient oxygen is inhaled and skeletal muscle becomes temporarily deprived of oxygen. Glucose cannot therefore be oxidized, and is degraded to lactic acid instead. Lactic acid is secreted into the blood-stream from which it is removed by the liver. There it is eventually oxidized via pyruvate and the tricarboxylic acid cycle to CO_2 and H_2O, or converted back to glucose as indicated in Fig. 2.11.

2.2.1.3 Metabolism of disaccharides

The two most important dietary disaccharides are **sucrose** or common sugar (α-D-glucosyl-β-D-fructoside) and **lactose** or milk sugar (β-D-galactosyl-1,4-D-glucose); maltose (α-D-glucosyl-1,4-D-glucose) occurs as a partial breakdown product of starch. Each disaccharide is degraded into its component sugars by intestinal enzymes and absorbed into the blood-stream (section 8.3.2.2). The further metabolism of fructose (from sucrose) and galactose (from lactose), into products capable of entering the pathway of glycolysis, will now be considered.

In the case of **fructose**, a kinase enzyme converts the sugar to fructose 1-P (Fig. 2.6). The aldolase B of liver splits fructose 1-P to glyceraldehyde and dihydroxyacetone-P (Fig. 2.12). The latter compound then enters the glycolytic sequence; glyceraldehyde is reduced to glycerol. In a hereditary disorder known as fructose intolerance, the symptoms of which are a lack of glucose in the blood-stream and a failure to gain weight which can result in death, aldolase B is missing. Once diagnosed, the treatment is simple: replacement

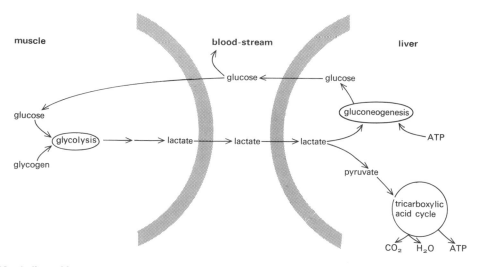

Fig. 2.11 Metabolism of lactate

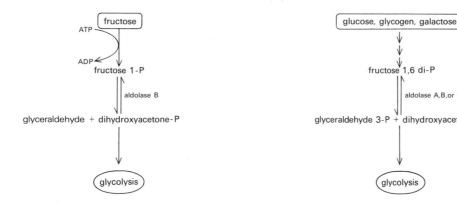

Fig. 2.12 Metabolism of fructose

Aldolase B is involved; the function of the other aldolase iso-enzymes is shown for comparison.

of the sucrose in the diet by lactose or glucose. Because aldolase A, of which a certain amount is present in liver, is normal, glycolysis through fructose 1,6-diP is unaffected by the disease.

Although only two stages are involved in the entry of fructose into the glycolytic pathway, a striking difference in the outcome of eating sucrose rather than starch has been postulated. This is that the consumption of sucrose leads to abnormal gains in weight (by conversion into fat, rather than degradation to CO_2 and H_2O), and to a greater predisposition towards atherosclerosis and resulting ailments. While it is accepted that the metabolic fate of sucrose differs from that of glucose or starch (see above), experimental results do not confirm the postulate that sucrose leads to a greater gain in weight than an equivalent amount of starch. In short, people who consume much sugar gain weight simply because their total caloric intake is greater.

The metabolism of **galactose** is more complicated (Fig. 2.13), despite the fact that it differs from glucose only in the position of the hydroxyl group at C-4. Galactose is first converted to galactose 1-P (Fig. 2.6) which is then metabolized to a nucleotide derivative, uridine 5′-pyrophosphate galactose, UDP galactose (Fig. 2.13). It is this that is the substrate for an epimerase enzyme, converting the galactose moiety to glucose. The next stage yields glucose 1-P which enters the glycolytic sequence (Fig. 2.5). In a hereditary disorder known as galactosaemia, the second enzyme is missing. As with fructose intolerance patients are generally children, since the disorder is fatal by an early age if not recognized and prevented (by substituting lactose-free diets for milk). The inability to metabolize galactose results in abnormally high concentrations of galactose and galactose 1-P in blood and tissues. These are toxic, presumably by interfering with some normal glucose-metabolizing enzymes.

Note that the ability to *synthesize* galactose-containing compounds such as glycoproteins and glycolipids, by reactions (4) and (3) followed by appropriate transfer reactions (section 2.5.1), is unimpaired.

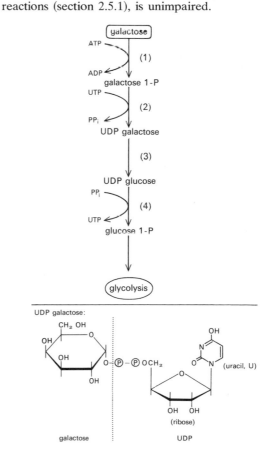

Fig. 2.13 Metabolism of galactose

Reactions 2, 3, and 4 are reversible. Note that reaction 4 is essentially the same as reaction 2 in reverse.

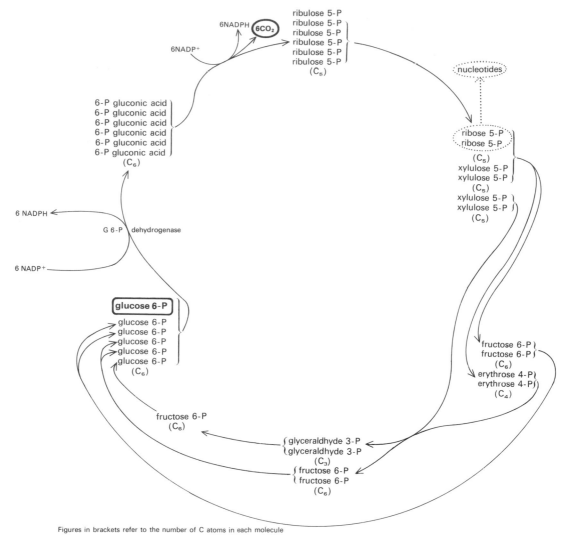

Figures in brackets refer to the number of C atoms in each molecule

Fig. 2.14 Pentose phosphate cycle

During each turn of the cycle one glucose 6-P molecule is oxidised to 6 molecules of CO_2 (and 5 molecules of glucose 6-P are regenerated concomitantly). Alternatively glucose 6-P may be oxidized only as far as ribose 5-P (and one molecule of CO_2), which is a precursor of nucleotides (Figs 2.29, 2.36, and 2.41).

2.2.1.4 Pentose phosphate cycle

An alternative pathway for the degradation of glucose is known as the **pentose phosphate cycle** or hexose monophosphate (HMP) shunt. In this process glucose is oxidized all the way to CO_2 and H_2O by enzymes located entirely in the cytoplasm of cells (Fig. 2.14). Two points to note about this rather complicated sequence are the following. First, the oxidizing power is not molecular oxygen, as in the mitochondrial electron transport chain (Fig. 3.9). Instead it is $NADP^+$, regenerated within the cytoplasm by the operation of reductive syntheses such as the formation of long-chain fatty acids and cholesterol from acetyl CoA

(section 2.3.2). Indeed it is in just those tissues that have high rates of fatty acid and cholesterol synthesis such as fat cells, liver, or mammary glands during lactation, that the HMP shunt operates to a significant degree. Secondly, the process does not need to go to completion. Operation of the first three steps of the process results in the removal of one carbon atom only and the production of ribose 5-P, which is the starting-point for the *de novo* synthesis of nucleotides (section 2.3.4). This is in fact the major mechanism for the synthesis of **ribose** (and hence of deoxyribose) in animals.

In an hereditary disorder known as glucose 6-P

dehydrogenase (G6PDH) deficiency, primaquine sensitivity, or favism, the first enzyme of the hexose monophosphate shunt appears to be absent, at least from red cells. The reason is not that the gene specifying G6PDH is missing. The gene is merely altered in such a way that the enzyme that is produced is unstable, and is degraded within cells. Catalytically, the enzyme is unaltered. In cells that are able to replace degraded enzyme by newly synthesized enzyme, the defect is not serious. In red cells, which lack the machinery for protein synthesis, the defect leads to anaemia. The reason is as follows: lack of G6PDH leads to lack of NADPH (Fig. 2.14). This in turn leads to a decreased concentration of an —SH compound called glutathione (GSH; see Fig. 7.17). GSH is required for red cell stability; in its absence the life span of red cells, and hence their number at any one time, is reduced. Another symptom is an intolerance to antimalarials such as primaquine and to the ingestion of Fava beans (hence the names for the disorder); in this case reduced levels of cytoplasmic NADPH prevent the NADPH-driven detoxication reactions of liver (section 8.4.4).

2.2.2 Fats

2.2.2.1 Phospholipids and triglycerides

Dietary carbohydrates and proteins are broken down to low molecular weight products, that is, sugars and amino acids, before being absorbed and transported by the blood-stream. Fats (mol. wt less than 1000) are absorbed and transported in virtually the same form as that in which they exist in food or animal cells. The only metabolic alteration is **hydrolysis** to yield diglyceride, monoglyceride, and free fatty acid. Of course it is also merely hydrolysis that degrades proteins to amino acids and carbohydrates to sugars. The difference is that a protein or carbohydrate contains hundreds of hydrolysable bonds (peptide bonds and glycoside bonds respectively), whereas a triglyceride or phospholipid contains only three (each an ester bond).

The most common sequence of hydrolytic reactions, in the gut as well as in cells, is shown in Fig. 2.15. Others are possible, depending on the sequence in which the ester bonds are hydrolysed. In the case of inositol-containing phospholipids, the main reaction is

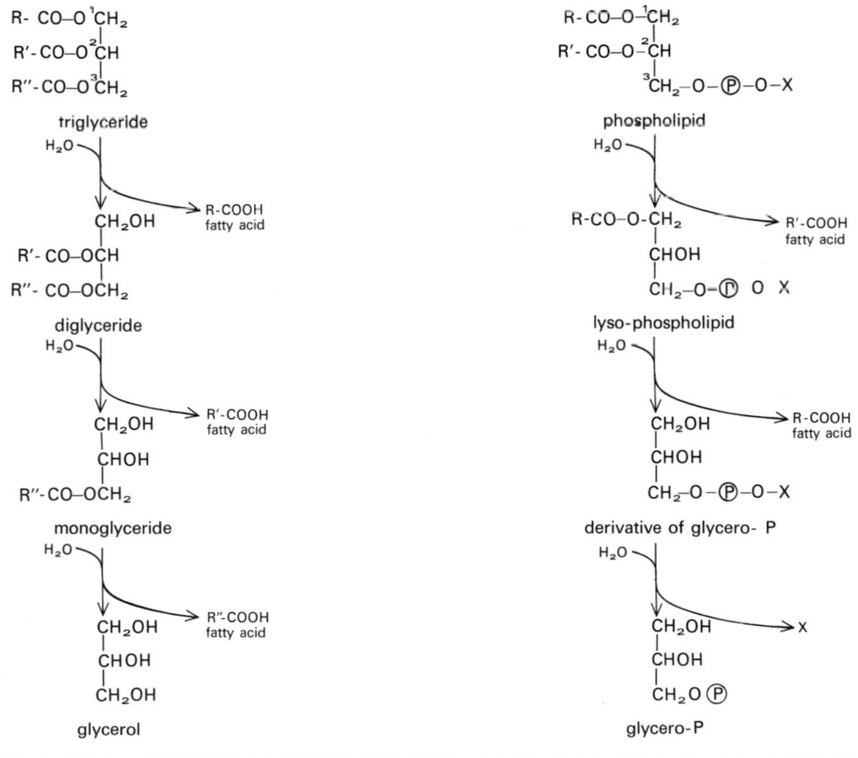

HOX represents a substituent such as choline, ethanolamine, inositol, etc.
Note that diglyceride, monoglyceride, and lyso-phospholipid are mixtures of the 1-, 2-, and 3- substituted derivatives

Fig. 2.15 Degradation of fats

hydrolysis to diglyceride and phosphoryl inositol (actually released as cyclic 2,3-phosphoryl inositol). A similar reaction degrades sphingomyelin to sphingosine and phosphoryl choline; the enzyme is missing in a hereditary disorder known as Niemann–Pick syndrome. The hydrolytic products are either resynthesized into triglyceride and phospholipid (section 2.5.2) or degraded further to CO_2 and H_2O (Fig. 2.16); degradation of fatty acids to acetyl CoA occurs within mitochondria and is described in the next chapter. No significant conversion of fat to sugar or amino acids is possible. The amount of glycerol which *is* convertible to sugar by gluconeogenesis (Fig. 2.27), or to certain amino acids such as alanine and serine, represents less than 10 per cent by weight of phospholipid or triglyceride.

2.2.2.2 Cholesterol

Cholesterol, absorbed largely from the diet but synthesized also by tissues such as liver and the gastrointestinal tract, is metabolized in a variety of ways (Fig. 2.17). Each is degradative in the sense that carbon atoms are removed by oxidation. On the other hand the pathways lead to compounds that have important functions as hormones, and may therefore equally well be considered as synthetic. Even metabolism to cholic acid and other bile acids (which are excreted) is not a fully degradative pathway, in that further oxidation of bile acids does not occur. Moreover bile acids are very effectively reabsorbed in the gut and are therefore eliminated only slowly from the circulation. Cholesterol, in fact, is the only foodstuff, and one of the few cell constituents for that matter, that is not capable of complete oxidation to CO_2 and H_2O by animals. Once absorbed into the circulation, it cannot easily be eliminated; metabolism to the bile acid 'pool' soon becomes saturated, and what cholesterol is excreted unchanged (in the bile) is largely reabsorbed in the lower intestine. Other major cell constituents that are not oxidized are the haem portion (Fig. 3.10) of haemoglobin and cytochromes, and creatine (Fig. 10.3). In contrast to cholesterol, however, each is effectively excreted without reabsorption: haem in faeces and creatine in urine. An increased concentration of cholesterol in plasma has been implicated in the aetiology of atherosclerosis, and a high intake of cholesterol-containing foods, such as animal fat, meat, or eggs, is often discouraged. The connection between cholesterol and atherosclerosis, however, is by no means proved.

2.2.2.3 Ketone bodies

In liver, the acetyl CoA that is produced by the oxidation of fatty acids is not entirely oxidized to CO_2 and H_2O. Part of it is converted into the ketone bodies, acetoacetate, and β-hydroxybutyrate (Fig. 8.17). These are secreted into the blood-stream and utilized by other tissues. That is, the process of fatty acid oxidation that is begun in liver is completed in muscle and other cell types. In addition to circulating nutrients such as glucose, fatty acids, and amino acids, then, there is a fourth type of 'nutrient' that is derived not from food, but formed only by the liver: the ketone bodies.

In certain pathological situations such as diabetes, the production of ketone bodies (ketogenesis) is increased to the point where supply exceeds demand. The plasma concentration of ketone bodies (which are acids) therefore rises, and this leads to a metabolic acidosis.

2.2.3 Proteins and amino acids

The mechanism of intracellular protein degradation is poorly understood. The enzymes (**proteases**) that are involved have not been characterized as well either as

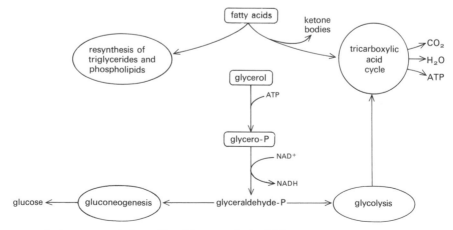

Fig. 2.16 Further metabolism of breakdown products of fats

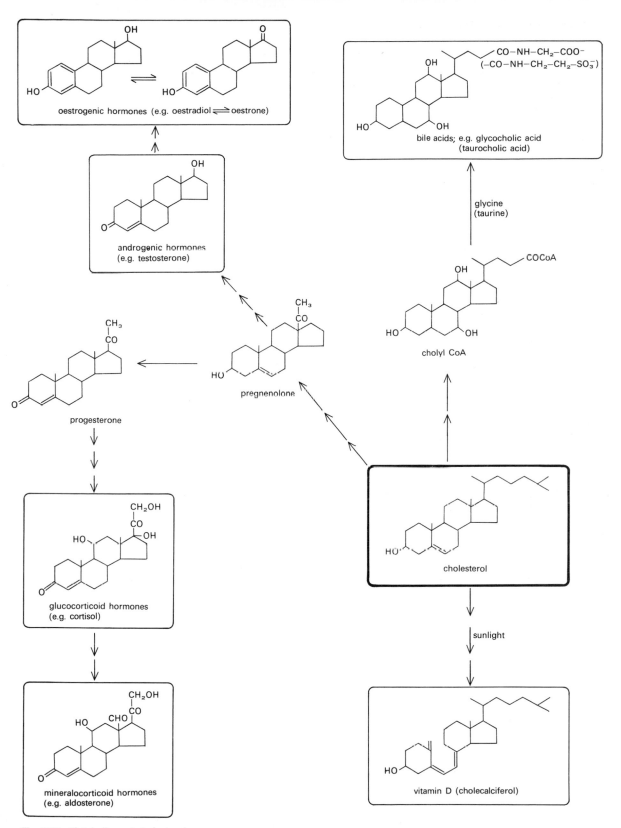

oestrogenic hormones (e.g. oestradiol ⇌ oestrone)

bile acids; e.g. glycocholic acid
(taurocholic acid)

androgenic hormones
(e.g. testosterone)

glycine
(taurine)

cholyl CoA

progesterone

pregnenolone

glucocorticoid hormones
(e.g. cortisol)

cholesterol

sunlight

mineralocorticoid hormones
(e.g. aldosterone)

vitamin D (cholecalciferol)

Fig. 2.17 Metabolism of cholesterol

phosphorylase that degrades intracellular glycogen, or as the proteases that degrade dietary proteins (Fig. 8.11). What is clear is that tissue proteins are degraded and resynthesized all the time. Compared with a dietary intake of some 50–100 g of protein per day, a 70 kg man will degrade and resynthesize ('turn over') some 200–300 g of protein a day. This figure represents an average of all proteins: some turn over very fast, others hardly at all. The rate of turnover falls to approximately half if protein is excluded from the diet. The decrease is due largely to a decrease in synthesis, rather than to an increase in degradation. If the caloric content of the diet, as well as the nitrogen content, is reduced to zero, and starvation sets in, the difference between degradation and synthesis becomes more marked. Much of the protein that is degraded under these conditions is metabolized to glucose (see below), rather than to CO_2 and H_2O. Muscle proteins and plasma proteins such as albumin are the chief sources for such degradation.

The breakdown of proteins in the body is complete; no peptides accumulate. The breakdown of the amino acids that are formed occurs as follows. The first reaction in the breakdown of amino acids, which normally occurs in liver, is removal of the amino group by a process known as **transamination**. Transamination does not in itself cause a net breakdown of amino groups. Instead an amino group is transferred from one of the common amino acids (except proline) to one of the two α-keto acids, oxaloacetate or α-ketoglutarate, yielding aspartate and glutamate respectively, plus the α-ketoacid corresponding to the original amino acid (Fig. 2.18). The enzymes have an absolute requirement for pyridoxal phosphate, which acts as an intermediate in the reaction (Fig. 2.52). The oxaloacetate and α-ketoglutarate are regenerated, so that only a catalytic amount of each is necessary.

Regeneration of oxaloacetate occurs in one of two ways. Each involves condensation between the amino group of aspartate and the keto group of another molecule (Fig. 2.18). The resulting compound (Schiff base) is split (not hydrolysed) to yield fumarate. This is then oxidized, via malate, back to oxaloacetate.

The keto compound is either citrulline or inosine monophosphate (IMP). In the case of citrulline, the product is arginine. Arginine is itself an intermediate of the urea cycle (Fig. 3.17), by which it is reconverted back to citrulline. In the case of IMP, the product is adenosine monophosphate (AMP). AMP is then deaminated to regenerate IMP; the ammonia which is liberated is metabolized to urea by way of the urea cycle (Fig. 2.18).

Regeneration of α-ketoglutarate occurs in a different manner. A specific enzyme, glutamate dehydrogenase, catalyses the oxidative deamination of gluta-

mate to yield α-ketoglutarate and free ammonia (Fig. 2.19). The ammonia then enters the urea cycle.

In short, the amino group of an amino acid is converted into the two amino groups of urea by one of two pathways: by way of free ammonia, and by way of aspartate (Fig. 2.18). Note that the urea cycle enzymes are in mitochondria, whereas transamination occurs partly in cytoplasm and partly in mitochondria.

During starvation, muscle proteins begin to break down. The amino acids that are formed are transaminated in muscle, not liver. (The 'branched-chain' amino acids (leucine, iso-leucine, and valine) are transaminated in muscle irrespective of whether the source is dietary protein or muscle protein). The mechanism of transamination is the same, except that pyruvate, not oxaloacetate or α-ketoglutarate, is the main keto group acceptor; alanine is formed (Fig. 2.20).

Alanine is not metabolized back to pyruvate in muscle. Instead, it passes into the circulation and reaches the liver. Here the ammonia is removed by the liver transaminase enzymes, and fed into the urea cycle. The pyruvate that results is then metabolized to glucose, by the pathway of gluconeogenesis (section 2.3.1), and released into the blood-stream.

The origin of the pyruvate in muscle is two-fold: from glucogenic amino acids (Fig. 2.22) and from glycogen (Fig. 2.5). What the muscle–liver 'pyruvate–alanine' cycle (Fig. 2.20) achieves is the following. First, the conversion of muscle glycogen and protein into blood glucose; muscle lacks the enzymes of gluconeogenesis, including the enzyme (glucose 6-phosphatase) necessary for the secretion of glucose into the blood-stream. Secondly, the detoxication of ammonia formed by protein breakdown (muscle lacks the urea cycle enzymes). In short, the pathway operates to produce glucose, that is, metabolic 'fuel', from non-dietary sources such as muscle glycogen and muscle protein.

L-proline (Fig. 1.2) is the only amino acid that cannot be transaminated. Its nitrogen atom is released, as ammonia, only after it has been converted by a series of enzymic steps to L-glutamate. Note that some of the ammonia that is released by transamination is used to form glutamine (Fig. 2.21). The breakdown of glutamine back to ammonia is part of the mechanism by which the pH of plasma is controlled (section 7.2.2.2).

The further metabolism of the α-keto acids formed by transamination results in their degradation to pyruvate, to acetyl CoA or to intermediates of the tricarboxylic acid cycle (α-ketoglutarate and oxaloacetate), as depicted in Fig. 2.22. Like pyruvate, the intermediates of the tricarboxylic acid cycle can enter the pathway of gluconeogenesis, as well as being broken down to CO_2 and H_2O.

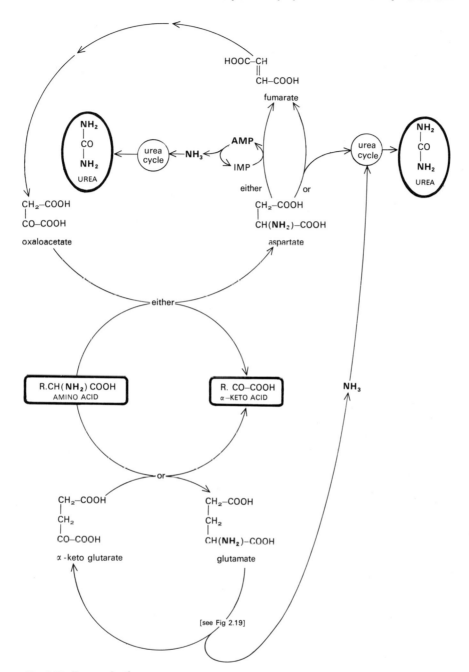

Fig. 2.18 Transamination

Since degradation of any one amino acid is not crucial to the viability of cells, several hereditary disorders in which an enzyme is missing are known. The most common disorder of amino acid catabolism is phenylketonuria (somewhere between 1 in 10 000 to 100 000 affected at birth). In this instance, the missing enzyme is actually one that converts one amino acid into another, namely phenylalanine to tyrosine, the major catabolic route of phenylalanine being via tyrosine. As a result of the deficiency, phenylalanine is degraded by an alternative pathway which results in the formation of phenylpyruvic acid and other breakdown products, which are excreted in the urine. The enzymes of this pathway are insufficient to deal with the amounts of phenylalanine present in dietary protein however, and high levels of phenylalanine and its

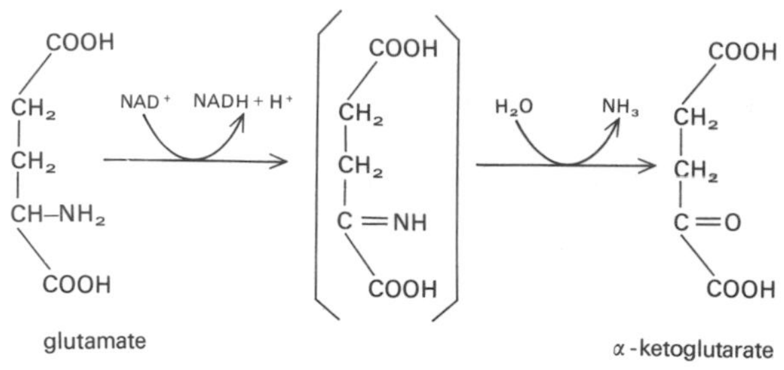

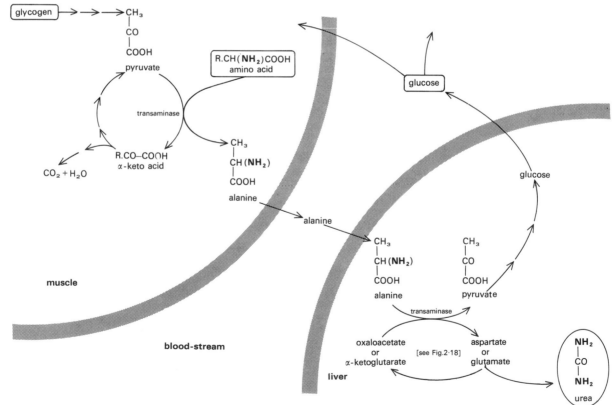

Fig. 2.19   Glutamate dehydrogenase

degradation products accumulate in blood and tissues. These are toxic, presumably by acting as antagonists of some essential metabolic process; the brain is particularly affected, the main symptoms being mental retardation and convulsions. The remedy, provided the disease is recognized early enough after birth, is theoretically as straightforward as that for fructose intolerance or galactosaemia (section 2.2.1.3): removal of phenylalanine from the diet. It is however considerably more difficult and expensive to substitute amino acid mixtures, lacking only phenylalanine, for natural protein, than it is to substitute starch or sugar for sucrose or lactose. By the age of 13–15 years, the extra phenylalanine 'load' is not so detrimental, and normal life can continue.

Specific amino acids participate in reactions *other* than transamination (leading predominantly to degradation to CO_2, H_2O, and urea) or protein synthesis. The most important are illustrated in Fig. 2.23. The synthesis of the porphyrin (haem) ring of haemoglobin and cytochromes is discussed in section 7.3.3; the synthesis of purines and pyrimidines, in section 2.3.4. Amino acids are also the precursors of certain hormones (section 9.3.2), neurotransmitters (section 12.3.2), and other pharmacologically active compounds, including the catecholamines (noradrenalin and adrenalin), histamine, serotonin, and γ-amino butyrate. Transfer of methyl groups (derived from the *S*-methyl group of methionine) is involved in the biosynthesis of adrenalin (Fig. 9.19), choline and carnitine.

2.2.4 Cellular turnover: lysosomal degradation

As indicated above, cellular constituents are in a continuous state of synthesis and degradation, or turnover.

Fig. 2.20   Pyruvate–alanine transaminases

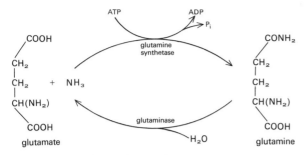

Fig. 2.21 Synthesis and degradation of glutamine

Similar enzymes convert aspartate to asparagine and vice versa.

Turnover allows the concentration of a molecule to be varied, by *temporal* separation of synthesis and degradation. The turnover of messenger RNA (Chapter 5) for example, is a means of controlling the lifetime, and hence the translational activity, of individual RNA templates. Because the polymer is degraded entirely, the constituent mononucleotides are utilized again and again for different templates. The same is true for many proteins, such as some of the transaminases and other enzymes involved in amino acid metabolism, that have turnover times of a few hours. It also applies to more stable proteins, such as haemoglobin, that serve a function for a limited period, albeit some 120–130 days, only.

Turnover also allows the disposition of molecules that are structurally fixed, to be varied by *spatial* separation of synthesis and degradation. Keratins of hair, for example, are synthesized at the follicular end and degraded at the non-growing tip. Cellular membranes, which are assembled at one site and degraded at another, fall into this category (see below). By separating degradation and resynthesis in space and time, the cell achieves a flexibility of function and repair it would not otherwise possess.

Of course, not all molecules are subject to this type of turnover. In the case of many metabolic intermediates, such as those of the tricarboxylic acid cycle, the concentration remains fairly constant apart from oscillations due to feedback control; synthesis and degradation are rather closely matched and a change in their rate leads merely to an altered 'flux' through the intermediates. And macromolecules with rather different functions, such as DNA, do not turn over at all.

The synthesis of cell constituents takes place in specific structures. Nucleic acids are synthesized in the nucleus, proteins and phospholipids on the endoplasmic reticulum, and glycogen by soluble enzymes within the cytoplasm. Where are these macromolecules degraded? In the case of soluble constituents such as cytoplasmic proteins, or cytoplasmic granules of glycogen, degradation appears to occur *in situ*. In the case of other constituents, degradation occurs largely in organelles known as **lysosomes**. Lysosomes are approximately the same size as mitochondria and in crude cell fractionation methods based on differential

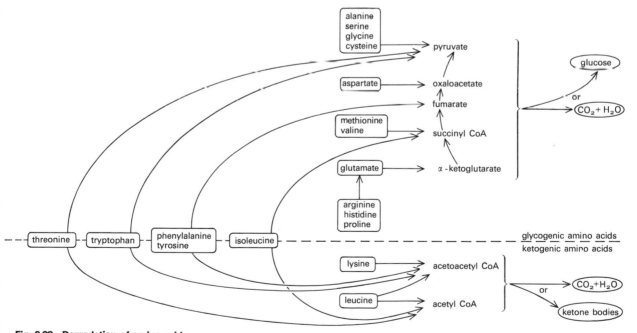

Fig. 2.22 Degradation of amino acids

All amino acids are capable of being broken down to CO_2 and H_2O. Note that some amino acids are both glycogenic and ketogenic.

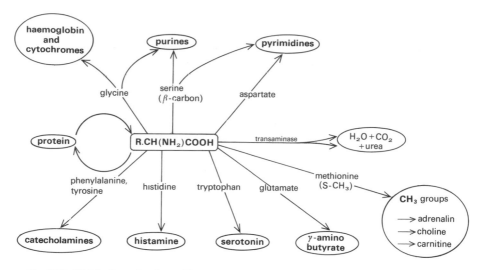

Fig. 2.23 Metabolism of amino acids

Details of the pathways are shown in subsequent figures (see Index).

centrifugation, lysosomes and mitochondria sediment together. They can be separated by further centrifugation through gradients of concentrated sucrose or other solution of high density.

The structure of lysosomes is simple: a bag-shaped membrane surrounding an aqueous solution of degradative enzymes. The pH within lysosomes is low, and most lysosomal enzymes have pH optima around 4–5. The enzymes include phosphatases, lipases, glycosidases, and other enzymes capable of hydrolysing low molecular weight compounds (such as lysozyme), and proteases and RNAases capable of

hydrolysing proteins and RNA. The manner in which cell constituents enter lysosomes is largely by membrane fusion. That is to say vesicles of plasma membrane or other membrane fuse with the lysosomal membrane such that both contents and membranes mix. The degradative enzymes then digest the contents, as well as any membrane components 'foreign' to the lysosome (Fig. 2.24). As a result lysosomes degrade not only intracellular constituents but also extracellular components such as microbes and viruses. The latter process is particularly marked in scavenging cells such as neutrophils and macrophages.

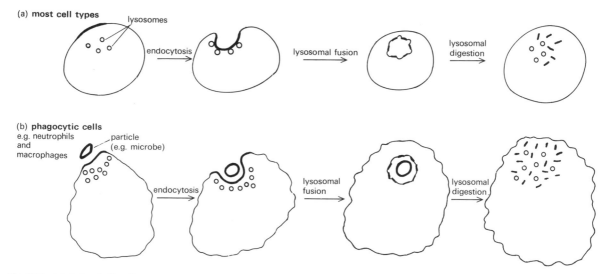

Fig. 2.24 Lysosomal digestion

(a) Digestion of cellular constituents, such as plasma membrane (see also Fig. 2.64). (b) Digestion of extracellular material, such as a microbe; note that a portion of the plasma membrane is digested at the same time.

Fig. 2.25 Some lysosomal storage diseases

Name	Type of defect	Missing enzyme	Substance accumulating	Symptoms
Tay Sachs disease	Sphingolipidosis	Hexosaminidase	Ganglioside GM$_2$: Cer-Glc-Gal$\downarrow$Gal-NAc NANA	Visual lesions; enlarged brain. Average age at death, 2–3 years.
Fabry's disease	Sphingolipidosis	Galactosidase	Trihexosyl ceramide: Cer-Glc-Gal$\downarrow$Gal	Skin lesions; corneal damage. Renal failure.
Gaucher's disease	Sphingolipidosis	Glucosidase	Glucosyl ceramide: Cer$\downarrow$Glc	Enlargement of liver and spleen. Central nervous system sometimes involved.
Niemann–Pick disease	Sphingolipidosis	Sphingomyelinase (Phospholipase C)	Sphingomyelin: Cer$\downarrow$(P)-choline	Enlargement of spleen and other organs. Central nervous system often involved
Hurler's syndrome	Mucopolysaccharidosis	Iduronidase	Dermatan sulphate: see Fig. 10.12	Skeletal deformities. Severe mental retardation.
Pompe's disease	Glycogen storage disease (Type II)	Glucosidase	Glycogen: see Fig. 2.57	Enlargement of heart; cardio-respiratory failure. Average age at death, 6 months.

Cer = ceramide, Figure 1.13. The arrow indicates the position at which the missing enzyme would act if present.

The outcome of lysosomal fusion and digestion of membranes, followed by resynthesis and reassembly by the endoplasmic reticulum, is that cellular membranes are in a continual state of flux. As might be expected, the components of membranes, that is phospholipids, glycolipids, membrane proteins, and glycoproteins, have rather similar rates of turnover. Superimposed on this flow of membrane may be excision and insertion of specific components. Phospholipids, for example, appear to be exchanged between membranes at rates exceeding membrane flow.

Whereas the degradation of 'bulk' protein, RNA, or phospholipid is a process essential for the viability of cells, this is not true of minor components such as sphingomyelin, glycolipids, and other carbohydrate-containing cell constituents. Several hereditary disorders have been described, in each of which a particular lysosomal degradative enzyme is missing. As a result, the substrate of that enzyme accumulates (Fig. 2.25); where it is a lipid, it forms droplets of fat in and around lysosomes. In most cases the outcome of such 'lysosomal storage disorders' is serious and the patient does not live beyond childhood. As in other hereditary disorders in which the defect is a missing enzyme, this applies to the homozygous state only. Because brain cells are particularly rich in sphingomyelin and glycolipids, this is where most of the tissue damage occurs, with the result that mental retardation forms one of the characteristic symptoms of the diseases listed in Fig. 2.25. In other cases (e.g. Gaucher's disease) liver and spleen are predominantly affected, presumably because the substrate of the missing enzyme is high in those tissues. In no case does the localization of damage apper to result from absence of a *cell*-specific enzyme. In fact several of the disorders can be detected in the affected foetus prior to birth, by analysis of foetal cells contained in amniotic fluid. As a result pregnancies can be terminated and much emotional and economic misery averted.

2.3 Synthesis of cellular building-blocks

2.3.1 Sugars

2.3.1.1 Gluconeogenesis

The synthesis of glucose from non-sugar precursors is a process known as **gluconeogenesis**. Like ketogenesis it occurs to a significant extent only in the liver; unlike ketogenesis, the enzymes are largely cytoplasmic. The non-sugar precursors are lactate, glycerol, or the glucogenic amino acids. In each case the precursor is a compound containing three or more carbon atoms. Compounds containing only two carbon atoms, such as acetyl CoA, cannot be converted to anything other than fatty acids, cholesterol, or ketone bodies. Compounds containing only one carbon atom, such as CO_2, are not converted in net amount to any cell constituent by animals (Fig. 2.26). Most of the steps of gluconeogenesis are the reverse of stages in glycolysis, and are catalysed by the same enzymes (Fig. 2.5). Only where the glycolytic reaction is irreversible is an alternative route followed (Fig. 2.27). This occurs at three places.

The first is the formation of phosphoenolpyruvate from pyruvate. As indicated in Fig. 2.9, the energy is too great to allow simple reversal. Instead, the reverse

Fig. 2.26 Gluconeogenic precursors

C_3 e.g. lactate, pyruvate, glycerol, alanine, serine

C_4 e.g. oxaloacetate, malate, aspartate

C_5 e.g. α-ketoglutarate, glutamate

N.B.

C_2 (e.g. acetyl CoA)→fatty acids, cholesterol, and ketone bodies only

C_1 (e.g. CO_2)→no net synthesis of any constituents (contributes to synthesis of urea, pyrimidines, and purines; acts *catalytically* in synthesis of fatty acids, gluconeogenesis, etc.).

reaction is split into two parts, each requiring ATP. The first is a CO_2-fixation reaction leading to ox-aloacetate; ATP is required for this biotin-catalysed formation of a C–C bond (Fig. 2.53). The second reaction is a CO_2-releasing reaction, coupled to a phosphorylation. The nucleotide involved is guanosine 5′-triphosphate, GTP, rather than ATP. Both reactions occur largely within mitochondria. The overall sequence from pyruvate to phosphoenolpyruvate therefore involves passage of intermediates into and out of mitochondria. The means whereby this is

achieved in the case of oxaloacetate is outlined in Fig. 2.32.

The other two places where gluconeogenesis is not the reverse of glycolysis are two other reactions involving ATP. In each case reversal is achieved by a simple phosphatase (Fig. 2.27). Thus it is seen that the energy needed to reverse the exergonic reaction of glucose to lactate, is provided by the hydrolysis of 6 molecules of ATP: 2 for each molecule of pyruvate converted into phosphoenolpyruvate (that is, a total of 4) and 1 for the formation of each molecule of 1,3-diphosphoglyceric acid (that is, a total of 2). Six molecules of ATP are just about equivalent to the overall energy change between lactate and glucose, so that the sequence is virtually 100 per cent efficient in terms of ATP (Fig. 2.28). It should be noted that when pyruvate or an amino acid is the gluconeogenic precursor, reducing power, in the form of cytoplasmic NADH, is required in addition to ATP.

The reader may wonder how the two opposing reactions of glycolysis and gluconeogenesis, catalysed largely by the same cytoplasmic enzymes, are prevented from occurring simultaneously. The answer is that the activity of the enzymes that catalyse the reactions that are specific to glycolysis (glucose→ glucose 6-P, fructose 6-P→fructose 1,6 di-P and phosphoenolpyruvate→pyruvate) and those that are specific to gluconeogenesis (pyruvate→ oxaloacetate, oxaloacetate→phosphoenolpyruvate, fructose 1,6 di-P→fructose 6-P, and glucose 6-P→ glucose) are under separate control by cytoplasmic factors (section 2.6). These reflect the need to degrade, or to synthesize, glucose. When intracellular sugar is high, the enzymes specific to glycolysis are activated, while the enzymes specific to gluconeogenesis are inhibited. When intracellular sugar is low, the enzymes specific to glycolysis are inhibited, while the enzymes specific to gluconeogenesis are activated.

2.3.1.2 Conversion of glucose to other sugars

Apart from glucose, cell constituents contain ribose (in RNA), deoxyribose (in DNA) and galactose, mannose,

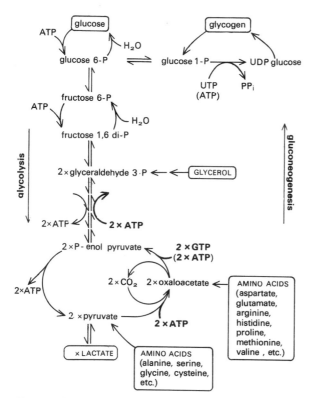

Fig. 2.27 Pathway of gluconeogenesis

The pathway of glycolysis is included for comparison.

Fig. 2.28 Energetics of glycolysis and gluconeogenesis

		$\Delta G^{\ominus}$
glycolysis	glucose	→ 2 lactate − 200 kJ
	2ADP + 2P_i → 2 ATP	+ 60 kJ
	(net change − 140 kJ)	
	efficiency in terms of ATP produced: 30%	

		$\Delta G^{\ominus}$
gluconeogenesis	2 lactate	→ glucose + 200 kJ
	6ATP → 6ADP + 6P_i	− 180 kJ
	(net change + 20 kJ)	
	efficiency in terms of ATP utilized ~100%	

fucose, xylose, glucuronic acid, *N*-acetylglucosamine, *N*-acetylgalactosamine, and *N*-acetylneuraminic acid (in glycolipids, glycoproteins and proteoglycans), as well as small amounts of other sugars. The formation of ribose, which occurs by way of the hexose-monophosphate shunt, has been mentioned; reduction of ribose to deoxyribose is discussed in section 2.3.4. The other sugars are derived from glucose 6-P or glucose 1-P by a series of reactions involving isomerization (mannose, galactose), oxidation (glucuronic acid), reduction (fucose), decarboxylation (xylose), and amination (*N*-acetylglucosamine, *N*-acetylgalactosamine, and *N*-acetylneuraminic acid). The overall pathway (which occurs in most cell types) is shown in Fig. 2.29. It will be noticed that in each case the **sugar nucleotide** (generally a derivative of uridine 5′-pyrophosphate, UDP or guanosine 5′-pyrophosphate, GDP) is the final low molecular weight product from which the sugar becomes incorporated into the respective polymer. In several cases nucleotides are also intermediates in the conversion reactions.

Galactose is synthesized by reversal of the degradative steps (Fig. 2.13). That is, glucose 1-P is first converted into the UDP derivative (reaction (4)). Isomerization of the hydroxyl groups at C-4 next occurs (reaction (3)). UDP galactose is the substrate for incorporation of galactose into oligosaccharides; in the case of lactose synthesis by mammary gland, condensation of UDP galactose with glucose leads directly to lactose.

Isomerization of glucose to mannose occurs in a different manner. Fructose 6-P is first formed and is then converted directly to mannose 6-P.

Glucose is oxidized by cells in two ways (in addition to the formation of pyruvate and subsequent oxidations). One is by oxidation of the C-1 (aldehydic) carbon atom to the corresponding carboxylic acid, known as gluconic acid. This is the first reaction of the pentose phosphate pathway (Fig. 2.14). The other is by oxidation of the C-6 (primary alcohol) carbon atom. This also results, via an intermediate aldehyde, in a carboxylic acid, known as glucuronic acid. In this case, the reducing function at C-1 remains, so that the possibility for glycoside formation is retained. In fact the formation of glucuronic acid from glucose occurs while the sugar is glycosidically linked through pyrophosphate to uridine (Fig. 2.29). UDP glucuronic acid is then incorporated into proteoglycans. It is also used in the detoxication of foreign compounds by the liver (section 8.4.4).

Reduction of the primary alcoholic group at C-6 of glucose to a methyl group (following a series of isomerizations) results in the formation of L-fucose. This unusual sugar can be considered either as a deoxyhexose, or as a methyl derivative of a pentose. The point to note is that the interconversions are carried out as the GDP derivative, which is also the form from which fucose is incorporated into glycoproteins and glycolipids.

Decarboxylation of a 6-carbon (C_6) uronic acid (UDP glucuronic acid) results in the formation of a C_5

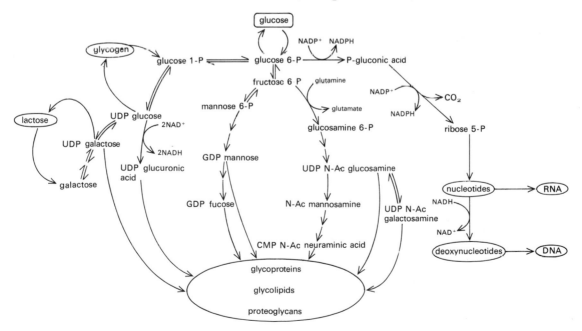

Fig. 2.29 Conversion of glucose to other sugars

sugar (xylose). Incorporation is chiefly into proteoglycans.

The amination of sugars takes place by reaction of fructose 6-P with L-glutamine. The amide group of glutamine is transferred to the C-2 carbon of fructose 6-P, resulting in the formation of glucosamine 6-P and glutamic acid. Glucosamine 6-P is the starting-point for the synthesis of all other amino sugars. The first reaction is acetylation of the amino group by acetyl CoA; almost all amino sugars occur as the N-acetylated derivative in animals. The further reactions of N-acetylglucosamine 6-P are analogous to the reactions of glucose 6-P (Fig. 2.29): isomerization to the 1-P, followed by reaction with UTP to yield the UDP derivative. An isomerase enzyme analogous to that catalysing reaction (3) of Figure 2.13 converts UDP N-acetylglucosamine to UDP N-acetylgalactosamine.

The formation of N-acetylneuraminic acid (NANA), a C_9 amino sugar, is more complex. The starting-point is N-acetylmannosamine 6-P, formed by simultaneous isomerization (at C-2) and hydrolysis (C-1) of UDP N-acetylglucosamine. This C_6 amino sugar then condenses with a C_3 acid (phosphoenolpyruvate) to yield N-acetylneuraminic acid. N-acetylneuraminic acid is converted into the cytidine 5'phosphate (CMP) derivative by reaction with cytidine 5'-triphosphate, CTP. Note that unlike the reactions leading to nucleoside diphosphate derivatives (for example Fig. 2.13 reactions (2) and (4), the formation of CMP NANA is not reversible.

2.3.2 Fatty acids and cholesterol

The formation of long-chain fatty acids and cholesterol will be considered in this section. The formation of lipids (lipogenesis) occurs chiefly in liver, adipose tissue, lactating mammary gland, and developing brain (for the formation of myelin, see section 12.2). The pathways leading to the synthesis of palmitate and cholesterol from acetyl CoA are in a sense analogous to the formation of glucose from lactate. Conversion of palmitate to other fatty acids is analogous to the conversion of glucose to other sugars. The final stage of fat synthesis, that is, the formation of phospholipids and triglycerides, is analogous to the formation of polysaccharides; each involves condensation reactions to form hydrolysable bonds (ester and

glycoside bonds respectively). These latter processes are discussed in section 2.5.

2.3.2.1 Fatty acid synthesis

The synthesis of long-chain fatty acids from acetyl CoA used to be thought to be the reverse of the degradation of fatty acids to acetyl CoA (Fig. 3.14). It is known not to be so, but to differ in several respects (Fig. 2.30). Apart from the site of the reaction sequence, and the fact that hydrogen is transferred by NADP in the cytoplasm, but by NAD^+ and FAD in mitochondria, the main difference lies in the requirement for biotin and CO_2. Whereas degradation occurs by successive removal of two carbon fragments (as acetyl CoA), synthesis occurs by successive addition of a C_3 fragment (as malonyl CoA), with subsequent loss of a C_1 fragment (as CO_2). It is the synthesis of malonyl CoA (from acetyl CoA) that requires CO_2 and biotin. The stages are illustrated in Fig. 2.31.

Before describing the reaction sequence, the origin of **cytoplasmic acetyl CoA** has to be considered. As outlined in Fig. 2.2, acetyl CoA is formed, by oxidative degradation of amino acids, sugars, and fatty acids, within mitochondria. Since the bulky and hydrophilic nature of the nucleotide portion of the CoA molecule prevents acetyl CoA from diffusing easily across biological membranes, acetyl CoA leaves mitochondria by the following mechanism (Fig. 2.32). It is in fact citrate that passes into the cytoplasm. Here citrate is cleaved to yield acetyl CoA and oxaloacetate. The reaction is not the reverse of the mitochondrial citrate-synthesizing enzyme (Fig. 3.16) which is irreversible, but a more complicated reaction requiring a molecule of ATP. The oxaloacetate so formed is returned to mitochondria as malate (formed by reduction of oxaloacetate with NADH) or as aspartate (formed from oxaloacetate by transamination with an amino acid). Within the mitochondria oxaloacetate is re-formed. The net result of the sequence of reactions outlined in Fig. 2.32 is to move acetyl CoA units from mitochondria to cytoplasm, and to return hydrogen atoms (as NADH) or amino groups to mitochondria. The processes are freely reversible and are used in several 'shuttle' reactions.

The first reaction of fatty acid synthesis is the carboxylation of acetyl CoA to yield **malonyl CoA**. As in the carboxylation of pyruvate to yield oxaloacetate,

Fig. 2.30 Differences between fatty acid synthesis and fatty acid degradation

	Synthesis (acetyl CoA → palmitoyl CoA)	Degradation (fatty acyl CoA → acetyl CoA)
Site	cytoplasm	mitochondria
Cofactors	NADPH, biotin, CO_2	NAD^+; FAD
Intermediates	protein-bound thioesters	CoA-bound thioesters

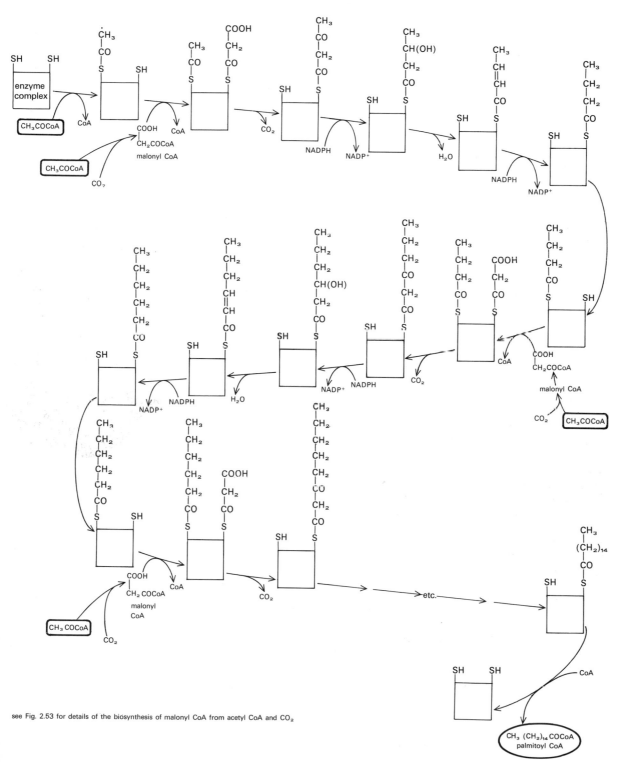

see Fig. 2.53 for details of the biosynthesis of malonyl CoA from acetyl CoA and CO_2

Fig. 2.31 Pathway of fatty acid synthesis

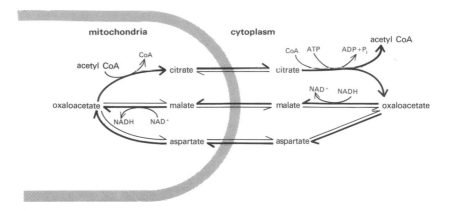

Fig. 2.32   Movement of carboxylic acids between mitochondria and cytoplasm
The predominant movements are shown in bold.

ATP and biotin are required (Fig. 2.53). The malonyl group is transferred from malonyl CoA to a cysteine residue on a protein 'carrier' that is part of the fatty acid synthetase enzyme complex; during the process, one thioester bond is broken and another is made; the free energy change is negligible. The acetyl group of a molecule of acetyl CoA is transferred by a similar mechanism to a site adjacent to the malonyl group. The two groups are now so placed that formation of a C–C bond between them — at the expense of one of the thioester bonds — is facilitated. The intermediate that is formed is unstable and decomposes spontaneously to give –S-linked acetoacetate and CO_2. The net result is that the energetically unfavourable condensation between two molecules of acetyl CoA is replaced (at the expense of ATP) by condensation via malonyl CoA (Fig. 2.33).

Successive malonyl CoA units are added and the intermediates reduced (Fig. 2.31), until palmitate (C_{16}) is reached. This is released from the enzyme complex as palmitoyl CoA. Palmitoyl CoA then either condenses with glycerophosphate to form triglyceride and phospholipids (section 2.5.2), or it enters mitochondria via carnitine (Fig. 3.14). Here acetyl CoA is added, albeit by an energetically unfavourable reaction (cf. Fig. 2.33 reaction a), to form stearyl CoA. This is either returned to the cytoplasm via carnitine for the synthesis of triglyceride and phospholipid, or it is first dehydrogenated at the 9–10 position to yield oleyl CoA.

The reactions just described account for the synthesis of the three major long-chain fatty acids (palmitate, stearate, and oleate) found in triglyceride and phospholipids. But it must be emphasized that triglycerides and phospholipids (with the exception of the pulmonary surfactant dipalmitoylphosphatidylcholine) contain in addition lesser amounts of other fatty acids, some longer than stearate, some shorter than palmitate, including the essential (that is required in the diet) highly unsaturated fatty acids linoleic, linolenic, and arachidonic acids. The extent to which the minor fatty acids are synthesized in cytoplasm or mitochondria is not clear. What does seem to be established is that the reason why palmitate is the major product of the cytoplasmic fatty acid synthesizing system is that its affinity for the enzyme complex is less than that of the preceding C_{14} acid, so that it 'drops off' and forms palmitoyl CoA rather than condensing with a further molecule of malonyl CoA. In other words the spectrum of fatty acids synthesized within the cytoplasm (e.g. see Fig. 2.63) is largely a measure of their relative affinity (that is, K_M) for the enzyme complex.

The amount of ATP required to synthesize a long-chain fatty acid from acetyl CoA appears at first sight to be rather low. One molecule of ATP is required for every molecule of malonyl CoA synthesized. That is, 7 molecules of ATP are required for the synthesis of the C_{16} fatty acid palmitate (the 'end' two carbon atoms, C-16 and C-15, are derived directly from acetyl CoA; all others via malonyl CoA). However it must be remembered (i) that acetyl CoA is in itself an activated compound (its synthesis from acetate, for example, would require one molecule of ATP) and (ii) that the reducing power used for the synthesis of palmitate (16-NADPH) is in a sense a form of ATP, in that if it were oxidized by the mitochondrial electron transport

Fig. 2.33   Use of CO_2 and ATP to 'drive' fatty acid synthesis

	$\Delta G^{\ominus}$
(a) 2 acetyl CoA $\rightleftharpoons$ acetoacetyl CoA	+ve
(b) acetyl CoA + CO_2 + ATP $\rightarrow$ malonyl CoA + ADP + P_1	−ve
malonyl CoA + acetyl CoA $\rightarrow$ acetoacetyl CoA + CO_2 + CoA	−ve

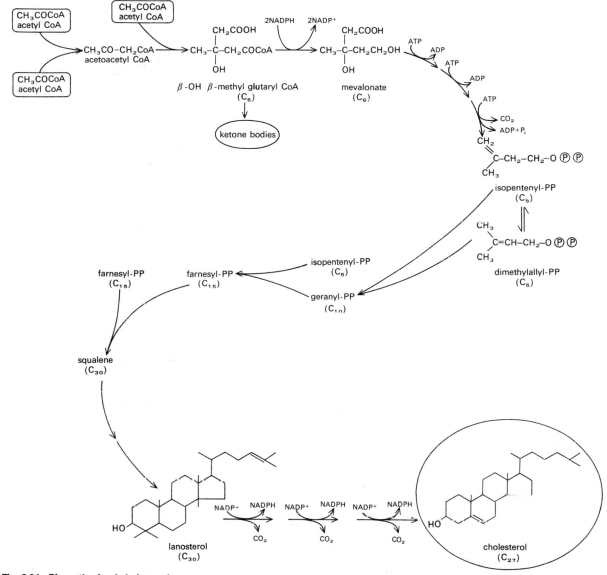

Fig. 2.34   Biosynthesis of cholesterol

chain, 48 molecules of ATP (3 per NADPH) would be generated.

2.3.2.2 Cholesterol synthesis

The biosynthesis of cholesterol is similar to that of long-chain fatty acids in that it also involves condensation and reduction of acetyl CoA units. While the reducing agent, NADPH, is also the same, the mechanism of the condensation reactions is entirely different (Fig. 2.34). In only one respect can an analogy be drawn. Just as in fatty acid synthesis a C_4 compound is synthesized via a transient C_5 intermediate, so in cholesterol biosynthesis a C_5 compound

is synthesized via a C_6 intermediate. In each case the rupture of a C–C bond with release of CO_2 'drives' the reaction sequence towards synthesis of the end-product.

In cholesterol synthesis, the reducing steps are carried out on a C_6 compound which, after decarboxylation to C_5 and activation by ATP, becomes the basic 'unit': C_5 plus C_5 then yields C_{10}; C_{10} plus C_5 yields C_{15}; and C_{15} plus C_{15} yields C_{30}. The other major difference between cholesterol and fatty acid synthesis is in the nature of the 'activating group' (Fig. 3.4). The formation of C–C bonds is a highly endergonic process. In the case of fatty acid synthesis the activating group is CoA; in the case of cholesterol synthesis it is pyrophosphate.

Other ways of synthesizing C–C bonds may be noted (Fig. 3.4). In the case of citrate synthesis in the tricarboxylic acid cycle (C_6 from C_4 plus C_2), the activating group is CoA. In the case of CO_2-fixation reactions, the activating group is ATP plus biotin. In the case of NANA synthesis (C_9 from C_6 plus C_3) activation is by means of the phosphoenol group of phosphoenolpyruvate. In the case of methylation of deoxyuridine 5'-phosphate, dUMP, a folate derivative is involved. In the case of fructose 1,6-diphosphate synthesis (C_6 from C_3 plus C_3) the reaction is exergonic.

The total amount of ATP required to synthesize a molecule of cholesterol from 18 molecules of acetyl CoA is 18 (3 for every molecule of β-hydroxy-β-methylglutaryl CoA converted to isopentenyl pyrophosphate). In addition 15 molecules of NADPH (equivalent to 45 molecules of ATP, if oxidized by the mitochondrial electron transport chain) are used. Thus a molecule of cholesterol (containing 27 carbon atoms) requires 18 molecules of ATP, and 15 molecules of NADPH. This may be compared with the requirement to synthesize 2 saturated fatty acid molecules, each of 14 carbon atoms. These would require 12 molecules of ATP and 28 molecules of NADPH. In other words, the ring structure of cholesterol requires somewhat more ATP per carbon atom than an equivalent amount of fatty acid, but since it is less saturated, it requires less NADPH.

Some of the enzymes involved in the biosynthesis of cholesterol are not truly in the cytoplasm, but are bound to the endoplasmic reticulum. The same is true of certain of the enzymes of phospholipid biosynthesis. On the other hand both pathways are more appropriately discussed in this chapter than in Chapter 5, which is best devoted to the more important function of the endoplasmic reticulum, namely the synthesis of proteins.

2.3.3 Amino acids

2.3.3.1 Non-essential amino acids

In the case of non-essential amino acids the pathways of degradation, via transamination and catabolism of the respective α-keto acids, are reversible. Under normal conditions, however, little amino acid synthesis occurs. For one thing, the spectrum of amino acids required for cellular protein synthesis is the same as the spectrum of amino acids found in dietary proteins. Provided one is eating sufficient essential amino acids, one is automatically eating sufficient non-essential ones also. And if one is not, synthesis of the non-

essential amino acids does not help to prevent the breakdown of cellular proteins. For another thing, synthesis of non-essential amino acids from glucose requires a source of amino nitrogen. And protein is the main dietary source of this. Hence the distinction between essential and non-essential amino acids is largely theoretical. On the other hand, there *is* a significant pool of circulatory amino nitrogen, largely in the form of glutamine and asparagine (Fig. 2.21); this contributes to the synthesis of those non-essential amino acids that are intermediates in the synthesis of *other* cell constituents such as nucleic acids (aspartate and glycine) and the haem group of cytochromes and haemoglobin (glycine).

2.3.3.2 Amino acid activation and modification

The synthesis of the main hydrolysable bonds of cell constituents (namely peptide bonds in proteins, glycoside bonds in carbohydrates and nucleic acids, and ester bonds in triglycerides, phospholipids, and nucleic acids) is an endergonic process, requiring some 8–16 kJ per bond. The energy is supplied through the hydrolysis of ATP (releasing approximately −30 kJ per molecule), thus 'driving' each reaction towards bond synthesis. The mechanism by which hydrolysis of ATP is coupled to the synthetic reaction varies from bond to bond (Fig. 3.4). For the synthesis of the glycoside bonds of carbohydrates and nucleic acids, the hydrolysis of ATP is used to synthesis pyrophosphate derivatives of the respective sugar molecules. For the synthesis of the phosphate ester bonds of phospholipids and nucleic acids, it is again the pyrophosphate derivatives (that is, cytidine 5'-pyrophosphate choline, CDP choline, or CDP ethanolamine and CDP diglyceride in the case of phospholipids; nucleoside triphosphate in the case of nucleic acids) that are used. For the synthesis of carboxylic ester bonds of phospholipids and triglycerides, it is the CoA derivatives of long-chain fatty acids (formed at the expense of ATP) that are used. In the case of amino acids, activation is as follows.

Amino acids first react with ATP to form an aminoacyl AMP molecule (Fig. 2.35). The enzymes catalysing this reaction are specific for each amino acid; at least 20 discrete enzymes exist. The next stage, which is catalysed by the same set of enzymes, involves the transfer of the activated amino acid to a soluble, relatively low molecular weight RNA, termed transfer or tRNA. In this case more than 20 different types of molecule exist, since several different tRNA molecules code for the same amino acid. Reactions (i) and (ii) of Fig. 2.35 probably occur without the amino acid leaving the enzyme; that is, aminoacyl AMP remains enzyme-bound. Although 1 ATP molecule is hydrolysed per amino acid activated, the subsequent formation of a peptide bond (section 5.2.2) involves two further high-energy molecules (GTP), so that the overall synthesis of a peptide bond requires three molecules of ATP.

The 20 or so **amino acid activating enzymes** that catalyse reactions (i) and (ii) of Fig. 2.35 have two interesting features. First, they possess two quite distinct and specific binding sites, one for amino acid, the other for tRNA. In this respect they resemble enzymes such as the fatty acid synthetase complex that has one site for malonyl residues and another site for residues containing 4, 6, 8, etc. carbon atoms (Fig. 2.31). Secondly, the tRNA binding site is unique, for this is the only instance in which a specific triplet of bases within an RNA molecule is recognized by a protein. While other enzymes, such as RNA polymerase and certain nucleases, recognize particular regions of RNA or DNA, none does so with the specificity of amino acid activating enzyme. The fidelity of translation of mRNA to protein depends entirely on the accuracy with which activating enzyme binds an amino acid to its appropriate tRNA.

Several amino acids become modified during, or subsequent to, the process of activation. Proline and lysine are hydroxylated, the resulting hydroxy amino acid appearing specifically in collagen and in a similar molecule that forms one of the components of complement. The hydroxylation reaction is dependent on the presence of L-ascorbic acid (vitamin C; Fig. 8.4); failure to synthesize collagen properly during scurvy (that is, vitamin C deficiency) accounts for the malfunctioning of cartilage and joints. Methylation (of lysine, histidine, and arginine), acetylation (of aspartate), and phosphorylation (of serine and threonine) all occur subsequent to polypeptide formation (**'post-translational modification'**). Phosphorylation is a particularly important process (see section 9.4.1.1), both for the activity of particular enzymes and for the activity of certain histones in controlling nuclear events.

2.3.4 Nucleotides

Nucleotides are nitrogenous base–sugar–phosphate complexes (Fig. 1.17). They are synthesized not, as might be expected, from the nitrogenous base, by way of the corresponding nucleoside (base–sugar complex); instead, the intact nucleotide is synthesized from various precursors, as described below, without ever passing through the nucleoside stage.

The biosynthesis of nucleotides is a relatively minor process in adults, since (i) DNA does not turn over and (ii) RNA is degraded predominantly only as far as nucleotides, from which fresh RNA is resynthesized. In cells of the foetus and of young children, as well as in proliferating cells of adults such as bone marrow and intestinal epithelium, however, net DNA and RNA synthesis occur. Since nucleic acid synthesis in such cells is an integral part of their viability, and hence of the organism, hereditary disorders in which one or other enzyme is missing have not been found. This is somewhat surprising, in view of the existence of 'salvage' pathways by which preformed pyrimidines and purines are used for the biosynthesis of nucleic acids. Presumably cells have evolved not to 'rely' on these pathways for the biosynthesis of DNA and RNA; in fact it is along the salvage pathway that hereditary disorders due to missing enzymes, such as gout and Lesch–Nyhan syndrome (see below) occur. It may be pointed out that the *de novo* pathway for the biosynthesis of pyrimidines and purines appears to be essential in microbes also. No naturally occurring strain has yet been isolated that lacks an enzyme of the pathway, and hence has a requirement for preformed purine or pyrimidine.

2.3.4.1 Pyrimidines

The pyrimidine ring is built up from ammonia, carbon dioxide, and aspartate (Fig. 2.36). The first reaction

(i) $R.CH(NH_2) COOH + ATP \rightleftharpoons R.CH(NH_2) CO-AMP + PP_i$
  amino acid   aminoacyl AMP

(ii) $R.CH(NH_2) CO-AMP + $ specific tRNA $\longrightarrow R.CH(NH_2) CO-tRNA + AMP$
  aminoacyl AMP   aminoacyl tRNA

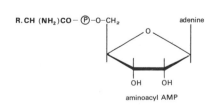

aminoacyl AMP

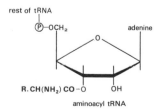

aminoacyl tRNA

Fig. 2.35 Activation of amino acids

In step (i) a mixed anhydride bond is formed at the expense of a pyrophosphate bond (see Fig. 3.5). In step (ii) an ester bond (i.e. not high energy of hydrolysis) is formed.

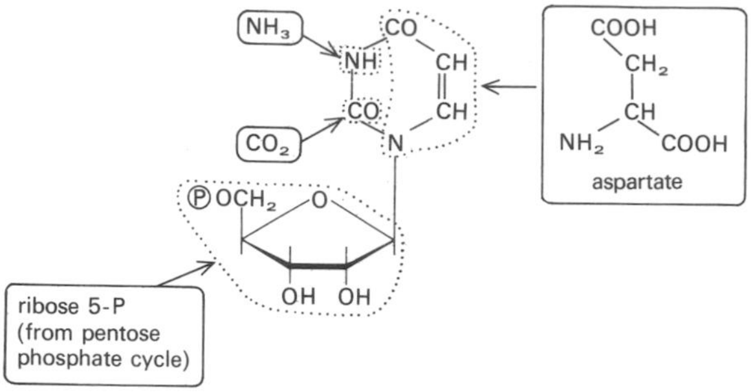

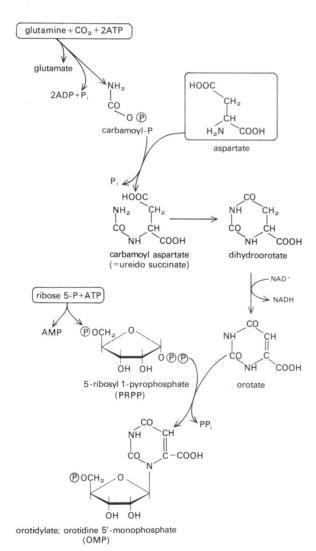

Fig. 2.36   Precursors of pyrimidine nucleotides

Fig. 2.37   Biosynthesis of pyrimidine nucleotides

involves condensation of ammonia (in the form of glutamine) with CO_2; ATP is required, as in other CO_2-fixation reactions (Fig. 3.4); a second ATP molecule is used, and the product is carbamoyl phosphate (Fig. 2.37). Carbamoyl phosphate is also the starting-point for the synthesis of urea in liver mitochondria (Fig. 3.17). The enzyme system catalysing carbamoyl phosphate synthesis in liver mitochondria is not, however, the same as the cytoplasmic enzyme. For example, in hereditary disorders which affect the mitochondrial enzyme, the cytoplasmic one, which is present in all cell types, is intact.

The next stage is condensation between carbamoyl phosphate and aspartate (Fig. 1.27), which is followed by ring closure to yield dihydro-orotate (Fig. 2.37). Dehydrogenation results in the formation of orotate, which has the aromatic pyrimidine structure. Note that isomerization (tautomerism) between keto and enol forms is possible. Such isomerization is the underlying cause of the ultraviolet absorption characteristics of all pyrimidines and purines (whether free or as DNA or RNA).

Orotate then condenses with 5-ribosyl 1-pyrophosphate (PRPP), itself formed from ribose 5-P Fig. 2.14) and ATP, to yield the nucleotide called orotidylate or **orotidine 5'-phosphate (OMP)**.

OMP is the precursor of all other pyrimidine nucleotides; it does not itself occur in nucleic acids. The reaction sequences are indicated in Fig. 2.38. It may be calculated from Figs 2.37 and 2.21 that the amount of ATP required to synthesize UMP from its precursors (ribose 5-P, CO_2, ammonia, and aspartate) is 4 molecules of ATP per molecule of UMP. The monophosphates of uridine and cytidine are converted into the corresponding triphosphates, which are the substrates for the synthesis of RNA (Fig. 4.20) and of sugar and lipid intermediates (Figs 2.29 and 2.62), by successive reactions with ATP (Fig. 2.39).

The 'salvage' pathway for the biosynthesis of pyrimidine nucleotides starts with the free pyrimidine base or its riboside (Fig. 2.40). These compounds are formed largely by breakdown of RNA and DNA in dead cells. A certain amount may be formed by turnover. Note that pathway (a) is essentially the same as the final stage in the biosynthesis of OMP (Fig. 2.37). The enzyme, however, is different. The salvage pathway for thymidine is discussed below (section 2.3.4.3).

2.3.4.2 Purines

The purine ring is built up from glycine, the amide nitrogen of glutamine, the amino nitrogen of aspartate, CO_2, and a 1-carbon unit derived from the serine (C_3) to glycine (C_2) interconversion, (Fig. 2.41). It is seen that although the 6-membered ring of a purine is a pyrimidine, the precursors of purines are completely

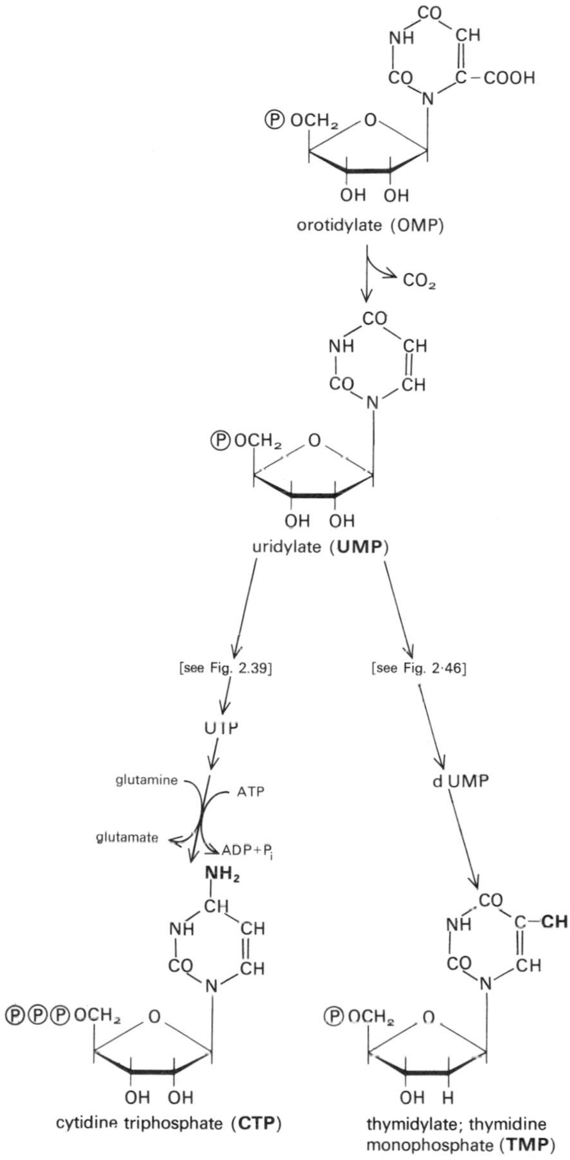

orotidylate (OMP)

CO_2

uridylate (**UMP**)

[see Fig. 2.39]   [see Fig. 2·46]

UTP

glutamine

ATP

glutamate

ADP+P$_i$

d UMP

cytidine triphosphate (**CTP**)

thymidylate; thymidine monophosphate (**TMP**)

Fig. 2.38 Interconversion of pyrimidine nucleotides

(i) UMP + ATP $\rightleftharpoons$ UDP + ADP
(CMP, AMP, GMP)   (CDP, etc.)

(ii) UDP + ATP $\rightleftharpoons$ UTP + ADP
(CDP, etc.)   (CTP, etc.)

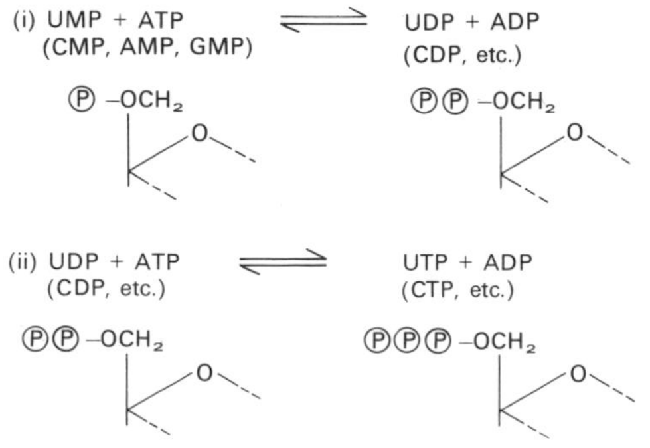

Fig. 2.39 Formation of nucleoside triphosphates

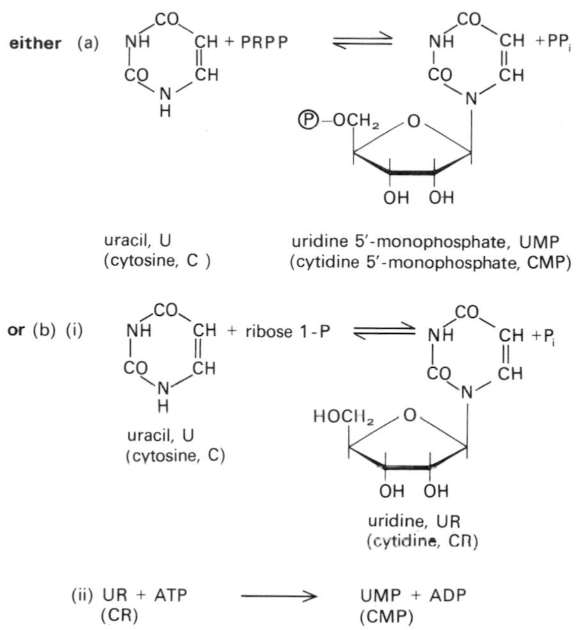

either (a)   uracil, U + PRPP $\rightleftharpoons$ uridine 5′-monophosphate, UMP + PP$_i$
(cytosine, C)   (cytidine 5′-monophosphate, CMP)

or (b) (i)   uracil, U + ribose 1-P $\rightleftharpoons$ uridine, UR + P$_i$
(cytosine, C)   (cytidine, CR)

(ii) UR + ATP $\longrightarrow$ UMP + ADP
(CR)   (CMP)

PRPP [Fig.2.37] and ribose 1–P are, like ribose 5–P, derived from the pentose phosphate cycle

Fig. 2.40 Salvage pathway for biosynthesis of pyrimidine nucleotides

different from those of pyrimidines (cf. Fig. 2.36). The biosynthetic pathways are different in another respect also. In purine biosynthesis the ring is built up on a ribose 5-P residue; that is, all intermediates are nucleotides of one sort or another. In pyrimidine biosynthesis, the formation of a nucleotide (that is, ribose 5-P derivative) is the last step of the sequence (Fig. 2.37).

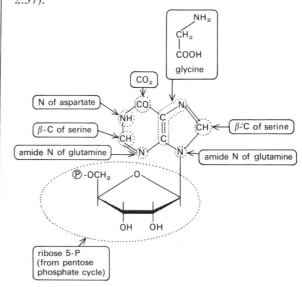

Fig. 2.41 Precursors of purine nucleotides

The sequence of reactions is indicated in Fig. 2.42. An amino group, in the form of L-glutamine, is first added to PRPP. The product reacts with glycine, followed by a folate-mediated 1-carbon transfer (Fig. 2.54). A second reaction with glutamine is followed by ring closure. CO_2 is then fixed, by a reaction *not* requiring biotin or ATP, followed by the insertion of an amino group derived from aspartate. The amino group transfer is analogous to that from aspartate in the urea cycle (Fig. 3.17). The final reaction of the sequence is another folate-catalysed 1-carbon transfer, followed by ring closure. The purine nucleotide so formed, **inosine 5′-phosphate (IMP)** (also called hypoxanthine ribotide), is, like OMP, not a constituent of nucleic acids. It is converted into adenine and guanine nucleotides by the steps indicated in Fig. 2.43. The enzyme catalysing the formation of AMP is particularly active, since it is responsible for the transamination of amino acids (Fig. 2.18) also.

The amount of ATP required to synthesize 1 molecule of IMP (4 nitrogen atoms) from its precursors is 7 molecules of ATP. This may be compared with 4 molecules of ATP required to synthesize a molecule of UMP (2 nitrogen atoms).

As with pyrimidine nucleotides, a **salvage pathway** for the biosynthesis of purine nucleotides exists. The reaction (Fig. 2.44) is analogous to reaction (a) of Fig. 2.40; reactions similar to sequence (b) of Fig. 2.40 do not appear to occur in the case of purines. The substrates for the salvage pathway are guanine or hypoxanthine. Hypoxanthine is derived from adenine by deamination; the nucleotide product (hypoxanthine ribotide or IMP) has therefore to be reaminated (Fig. 2.43) to form AMP.

The enzyme of the salvage pathway (called hypoxanthine–guanine phosphoribosyl transferase) (HGPRT) is faulty in certain cases of gout. In the disease known as the Lesch–Nyhan syndrome, the enzyme is missing. The severity of the disease (mental retardation, self-mutilation, and other symptoms) suggests an important function for the enzyme in brain metabolism.

2.3.4.3 Deoxyribonucleotides

The components of DNA differ from those of RNA in two respects. First, the sugar is 2-deoxyribose, not ribose. Secondly, the pyrimidine base thymine replaces uracil.

Deoxyribose is formed directly from ribose. The **reductase enzyme** acts not on free ribose, but on the nucleoside diphosphate (CDP, ADP, GDP) formed as indicated in Fig. 2.39. The reaction is complex and may require vitamin B_{12} as coenzyme (section 2.4.1). The resulting deoxyribonucleoside diphosphates are then converted to the respective triphosphates as indi-

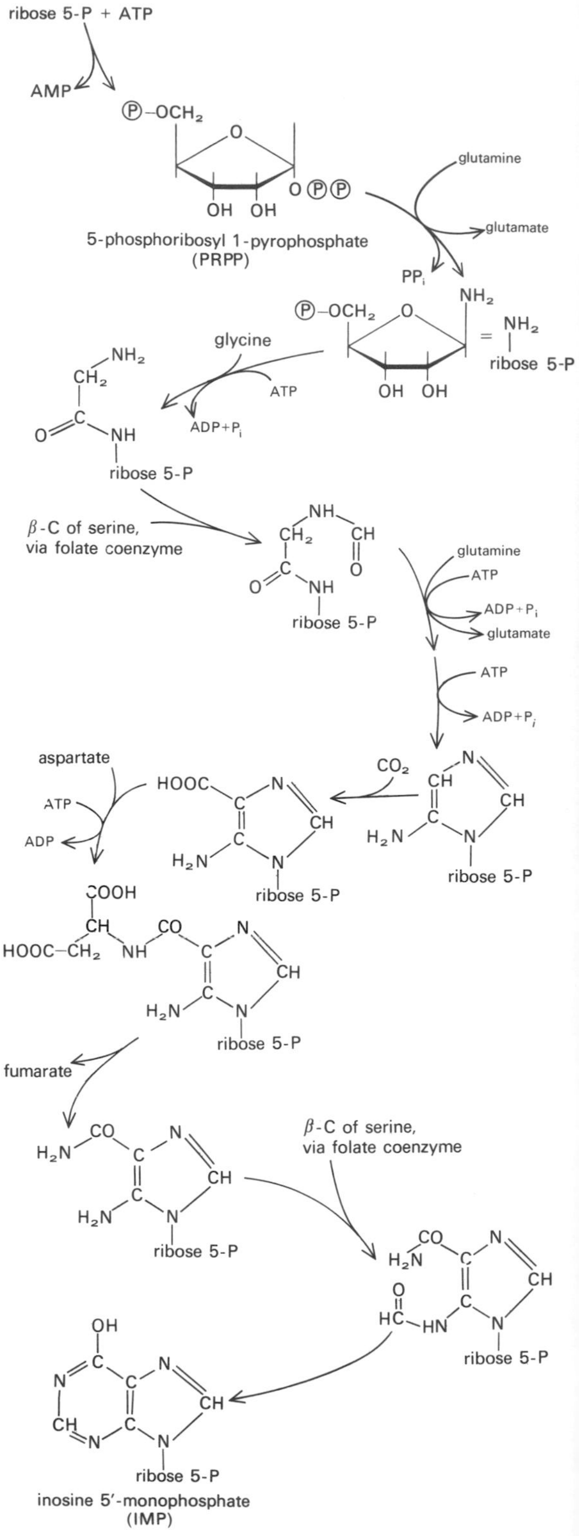

Fig. 2.42 Biosynthesis of purine nucleotides

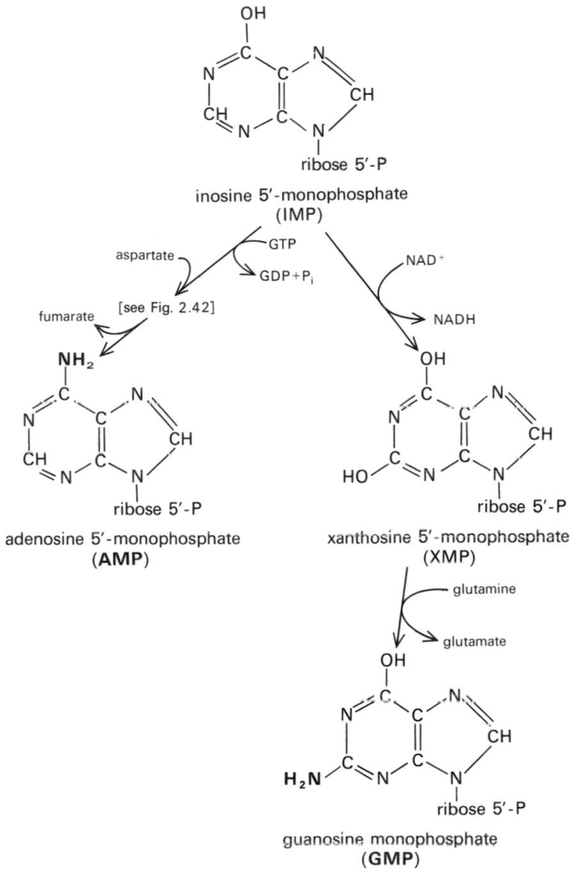

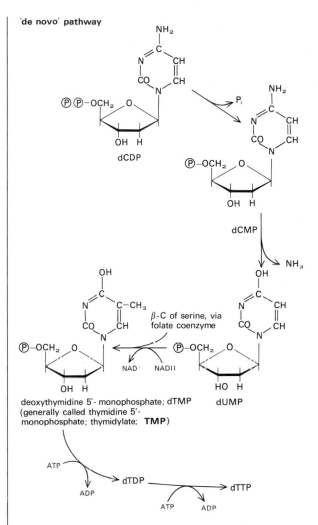

'de novo' pathway

deoxythymidine 5'- monophosphate; dTMP
(generally called thymidine 5'-
monophosphate; thymidylate; **TMP**)   dUMP

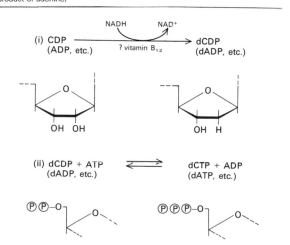

inosine 5'-monophosphate
(IMP)

adenosine 5'-monophosphate
(**AMP**)

xanthosine 5'-monophosphate
(**XMP**)

guanosine monophosphate
(**GMP**)

Fig. 2.43   Interconversion of purine nucleotides

**Fig. 2.44   Salvage pathway for biosynthesis of purine nuc-
leotides**

Guanine + PRPP ⇌ GMP + PP$_i$
(Hy)   (IMP)

(Hy = hypoxanthine, degradation
product of adenine)

(i) CDP → dCDP
(ADP, etc.)   ? vitamin B$_{12}$ (dADP, etc.)

(ii) dCDP + ATP ⇌ dCTP + ADP
(dADP, etc.)   (dATP, etc.)

Fig. 2.45   Formation of deoxynucleoside triphosphates

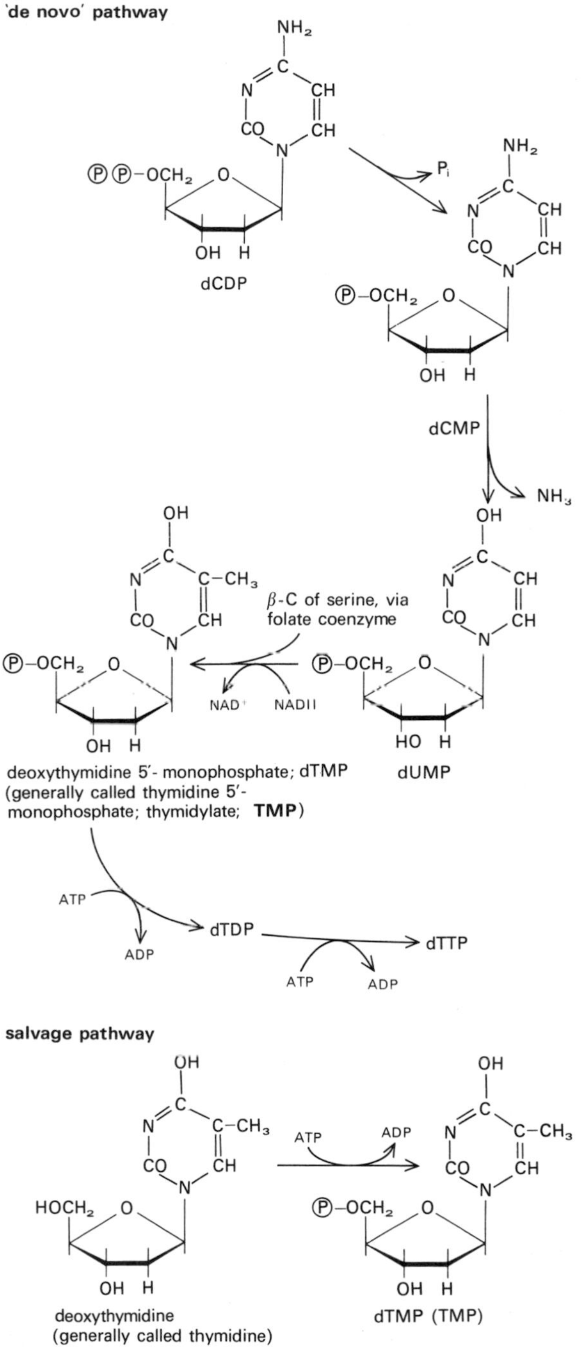

salvage pathway

deoxythymidine
(generally called thymidine)

dTMP (TMP)

Fig. 2.46   Formation of thymidine monophosphate

With thymine derivatives, it is generally assumed that the sugar is
deoxyribose, not ribose.

cated in Fig. 2.45. In the case of thymidine nucleotides the pathway is somewhat different.

The methylation of uracil to **thymine** occurs at the level of the deoxyriboside monophosphate. The reaction involves a folate-mediated 1-carbon transfer (Fig. 2.54) as in purine biosynthesis, except that a reduction occurs simultaneously with transfer (Fig. 2.46). Uracil deoxyriboside 5′-phosphate is first formed from the corresponding cytidine derivative by a deamination reaction. The thymidine deoxyriboside 5′-phosphate (usually referred to simply as thymidine monophosphate, the deoxy nature of the sugar being assumed) is metabolized to the triphosphate by reactions analogous to those of Fig. 2.39.

The **salvage pathway** for thymine nucleotides appears to be particularly important. The starting-point is thymidine (the deoxyriboside of thymine), which is phosphorylated by a specific kinase to form thymidine 5′-phosphate, TMP (Fig. 2.46). Thymidine kinase, like other salvage enzymes, is particularly high in some rapidly growing tumour cells having high rates of cell division and hence of DNA synthesis. Such cells scavenge whatever precursors are available. The enzyme is high in rapidly proliferating cells of normal tissues, such as bone marrow and intestinal epithelium, also. Indeed the enzyme is present in most cells capable of synthesizing DNA, and radioactive thymidine has come to be a most useful 'label' for newly synthesized DNA. The fact that the activity of thymidine kinase is under specific metabolic control suggests that the salvage pathway may have a biological significance greater than that of an alternative supply of biosynthetic precursors.

2.4 Synthesis and function of coenzymes

Many reactions of intermediary metabolism require the presence of a coenzyme. Coenzymes are low molecular weight compounds that participate in reactions involving several steps. The coenzyme is generally a substrate of the first step; it is resynthesized at a later step, and is therefore required in catalytic amounts only (Fig. 1.30). The same is true of metabolic intermediates, such as oxaloacetate in transamination (Fig. 2.18) or in the tricarboxylic acid cycle (Fig. 3.16). But oxaloacetate can also be degraded to CO_2 and energy, or be converted into sugar or amino acid; moreover it can be synthesized from sugar or amino acid. The coenzymes to be described in this section differ in that they do not enter any of the metabolic pathways of cells. Moreover it is a common feature of the coenzymes that each contains a component that cannot be synthesized *de novo* by animal cells — in other words one that needs to be provided, albeit in small amounts, in the diet. Such

components are the **vitamins**. The vitamins have been classified into water soluble and fat soluble. It happens to be largely the water-soluble vitamins (collectively known as the B group) that are components of coenzymes (Fig. 1.31). The fat-soluble vitamins and ascorbic acid (vitamin C) appear not to be coenzymes of general metabolic pathways; they are required only in specialized cells, such as connective tissue, bone, or retina, so that their action is more akin to that of hormones. It is not surprising, therefore, that the water-soluble vitamins, in the form of the coenzymes to be described below, play the same role in plant and microbial cells as they do in animal cells; plants and many microbes, of course, are able to synthesize the entire coenzyme molecule *de novo*. Cell-specific vitamins such as A, D, K or C, on the other hand, have no functions in plants, microbes, or the lower forms of animal life.

It might be thought that since they play a crucial role in intermediary metabolism, lack of the B vitamins would be fatal. In the strict sense this is true. However, few diets are *entirely* deficient of B vitamins. Hence death due directly to a B vitamin deficiency is rare. Instead various minor symptoms, involving the gastrointestinal tract, red blood cells, areas of skin, and the nervous system, as well as a proneness to infection, develop. Since the B vitamin-containing coenzymes play a role in all cells of the body, it is not clear why a vitamin deficiency manifests itself in one or other tissue. Presumably factors such as the relative importance of a particular metabolic pathway, the concentration of coenzyme required to maintain the pathway at sufficient rate (that is the K_M of coenzyme for its respective enzyme) and the retention of coenzyme within a cell and its elimination from the circulation, all play a part. It should be noted that while some coenzymes such as lipoic acid or biotin are covalently bound to their respective enzymes (forming a holoenzyme), others such as NAD^+, thiamin pyrophosphate, the folates or pyridoxal phosphate, are not. They are either loosely bound or are present in free solution. What is clear is that a correct intake of the various B vitamins is as important for the general wellbeing of an individual as the proper balance between his consumption of protein, fat, or carbohydrate (Fig. 8.2).

2.4.1 Oxido-reduction: nicotinamide and flavin nucleotides; vitamin B₁₂

The vitamin niacin, or nicotinic acid (B₅), is converted to the nicotinamide coenzymes by the pathway shown in Fig. 2.47. Several of the enzymes are particularly high in the nucleus of cells, where NAD^+ appears to play a role distinct from that of cytoplasmic and mitochondrial hydrogen carrier, namely one of controlling DNA transcription (section 4.2.2). The extent

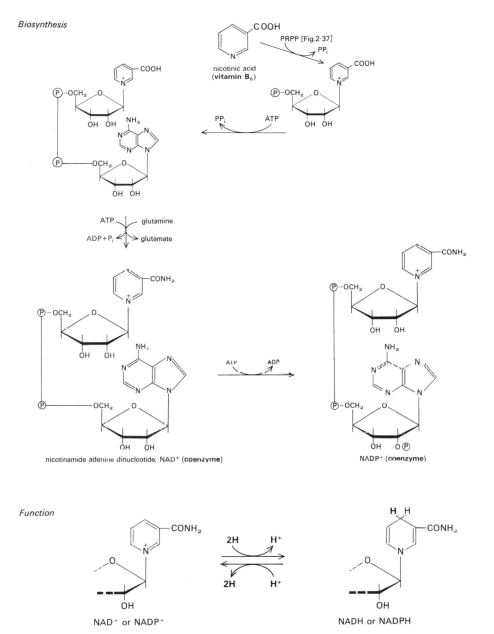

Fig. 2.47 Biosynthesis and function of nicotinamide coenzymes

to which cytoplasmic and mitochondrial NAD$^+$ is derived from nuclear NAD$^+$ is not clear. Certainly it does not readily diffuse between mitochondria and cytoplasm, or between cytoplasm and extracellular fluid. In addition to synthesis from dietary niacin, nicotinamide nucleotides are also synthesized from the amino acid tryptophan. Hence niacin deficiency (pellagra) is rarely seen except when dietary protein is insufficient, or is lacking in tryptophan.

The major function of NAD$^+$ is as hydrogen carrier

in mitochondria (section 3.4) or cytoplasm (for example, Figs 2.5 and 2.27). The main types of reaction in which nicotinamide nucleotides function are (i) **dehydrogenation** of an **alcohol** to an aldehyde or ketone, as in lactate→pyruvate, malate→oxaloacetate, or fatty acid oxidation, and (ii) **dehydrogenation** of an **aldehyde** to an acid, as in glyceraldehyde 3-P→3-P-glyceric acid (Fig. 2.48).

Mitochondrial NADH is reoxidized to NAD$^+$ with the generation of 3 molecules of ATP, and this is one

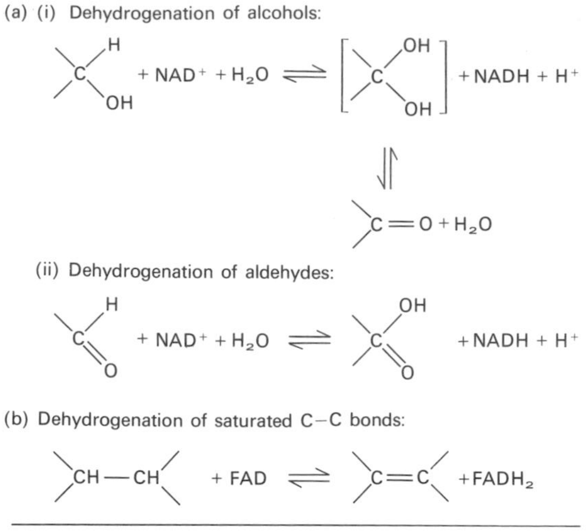

(a) (i) Dehydrogenation of alcohols:

(ii) Dehydrogenation of aldehydes:

(b) Dehydrogenation of saturated C–C bonds:

Note that reactions (a)(i) and (ii) are really the same, since each involves an oxidation of C–H to C–OH

Fig. 2.48 Reactions involving nicotinamide and flavin nucleotides

of the most important energy-yielding reactions of animal cells. Note that the pyridine ring in its reduced form has a specific ultraviolet adsorption maximum at 340 nm. Hence oxido-reduction reactions involving NAD^+ can be monitored by spectrophotometric assays of NADH formed or utilized. Although there are transhydrogenases capable of intercoverting $NADP^+$/NADPH and NAD^+/NADH, the major role of the $NADP^+$/NADPH coenzymes is in cytoplasmic reductive syntheses such as those of long-chain fatty acids and cholesterol (section 2.3.2).

The vitamin riboflavin (B_2) is converted to the coenzymes flavin mononucleotide (FMN) and flavin–adenine dinucleotide (FAD) by the pathways shown in Fig. 2.49. In this case, both hydrogen atoms in an oxido-reduction reaction become covalently attached to the coenzyme. Flavin nucleotides are in general more tightly bound to their respective enzymes than are the pyridine nucleotides. The major reaction in which flavin nucleotides participate is **dehydrogenation** of a **saturated C–C bond** to yield an unsaturated ethylenic

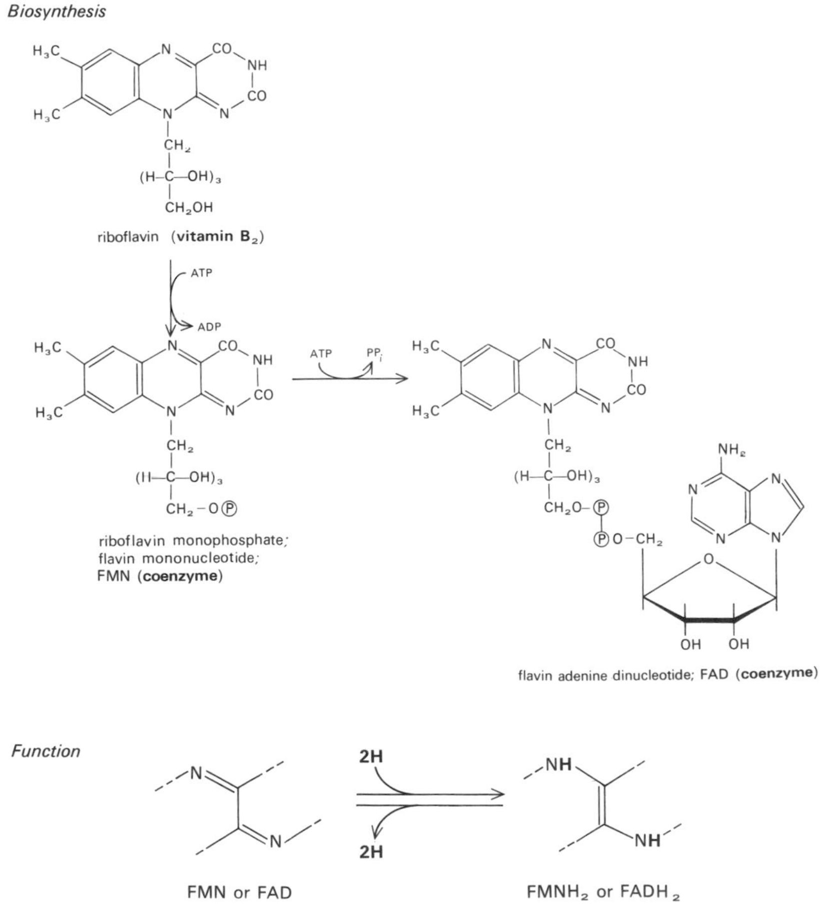

Fig. 2.49 Biosynthesis and function of flavin nucleotides

bond, as in succinate to fumarate, or in fatty acid oxidation (Fig. 2.48). Reoxidation of reduced flavin nucleotides by the mitochondrial electron transport chain results in the synthesis of 2 molecules of ATP.

Vitamin B_{12} is a complex derivative of cobalt, as shown in Fig. 2.50. Synthesis of the coenzyme involves no more than the replacement of the cyanide group by 5-deoxyadenosine. The coenzyme participates in a number of reactions, in each of which the cobalt–deoxyadenosine bond is involved. These are the biosynthesis of methionine from homocysteine and a methyl group donor, the isomerization of methyl-malonyl CoA to succinyl CoA, and, in certain species, possibly including man, the reduction of ribonuc-leotides to deoxyribonucleotides. The biosynthesis of methionine, which is an essential amino acid, is not of major importance in humans. Nor is the isomerization reaction, which is a step in the degradative sequence of the rather rare, odd-numbered fatty acids. It is the **reduction** of **nucleotides** to deoxynucleotides (Fig. 2.45) that is probably the most important vitamin B_{12}-requiring reaction in humans. The mechanism is not yet clear.

It should be pointed out that vitamin B_{12} (originally called 'extrinsic factor') requires a specific glycoprotein secreted by the stomach (called 'intrinsic factor') for its absorption in the small intestine. An apparent lack of vitamin B_{12} coenzyme function (which is manifest as pernicious anaemia) can therefore be due either to lack of dietary vitamin B_{12} or to lack of intrinsic factor in the intestine. The latter is the more common cause.

2.4.2 Oxidative decarboxylation: thiamin pyrophosphate, lipoic acid, and CoA

Three coenzymes, thiamin pyrophosphate, lipoic acid, and CoA function in the **oxidative decarboxylation** of **α-keto acids** and are conveniently discussed together. The biosynthesis of thiamin pyrophosphate and of CoA from their respective B vitamins thiamin (vita-min B_1) and pantothenic acid is shown in Fig. 2.51. The two α-keto acids that are of major significance to man are pyruvate and α-ketoglutarate. The products of oxidative decarboxylation are acetyl CoA and suc-cinyl CoA respectively. (Note that in oxaloacetate,

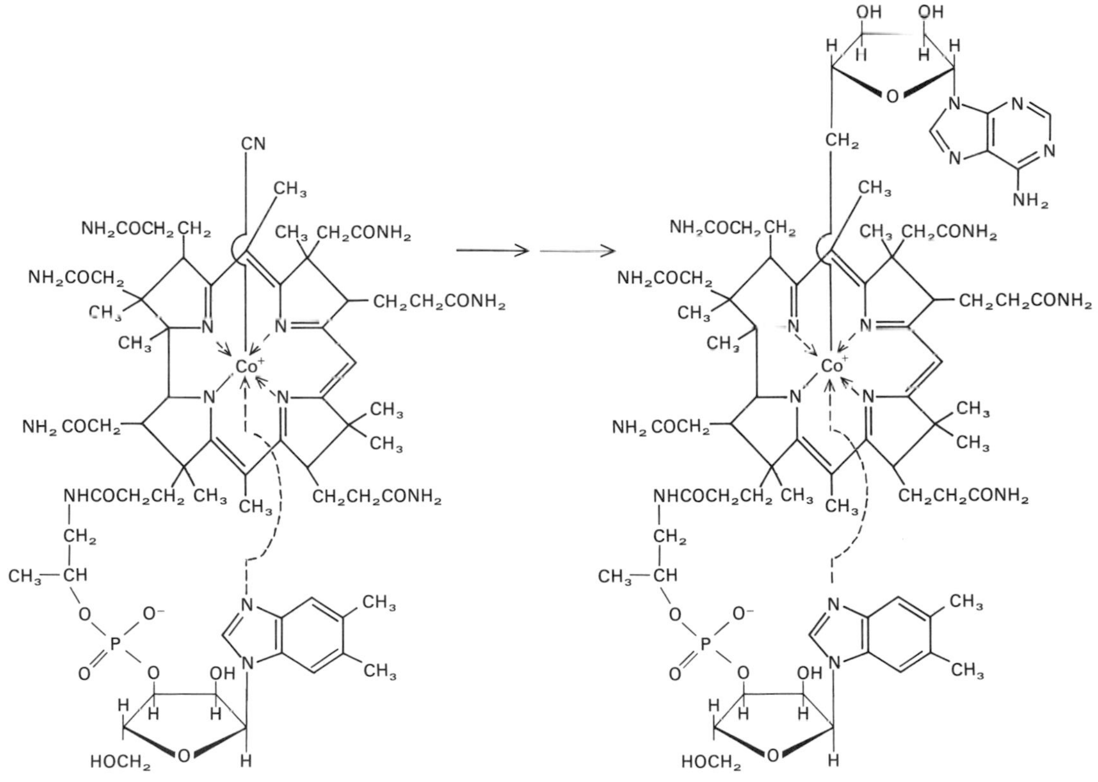

vitamin B_{12} B_{12} coenzyme

Fig. 2.50 Biosynthesis of B_{12} coenzyme

Vitamin B_{12} is generally isolated as the cyanide derivative; whether this is the predominant form in nature is not clear. Note that the covalent bond between a metal (Co) and a carbon atom (—CH_2—) in the coenzyme form is unique in biology.

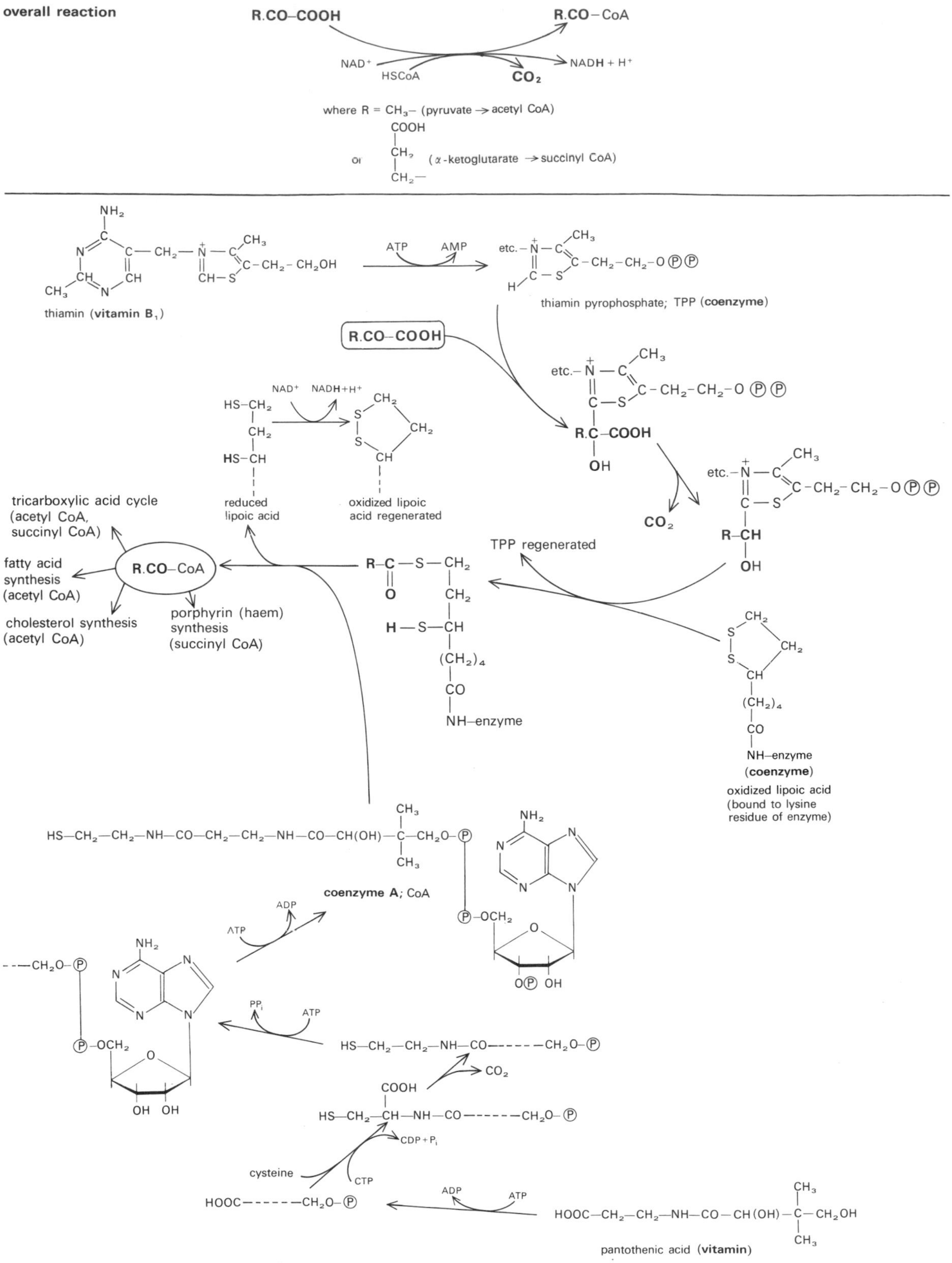

Fig. 2.51 Biosynthesis and function of thiamine pyrophosphate, lipoic acid and coenzyme A

which is both an α- and a β-ketoacid, the β-keto function predominates and oxaloacetate is decarboxylated (non-oxidatively) to pyruvate or phosphoenolpyruvate, and not to malonyl CoA).

The first stage in the oxidative decarboxylation of α-keto acids is condensation with thiamin pyrophosphate, as a result of which the terminal C–C bond is weakened and CO_2 is released. The next stage is an oxidation of the resulting aldehyde derivative to the corresponding acyl derivative, mediated by transfer to the oxidized form of lipoic acid. It should be noted that the 'vitamin status' of lipoic acid is not clear. Since it occurs covalently bound through its carboxyl group to the enzyme complex, only trace amounts are required. Note also that it is a lipid-soluble enzyme, providing an exception to the generalization that the coenzymes of intermediary metabolism are all derived from water-soluble vitamins. What is important is that by forming a thioester linkage, a high-energy bond is created (compare Fig. 2.51 with the oxidation of glyceraldehyde 3-P, depicted in Fig. 2.8). The thioester bond is retained in acetyl (or succinyl) CoA by transfer from lipoic acid to CoA. The various fates of acetyl CoA and succinyl CoA are indicated in Fig. 2.51.

CoA is also involved in the oxidative degradation of fatty acids. In this case a C–C β-keto acid bond is cleaved concomitantly with the formation of a thioester (Fig. 3.14). The degradation of fatty acids has therefore some similarity with the oxidative decarboxylation of α-keto acids, in that it is also an oxidative sequence leading to the formation of a CoA derivative; however the elimination is a β-process, not an α-process.

2.4.3 Transamination: pyridoxal phosphate

The biosynthetic sequence from pyridoxine (vitamin B_6) to pyridoxal phosphate is shown in Fig. 2.52. The major reaction in which pyridoxal phosphate participates is **transamination** (Fig. 2.18). The coenzyme condenses with an amino acid to form a Schiff base, which rearranges to another Schiff base, which is then hydrolysed to form pyridoxamine phosphate and the corresponding α-keto acid (Fig. 2.52). The coenzyme is regenerated by reversal of the reaction sequence with another α-keto acid (for example oxaloacetate or α-ketoglutarate, see Fig. 2.18).

Although transamination is quantitatively the most important reaction in which pyridoxal phosphate participates, two other reactions involving amino acids are of significance. In each case a Schiff base with the α-amino group of the amino acid is formed, as in transamination. In decarboxylation reactions the α-C–C bond is split (sequence b of Figure 2.52). In

β-elimination reactions, the β-C–C bond is split. Thus it is seen that condensation with pyridoxal phosphate has the general result of weakening an amino acid at its α-carbon atom. Whether the C–N bond, or one of the C–C bonds is then split depends in some way on the specificity of the enzyme involved. Decarboxylation reactions are important in the biosynthesis of neurotransmitters and other 'local' hormones such as the catecholamines, serotonin, and **histamine** (Figs 9.19 and 12.10). The serine$\rightarrow$glycine reaction is important in nucleotide biosynthesis, in that it generates the folate-linked 1-carbon compound that is a precursor of the purine ring (Fig. 2.42) and of the methyl group of thymine (Fig. 2.46).

In addition to these reactions pyridoxal phosphate is involved in other reactions of amino acid metabolism (such as the degradative elimination of hydrogen sulphide from cysteine), or the reaction of glycine with succinyl CoA during haem biosynthesis (Fig. 7.27). Pyridoxal phosphate also functions in some way in the glycogen phosphorylase enzyme system.

2.4.4 1-Carbon transfer: biotin and the folates

The coenzymic form of biotin (vitamin B_7) is as an enzyme-bound complex (Fig. 2.53). As with lipoic acid, only catalytic amounts are required, and a deficiency syndrome has been observed only by use of an inhibitor, namely the egg white protein avidin; this reacts with biotin and prevents its combination with the carboxylating enzyme. The reactions in which biotin participates are all **carboxylations**, requiring ATP for the synthesis of an enzyme–biotin–CO_2 intermediate (Fig. 2.53). The two most important carboxylation reactions in man are:

i. carboxylation of pyruvate to oxaloacetate, as part of the pathway of gluconeogenesis (Fig. 2.27)

ii. the carboxylation of acetyl CoA to malonyl CoA, during the synthesis of long-chain fatty acids (Fig. 2.31).

Folic acid (pteroylglutamic acid) is reduced to tetrahydrofolic acid (THF), which then accepts a 1-carbon fragment (generally derived from serine) to form an N^5, N^{10}, 1-carbon derivative (Fig. 2.54). The 1-carbon fragment is then transferred to the nitrogen atom of an acceptor molecule. Two of the most important acceptors are the two intermediates of **purine biosynthesis** (Fig. 2.42); another important reaction leads to the **formation** of **thymidine 5'-phosphate** (TMP) (Fig. 2.46). In this case the N^5, N^{10}, 1-carbon derivative undergoes an intramolecular rearrangement, such that the 1-carbon fragment is reduced (from the oxidation level of formaldehyde to that of hydroxymethyl) at the expense of an oxidation within

(a) transamination

overall reaction

$$R-\underset{\underset{NH_2}{|}}{\overset{\overset{H}{|}}{C}}-COOH + R'-\underset{\underset{O}{||}}{C}-COOH \rightleftharpoons R-\underset{\underset{O}{||}}{C}-COOH + R'-\underset{\underset{NH_2}{|}}{\overset{\overset{H}{|}}{C}}-COOH$$

Note that all stages shown below are reversible

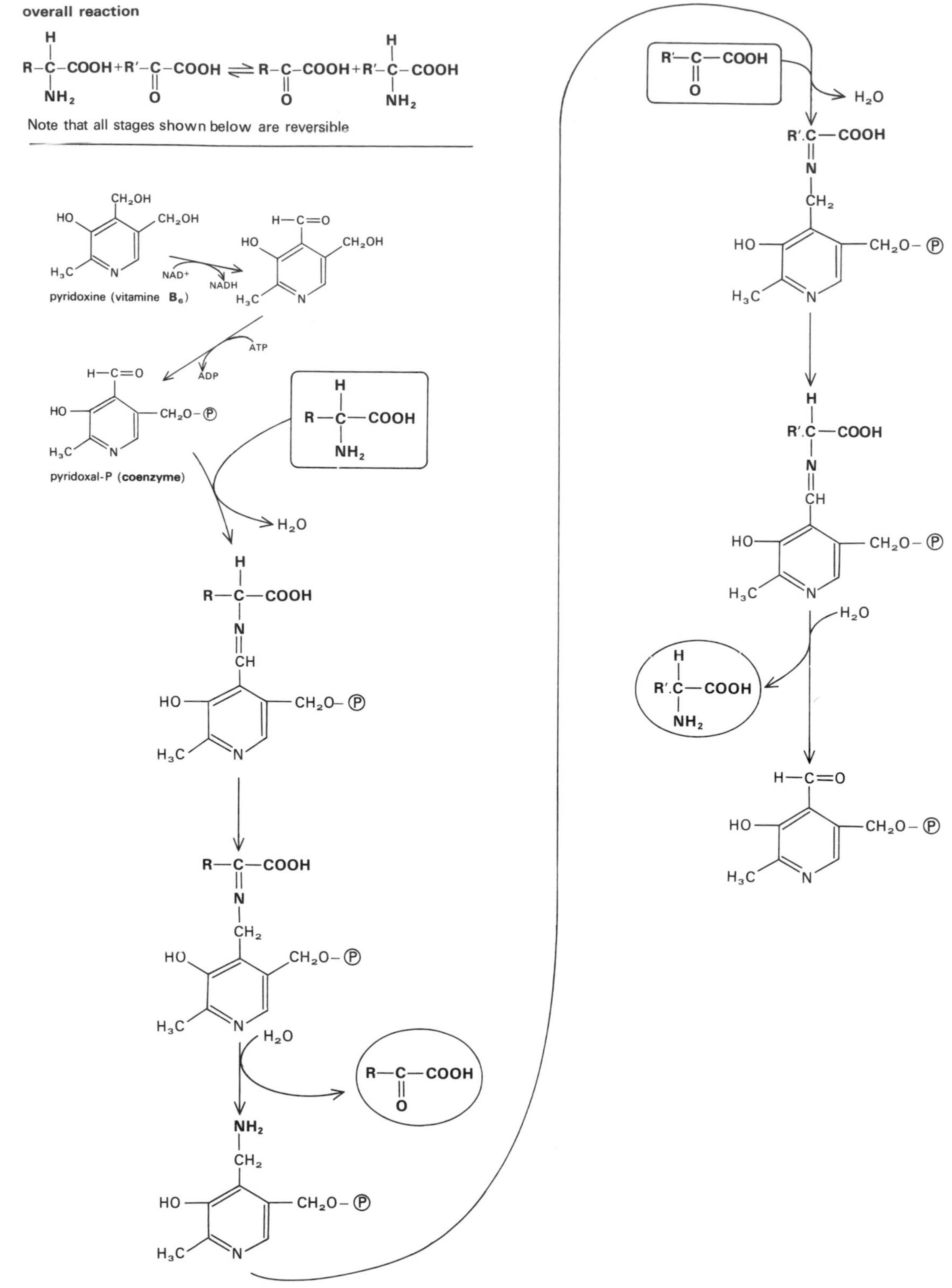

Fig. 2.52 Biosynthesis and function of pyridoxal phosphate

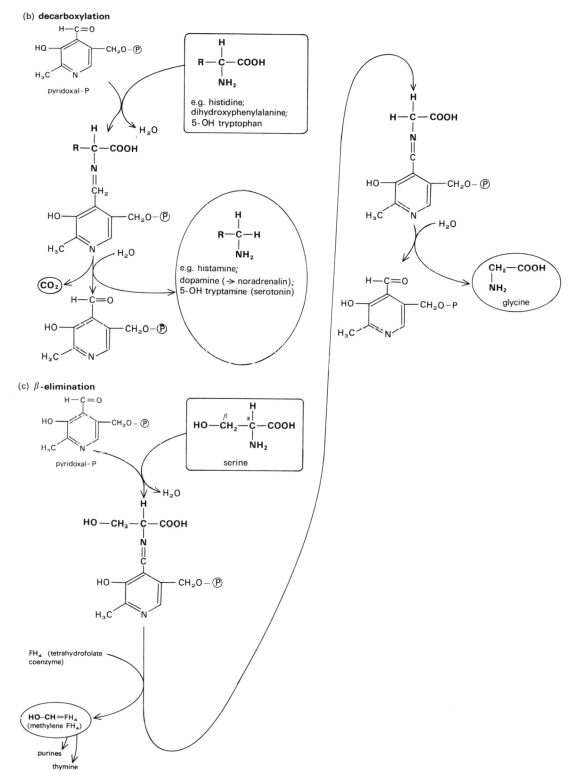

(b) **decarboxylation**

(c) **β-elimination**

Reactions (a), (b), and (c) represent the most important functions of pyridoxal-P in (animal) metabolism.

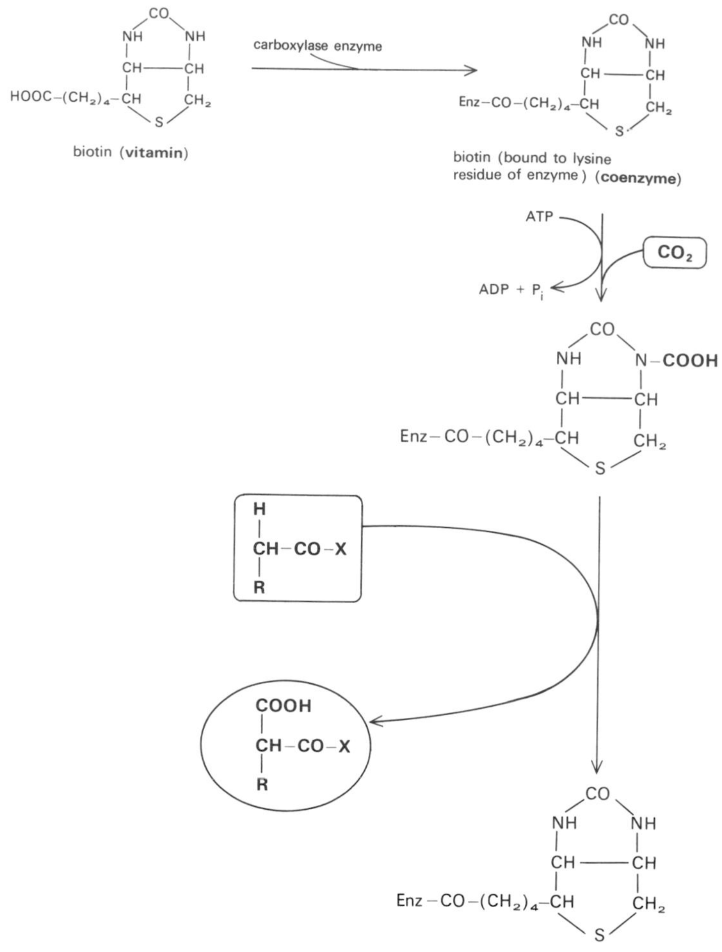

biotin (**vitamin**)

carboxylase enzyme

biotin (bound to lysine residue of enzyme) (**coenzyme**)

ATP

ADP + P$_i$

CO$_2$

reaction	R	X
acetyl CoA → malonyl CoA	H	CoA
pyruvate → oxaloacetate	H	COOH
propionyl CoA (end-product of β oxidation of odd-numbered fatty acids) → methyl malonyl CoA	CH$_3$	CoA

Fig. 2.53 Biosynthesis and function of biotin

Note that avidin, a protein of egg-white, competes with the carboxylase enzyme for binding to biotin, and thus prevents the function of biotin as a coenzyme; avidin is an 'anti-vitamin'.

the pteridine ring. The resulting dihydrofolate (DHF) is reduced back to THF by NADPH (Fig. 2.54).

Several important drugs act by interfering with the folate coenzymes (Fig. 2.55). The sulphonamides are analogues of *p*-aminobenzoic acid and antagonize the incorporation of *p*-aminobenzoic acid into the folate coenzyme. Since the enzymes for the incorporation are not present in animals (which require folate as a vitamin) the drug is relatively inert in humans. It is highly toxic to bacteria which do synthesize their own folates, and has proved to be one of the most useful bactericidal agents.

Another type of drug interferes with the *action* of the folate coenzymes. Amethopterin and aminopterin are inhibitors of the reductase step (Fig. 2.55) and therefore prevent the synthesis of THF from DHF. As a result the synthesis of purines, and particularly of TMP (Fig. 2.54), is prevented. DNA synthesis is affected more than RNA synthesis, because of the turnover and reutilization of ribonucleotides.

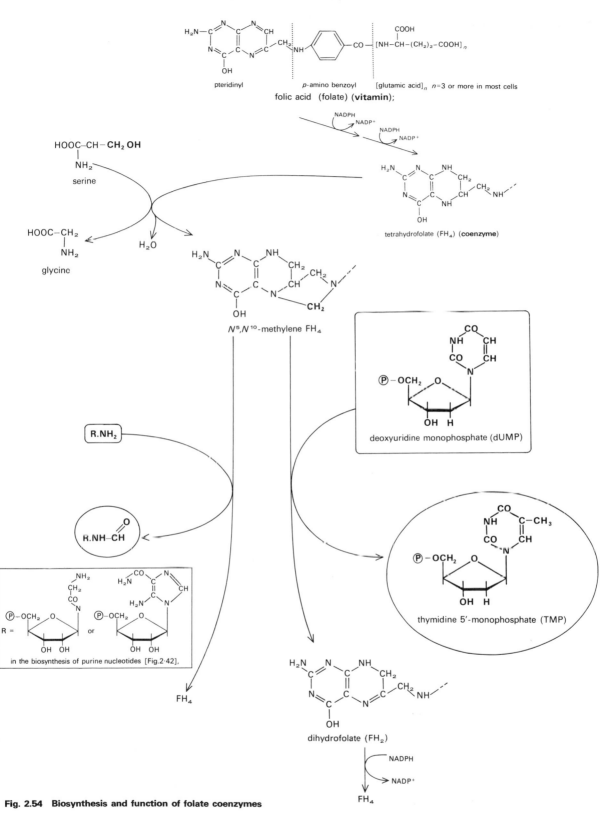

Fig. 2.54 Biosynthesis and function of folate coenzymes

During transfer of the —CH₂— group in the biosynthesis of purine nucleotides, its oxidation state remains the same (level of formaldehyde); in the biosynthesis of TMP, its oxidation state is reduced (level of 'hydroxymethyl'), at the expense of FH₄, as shown.

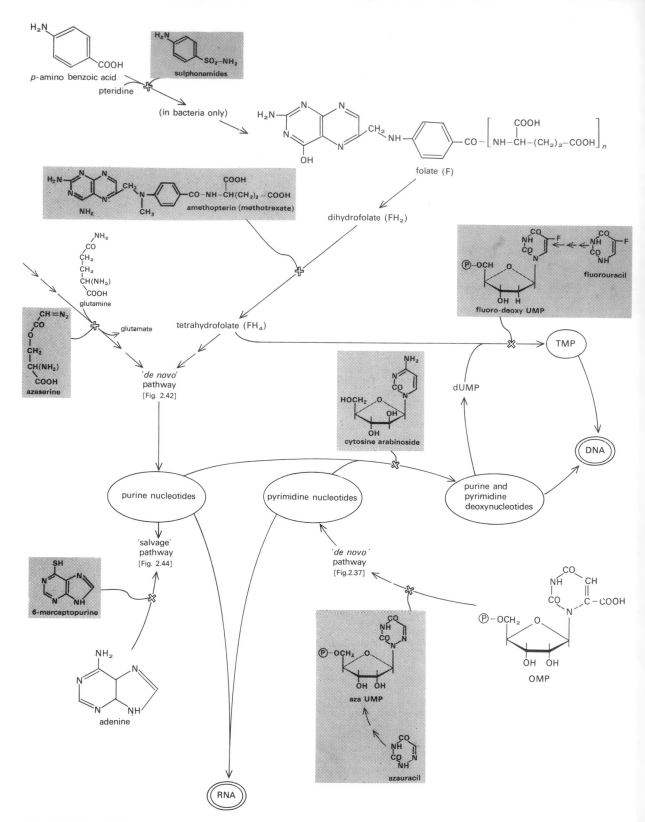

Fig. 2.55 Some inhibitors of nucleic acid synthesis

The inhibitors are shown shaded.

Amethopterin has proved to be a potent anti-cancer agent and is currently used in the treatment of several forms of leukaemia. Of course the drug inhibits all cells having high rates of DNA synthesis, such as intestinal epithelium and cells in the bone marrow, and patients receiving amethopterin suffer from side-effects at these sites. At present this drawback is an acceptable hazard in the treatment of cancer.

Another drug that has been used in the treatment of certain malignant tumours is fluorouracil. This is converted by the salvage enzymes of pyrimidine metabolism to fluorouridine 5′-phosphate and hence to fluorodeoxyuridine 5′-phosphate (FdUMP). The latter compound is a potent inhibitor of the thymidine 5′-phosphate synthetase system (Fig. 2.55).

In addition to these drugs, all of which act by interfering with the folate coenzymes, some other drugs that inhibit the synthesis of DNA and have found some use in the treatment of cancer, may be mentioned. One type are two analogues of L-glutamine, namely 6-diazo, 5-oxo L-norleucine (DON) and azaserine; note that, despite its name, azaserine is an analogue of glutamine not of serine. These compounds inhibit purine synthesis at the stage of amino group transfer from glutamine. As might be anticipated, the synthesis of amino sugar-containing compounds, which are all derived from glutamine (Fig. 2.29), is also affected. The second type of anti-cancer agents are analogues of purines, such as 6-mercapto purine. These act by inhibiting the salvage pathway, and have proved surprisingly effective.

It must be emphasized that no drug is as effective in preventing the growth of malignant tumours as are the sulphonamides in preventing the growth of bacteria. This is because the only difference between a tumour cell and a normal cell is a relatively greater dependence on DNA synthesis. Hence all anti-tumour drugs have toxic side-effects. Another point to note is that because of the high proliferative capacity of cancer cells, mutant clones that are resistant to inhibitory drugs readily develop and outgrow the sensitive parent cells. This property is, of course, exhibited by microbes also. In order to overcome the problem, combination therapy is used. In other words, a drug is given for a period of time and just as resistance develops, a different type of drug is substituted. By this means remarkable remissions in the case of acute lymphatic leukamia, for example, have been obtained.

2.5 Incorporation of building blocks into cell constituents

In this section the biosynthesis of carbohydrates and fats from their respective building blocks will be considered. The biosynthesis of nucleic acids and proteins is discussed in Chapters 4 and 5 respectively.

2.5.1 Carbohydrates

2.5.1.1 Glycogen

The synthesis of glycogen does not occur by reversal of the degradative pathway. Although the reaction catalysed by the enzyme phosphorylase (Fig. 2.4) is potentially reversible, the intracellular concentration of glucose 1-P is so low compared with that of inorganic phosphate as to make the reaction virtually irreversible. Reversal is achieved by the pathway depicted in Fig. 2.56. In other words a molecule of ATP (required for the resynthesis of UTP from UDP, Fig. 2.39) is required for every glucose unit that is incorporated.

The enzyme **glycogen synthetase** has several properties in common with glycogen phosphorylase. It too exists in two forms, which are interconverted in one direction by a cAMP-stimulated protein kinase, and in the other by a phosphatase. In this instance, however, the more active form is the unphosphorylated, not the phosphorylated, enzyme. Hence conditions leading to an increase in protein kinase, such as a hormonally stimulated increase in cAMP, results in simultaneous activation of phosphorylase (leading to glycogen breakdown) and inactivation of glycogen synthetase (leading to reduced glycogen synthesis). In other words the processes of degradation and synthesis of glycogen are kept mutually exclusive, just as the processes of glycolysis and gluconeogenesis are.

The inactive form of glycogen synthetase is not strictly speaking inactive relative to the active form. Rather it has an absolute requirement for the presence of glucose 6-P, which renders it *apparently* inactive in situations where the concentration of glucose 6-P is low. The active form is independent of the concentration of glucose 6-P. In other words, glucose 6-P is an additional controlling factor in glycogen synthesis. When the concentration of glucose 6-P is high, glycogen is synthesized, whether protein kinase is active or not. When the concentration of glucose 6-P is low, glycogen is synthesized only when protein kinase is inactive.

The 1:6 branches of glycogen are formed by a different enzyme, called branching enzyme. Like glycogen synthetase, the enzyme utilizes UDP glucose, not glucose 1-P. In a hereditary disorder known as Andersen's disease or type IV glycogen storage disease, the branching enzyme is absent. The disease affects muscle and liver equally, suggesting that separate enzymes (unlike the case of phosphorylase) do not exist. Symptoms of this kind of glycogen storage disease (general weakness, muscle failure) are similar to those of phosphorylase deficiency. The underlying reason, of course, is the opposite: failure to synthesize, not to degrade, glycogen. Other types of glycogen

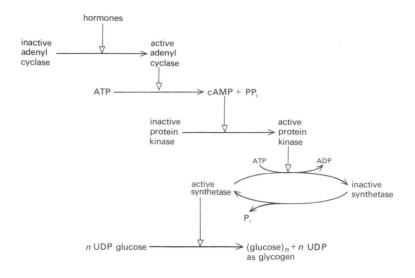

Fig. 2.56 Synthesis of glycogen. Note similarity to Fig. 2.4

storage disease are summarized in Fig. 2.57. It may be noted that type I (von Gierke's disease) is an indirect glycogen storage disease, in so far as it is the release of glucose, from liver to blood-stream, that is impaired. In type II, it is a lysosomal hydrolyase, not a cytoplasmic phosphorylase, that is absent.

2.5.1.2 Glycoproteins, glycolipids, and proteoglycans

The biosynthesis of glycoproteins is not clearly understood. What is known is that the protein chain is synthesized first, on the endoplasmic reticulum (section 5.2), and that the sugar components are added subsequently. The main site of this addition appears to be within an organelle known as the **Golgi apparatus**.

This is an as yet ill-defined series of tubules or vesicles that seem to be continuous with the endoplasmic reticulum. In addition to glycoproteins destined for incorporation into the surface membrane, proteins and glycoproteins destined for export from the cell, such as serum albumin in liver, pass through the Golgi system.

Sugars, in the form of their nucleotide derivatives (Fig. 2.29), are sequentially attached by **glycosyl transferases** to the protein chain. The amino acid to which the first sugar is added is generally a serine or a threonine residue, the sugar being linked as an *O*-glycoside (Fig. 2.58). In many glycoproteins that are secreted by cells, such as the immunoglobulins, and protein hormones, the amino acid–sugar linkage is an *N*-glycoside between asparagine and sugar. The addition of sugar residues takes place partly by way of lipid

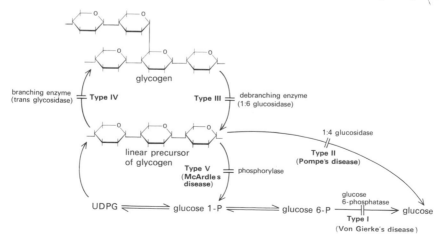

Fig. 2.57 Types of glycogen storage disease. Note that phosphorylase is absent in Type VI (Hers' disease) also.

intermediates such as the long-chain ester (80–100 carbon atoms), dolichol phosphate (Fig. 2.59).

Glycolipids are synthesized via the lipid intermediate ceramide (Fig. 2.62). Again sugars are added sequentially by successive glycosyl transferases. Galactose is often the first sugar that is attached.

Since the structure of proteoglycans is, like that of the glycoproteins, not yet fully known, it is not surprising that their biosynthesis is not well understood either. The general pathway is probably the same as that for glycoproteins, the predominant type of linkage being an O-glycosidic one between serine and xylose. Many proteoglycans are sulphated. In chondroitin sulphates, the sulphate is attached as an O-ester on the C-4 or C-6 of N-acetylgalactosamine. In heparin, the sulphate is attached to both C-6 and the 2-amino group of glucosamine, as well as to a uronic acid. Sulphation occurs after the carbohydrate chain has been completed. Inorganic sulphate is first activated by a two-step reaction involving ATP (Fig. 2.60).

Few hereditary disorders of the synthesis of glycoproteins, glycolipids, or proteoglycans have been de-scribed. On the other hand several disorders due to faulty turnover, by lysosomal degradation, exist. They are summarized in Fig. 2.25.

2.5.2 Fats

2.5.2.1 Triglycerides

As might be expected, triglyceride synthesis from glycerol and free fatty acids is not the reverse of its degradation (Fig. 2.15). Instead a high-energy bond is split for every ester bond synthesized. The high-energy bond is that of acyl CoA, rather than ATP, with the result that the **CoA derivatives** of **long-chain fatty acids** are condensed directly to **glycerol**. One extra molecule of ATP is required for every triglyceride synthesized from free glycerol, since phosphatidic acid is an intermediate (Fig. 2.61). The pathway from glycerol is however relatively minor and is confined largely to liver, which is the only tissue having significant levels of the kinase enzyme. The more common pathway, present in most cells, and especially high in adipose tissue, is by reduction of

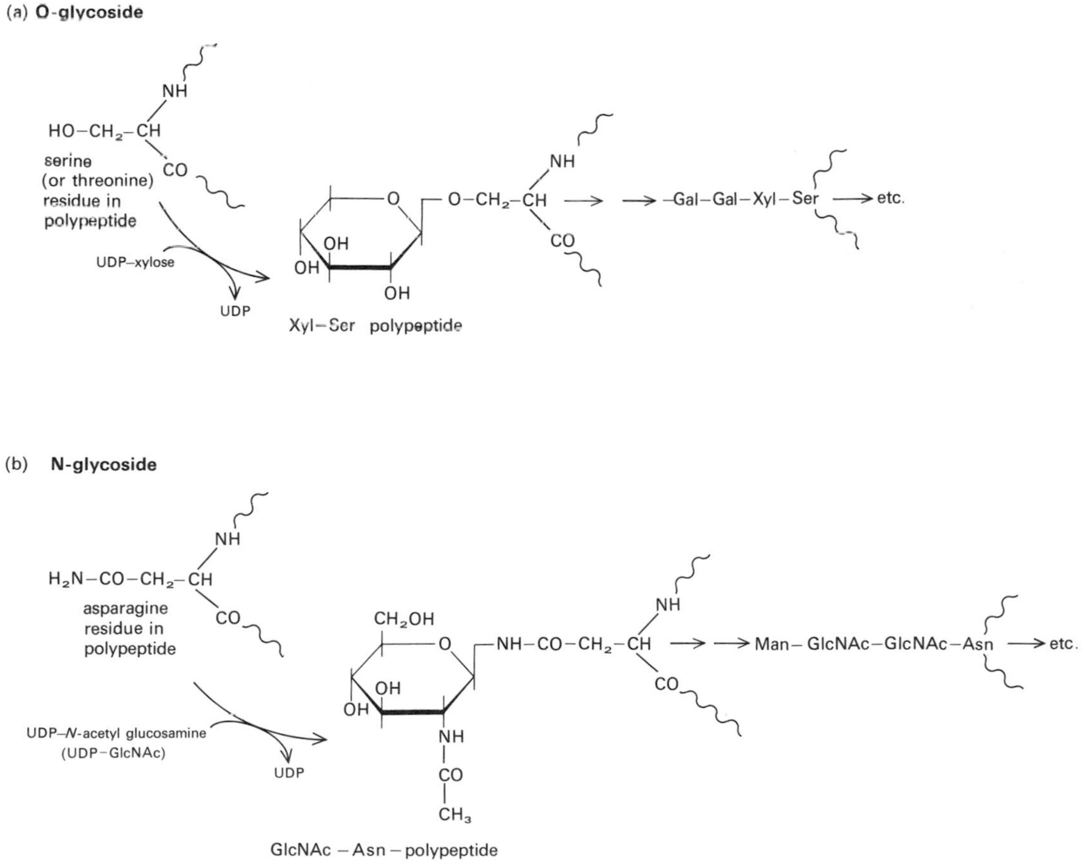

Fig. 2.58 Formation of 'linkage region' in glycoproteins and proteoglycans

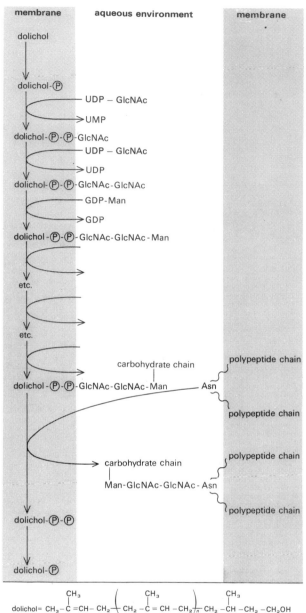

membrane   aqueous environment   membrane

dolichol

dolichol-(P)

— UDP – GlcNAc

→UMP

dolichol-(P)-(P)-GlcNAc

— UDP – GlcNAc

→UDP

dolichol-(P)-(P)-GlcNAc-GlcNAc

— GDP-Man

→GDP

dolichol -(P)-(P)-GlcNAc-GlcNAc-Man

etc.

etc.

carbohydrate chain   polypeptide chain

dolichol -(P)-(P)-GlcNAc-GlcNAc-Man ——— Asn

polypeptide chain

carbohydrate chain   polypeptide chain

Man-GlcNAc-GlcNAc-Asn

polypeptide chain

dolichol-(P)-(P)

dolichol-(P)

$$\text{dolichol} = CH_3 - \overset{\overset{CH_3}{|}}{C} = CH - CH_2 - \left(CH_2 - \overset{\overset{CH_3}{|}}{C} = CH - CH_2 \right)_n CH_2 - \overset{\overset{CH_3}{|}}{CH} - CH_2 - CH_2OH$$

n is generally 14−18; the degree of unsaturation also varies among different dolichols

Fig. 2.59 Lipid intermediates in glycoprotein synthesis

The function of dolichol is to act as a membrane-embedded acceptor for the build-up of the carbohydrate chain, which is then transferred to a membrane-embedded protein. The sugars shown are those typically found at the 'linkage region' (Fig. 2.58).

dihydroxyacetone phosphate, formed from glucose. The latter steps, dependent on absorption of glucose from the blood-stream, explain why synthesis of triglyceride by adipose tissue is an insulin-sensitive process.

The type of triglyceride found in tissues reflects largely the type of fatty acid in the diet. This is in contrast to phospholipids, in which a higher degree of specificity is exercised.

2.5.2.2 Phospholipids

The biosynthesis of phospholipids from **fatty acyl CoA** derivatives, **glycerophosphate** and **nitrogenous base** is depicted in Fig. 2.62. As for cholesterol biosynthesis, the enzymes are largely particulate, being attached to the endoplasmic reticulum rather than in free solution in the cytoplasm. The activating group for the synthesis of the phosphate ester is CDP. It is attached either to the nitrogenous base, as in the case of choline and ethanolamine, or to diglyceride as in the case of inositol. Serine-containing phospholipids are synthesized predominantly by an exchange reaction between phosphatidyl ethanolamine and free serine. Phosphatidyl serine may be decarboxylated (by a pyridoxal 5-P requiring enzyme (Fig. 2.52), back to phosphatidyl ethanolamine.

Phosphatidyl inositol is phosphorylated to mono- and diphosphoderivatives by ATP-dependent kinases. The significance of these phosphatidyl inositol polyphosphates is not clear. The rapid turnover of the phosphate groups on the inositol ring has led to the postulate that they are involved in nerve transmission.

The fatty acid composition of phospholipids is subject to less variation than that of triglycerides. Thus the C-2 position is generally esterified by an unsaturated fatty acid. The spectrum of fatty acids in either position is rather similar in phosphatidylcholine and phosphatidylethanolamine, and markedly different from that in phosphatidyl inositol. The distribution in phosphatidylcholine and phosphatidylethanolamine tends to reflect that of the intermediate phosphatidic acid, whereas the distribution in phosphatidyl inositol tends to reflect that of CDP diglyceride (Fig. 2.63). Clearly the synthesis of CDP diglyceride involves an enzyme of some specificity that controls the nature of cellular phospholipids and hence of cellular membranes.

In addition to acyl side-chains, ether-linked, α,β-unsaturated side-chains are also found in the 1-position of phospholipids to a minor extent. These so-called plasmalogens are synthesized by way of fatty acyl aldehydes. A phospholipid unique to mitochondria is cardiolipin (diphosphatidyl glycerol). Its biosynthesis is depicted in Fig. 2.62. Ceramide, which is the starting-point for the synthesis of sphingomyelin and glycolipids, is synthesized as shown in Fig. 2.62. Only one of the long-chain fatty acids residues is variable. The other is a saturated C_{18}, derived from palmitate.

Choline and inositol are sometimes classed as B vitamins. This is because neither is synthesized to a

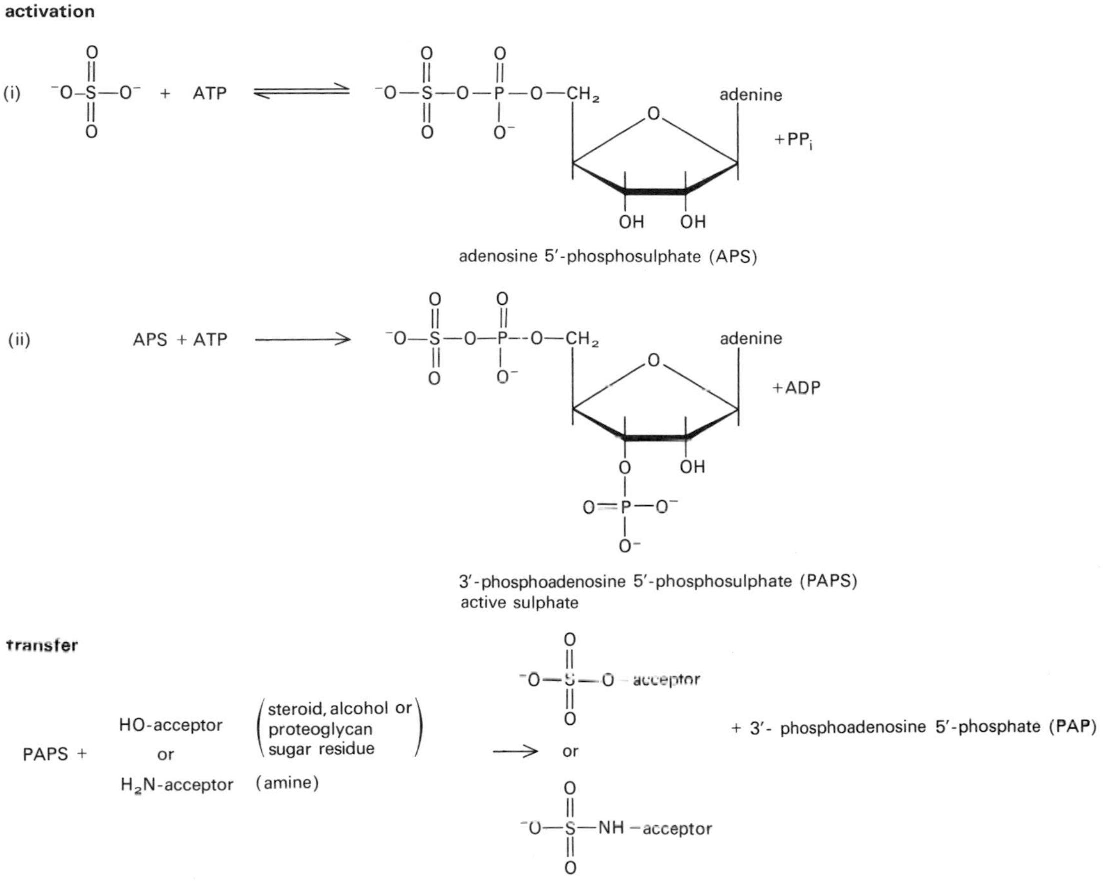

activation

(i) $^-O-S-O^-$ + ATP $\rightleftharpoons$ ... adenine $+PP_i$

adenosine 5′-phosphosulphate (APS)

(ii) APS + ATP $\longrightarrow$ ... adenine $+ADP$

$O=P-O^-$

3′-phosphoadenosine 5′-phosphosulphate (PAPS)
active sulphate

transfer

PAPS + HO-acceptor or H$_2$N-acceptor (steroid, alcohol or proteoglycan sugar residue) (amine) $\longrightarrow$ $^-O-S-O$ —acceptor or $^-O-S-NH$ —acceptor + 3′- phosphoadenosine 5′-phosphate (PAP)

Fig. 2.60 Activation and transfer of inorganic sulphate

sufficient extent by animal cells. In fact inositol is probably not synthesized at all (except by the microorganisms of the intestine, which contribute to the supply of several B vitamins). Choline can be synthesized from serine or glycine, provided that a sufficient supply of methyl groups, derived from methionine (Fig. 2.23), is available. In general, the dietary amounts of choline and inositol required are much greater than those of the B vitamins, and are nearer to the amounts of essential fatty acids required by man. All these three phospholipid building blocks (choline, inositol, and essential fatty acids) are therefore required in amounts approximately half way between a vitamin and a caloric nutrient such as protein, fat, or carbohydrate. Ethanolamine, of course, is not in the same category, as it can be derived by decarboxylation from serine.

2.5.3 Membrane biogenesis and turnover

In so far as phospholipids are integral components of biological membranes, their biosynthesis is intimately related to the biosynthesis of membranes. And membranes, together with chromatin and other constituents, are what cells are made of. In other words, the biosynthesis of cells is largely the biosynthesis of membranes. As yet, the details of membrane biogenesis are unknown. In particular is not clear how the specificity of membrane composition is achieved. What is known is that the components of membranes, phospholipids, cholesterol, proteins, glycoproteins, and glycolipids, are synthesized on the endoplasmic reticulum and its associated elements. From here they are somehow transferred to nuclear, mitochondrial, or surface membrane. The key question is how particular components become committed to particular membranes.

Two possibilities exist. The first is that the endoplasmic reticulum is itself heterogenous, and contains portions destined for nucleus, mitochondria, or surface membrane, as well as for endoplasmic reticulum itself. Such portions might then 'bud' off as vesicles and be transported to their site of insertion in a manner akin to the movement of membrane prior to lysosomal digestion (Fig. 2.24). The other possibility is that

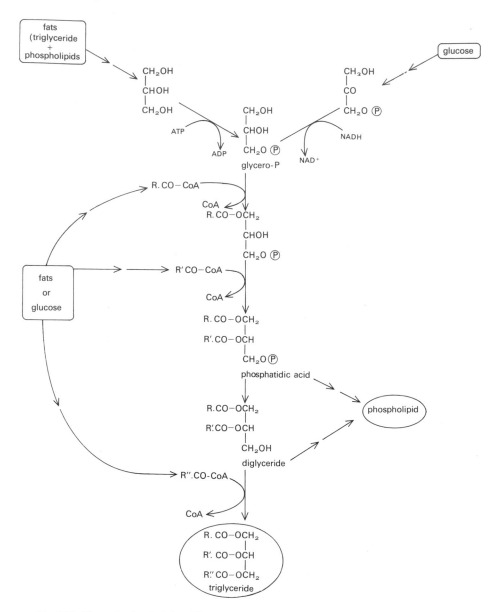

Fig. 2.61 Biosynthesis of triglycerides

discrete molecules of protein, phospholipid, and so forth are inserted directly into respective membranes, involving some kind of recognition process. Present evidence favours the first possibility, but the actual mechanism may well include both processes. It is also not known how the asymmetry of membranes, that is the different composition of the two halves of the bilayer (Fig. 6.2), is achieved. Once inserted, the components of biological membranes are free to diffuse within the plane of the membrane. Hence localized 'growth' regions are not likely to persist.

Because of the continuous movement of portions of

membrane from site of synthesis to site of insertion, and back to lysosomes for degradation, membrane components such as phospholipids and proteins have high rates of turnover, and the degradation products are used over again (Fig. 2.64). Just as RNA turnover allows the pattern of mRNA, and hence the spectrum of proteins, to be varied within a cell, so membrane turnover allows the composition of membranes to be continuously modified and adapted to specific needs. The concept that biological membranes are inert bags separating the contents of one organelle from another is no longer valid.

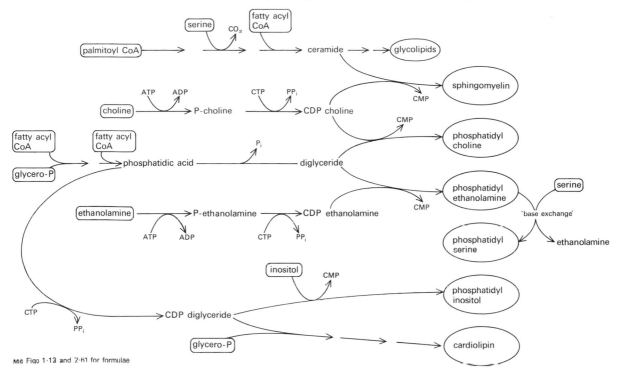

see Figs 1·13 and 2·61 for formulae

Fig. 2.62 Biosynthesis of phospholipids

2.6 Control of intermediary metabolism

The purpose of intermediary metabolism is two-fold: to degrade nutrients in order to obtain chemical energy in the form of ATP, and to synthesize cellular constituents and other end-products of metabolic pathways. The operation of any pathway fills a particular cellular need. That need, however, does not exist continuously. It varies according to the nutritional status of the individual, his energy expenditure, his hormonal balance, and so forth. There is thus a need to switch some metabolic pathways on and off, and to increase or decrease the rate at which other pathways operate. In short, to control intermediary metabolism.

Control of metabolism is achieved in one of two

Fig. 2.63 Fatty acid composition of some phospholipids (Approximate percentage by weight present in each phospholipid class)

	Length and degree of unsaturation of predominant fatty acids									
	16:0	16:1	18:0	18:1	18:2	20:4	22:0	22:5	22:6	24:0
Major phospholipids										
Phosphatidyl choline	33	3	35	10	8	8	—	—	2	—
Phosphatidyl ethanolamine	31	1	31	10	6	16	—	—	3	—
Phosphatidyl inositol	5	—	32	12	6	23	—	3	5	—
Sphingomyelin	36	4	26	2	1	18	3	—	—	10
Metabolic precursors										
Phosphatidic acid	16	1	30	22	6	13	—	—	1	—
CDP diglyceride	4	—	34	17	5	22	—	1	4	—

Data for liver (rat and ox) from a number of sources have been combined. 16:0 denotes a saturated fatty acid of 16 carbon atoms in length (i.e. palmitic acid); 18:1 denotes a fatty acid of 18 carbon atoms in length, having one double bond (i.e. oleic acid); and so forth. All phospholipids, of course, contains 2 fatty acids; the table indicates the *spectrum* of subclasses of each phospholipid that is present.

Note that the metabolic precursors of phospholipids are present in trace quantities only.

The general similarity in fatty acid spectrum of (a) phosphatidyl choline, phosphatidyl ethanolamine, and the metabolic precursor phosphatidic acid, and (b) phosphatidyl inositol and its immediate precursor CDP diglyceride is to be noted; the fatty acid spectrum of sphingomyelin falls into a third class. Quite unrelated is the fatty acid 'profile' of triglyceride.

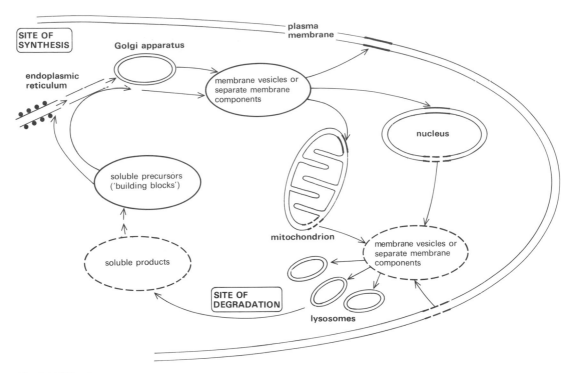

Fig. 2.64 Membrane turnover

ways. By alteration of the **amount** of enzymes present in a cell, and by alteration of their **activity**. In each case, the control may be positive or negative. That is, a substance that acts as a regulator may **induce** the synthesis of an enzyme, or it may **stimulate** the activity of an enzyme; alternatively, regulatory substances may **repress** the synthesis of an enzyme, or they may **inhibit** the activity of an enzyme.

Not all the enzymes of a pathway are controlled in this manner. Generally only the first one or two enzymes are sensitive to control; it is these that set the pace at which the pathway operates (Fig. 2.65). This assumes, in the case of positive control, that successive enzymes are present in sufficient amount to be able to metabolize substrates at an increased rate. This is generally true. In fact most enzymes are present in

cells in excessive amounts; only the enzymes that are sensitive to metabolic control are present in limiting amounts, or have their activity inhibited in order to become rate-limiting.

It will be appreciated that alteration of enzyme activity, by activation or by inhibition, leads to an immediate change in the rate at which a metabolic pathway operates. The effect lasts only for as long as the concentration of the regulator remains raised. As soon as the concentration of regulator falls below that required to activate or inhibit an enzyme, the effect is abolished. This type of control is sometimes referred to as 'fine' or 'short-term' control.

In the case where the inhibitor of a pathway is itself the product of that pathway, such control becomes self-regulating. If, for example, Z is an inhibitor of the pathway shown in Fig. 2.65, then a rise in Z will lead to an inhibition of the enzyme catalysing A to B. This results in a decreased formation of Z, as a consequence of which the concentration of Z falls below the inhibitory level.. The enzyme catalysing A→B therefore returns to full activity and a little more Z is formed. In short, the pathway A to Z fluctuates between a lower and an upper limit, set by the concentration of Z. If an inhibitory concentration of Z is maintained for some time, as might occur if Z enters cells from another source, the pathway may be shut down entirely. Such control, which is known as **negative feedback**, is quite common in biology; it operates

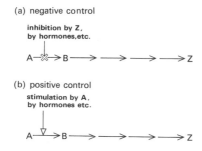

Fig. 2.65 Types of metabolic control

in the control of hormone secretion, for example, and is discussed again in section 9.1.

An example of negative feed-back in intermediary metabolism is in the synthesis of cholesterol (Fig. 2.66). The first stage of the biosynthetic pathway, which catalyses the conversion of β-OH, β-methyl-glutaryl CoA to mevalonic acid, is inhibited by cholesterol. When cholesterol is absent from the diet, the rate at which the pathway operates is set by the concentration of cholesterol, which is itself dependent on the rate at which it is secreted or further metabolized. When cholesterol is present in the diet in appreciable amounts, and high levels of circulating cholesterol are maintained, the pathway is virtually shut off most of the time. Another example of negative feedback is the inhibition of aspartate transcarbamylase—the first enzyme in pyrimidine biosynthesis—by CTP (Fig. 1.27). (Strictly speaking aspartate transcarbamylase is the second step in pyrimidine biosynthesis; the first step — formation of carbamoyl phosphate — is however not unique to pyrimidine biosynthesis, since it is the first step in the metabolism of ammonia by the urea cycle also.)

Unlike control by inhibition or activation of an enzyme, control by synthesis of an enzyme (induction) is longer-lasting. Once an enzyme has been synthesized, it remains intact for as long as the cell remains viable, or until the enzyme is degraded by turnover (section 2.2.4). Moreover synthesis of an enzyme takes time, albeit only a few minutes per protein molecule, whereas inhibition or activation is immediate. Inhibition of enzyme synthesis (repression) takes even longer to manifest itself. For the outcome becomes apparent only when existing enzyme has been degraded by turnover, and the lifetime of most proteins

is in the order of days rather than of minutes. For these reasons, alterations in enzyme synthesis are referred to as 'coarse' or 'long-term' control.

Some examples of metabolic control are listed in Fig. 2.67. The reader may find it difficult to discern 'initial enzymes of a metabolic sequence'. This is due to the fact that some sequences consist of only one enzyme or enzyme complex. The synthesis of glycogen from UDPG is an example. UDPG is the starting-point, since UDPG and compounds from which it is metabolized, such as glucose 6-P, lead to other pathways (Fig. 2.68). For this reason the synthesis of glucose 6-P from glucose should be considered as a single enzyme sequence. Conversely the breakdown of glycogen to glucose 1-P, or of triglyceride to glycerol and free fatty acids, which might be thought to be single enzyme sequences, are in effect the first steps of a long sequence leading to the complete oxidation of glycogen or triglyceride to carbon dioxide and water, with concomitant synthesis of ATP. Glycogen or triglyceride is degraded not to form metabolic intermediates, but to form ATP. As mentioned above, the purpose of intermediary metabolism is either to synthesize cellular constituents or to break down nutrients to obtain chemical energy in the form of ATP.

In the case of glycolysis, it is the second enzyme of the pathway (phosphofructokinase) that is subject to control (Fig. 2.68). This is because it is easier to influence the rate of a pathway through an irreversible reaction, such as that of $F6P + ATP \rightarrow F1,6\text{-diP} + ADP$, than through a reversible one such as that of $G6P \rightleftharpoons F6P$.

Several of the regulators listed in Fig. 2.67 are **hormones**. The main function of hormones is, in fact, to control metabolism. As is described in section 9.4, the hormone is generally not the regulator itself, but

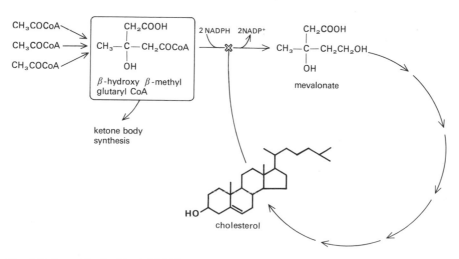

Fig. 2.66 Control by feed-back inhibition

The control of cholesterol synthesis is shown.

Fig. 2.67 Examples of metabolic control: (a) Alterations of enzyme activity

Pathway	Enzyme	Regulator	Alteration
Glycolysis	Hexokinase	Glucose 6-P	Inhibition
	Phosphofructokinase	Citrate; ATP	Inhibition
		AMP	Activation
	Pyruvate kinase	ATP; cyclic AMP	Inhibition
		Fructose di-P	Activitation
Gluconeogenesis	Pyruvate carboxylase	Acetyl CoA	Activitation
	Fructose 1,6-di-Pase	AMP	Inhibition
	Glucose 6-Pase	Glucose	Inhibition
Glycogen degradation	Phosphorylase	Cyclic AMP	Activation
		Glucose	Inhibition
Glycogen synthesis	Glycogen synthetase	Cyclic AMP	Inhibition
		Glucose, glucose 6-P	Activation
Triglyceride degradation	Lipase	Cyclic AMP	Activation
Fatty acid synthesis	Acetyl CoA carboxylase	Fatty acyl CoA	Inhibition
		Citrate	Activation
Tricarboxylic acid cycle and associated reactions	Pyruvate dehydrogenase	Acetyl CoA; ATP; NADH	Inhibition
		Pyruvate; ADP; Ca^{2+}	Activation
	Citrate synthetase	ATP; fatty acyl CoA	Inhibition
	Iso-citrate dehydrogenase	NAD^+; ADP; citrate	Activation
Cholesterol synthesis	Mevalonate synthetase	Cholesterol	Inhibition
Several	Adenyl cyclase	Various hormones (see Fig. 9.26)	Activation

Cyclic AMP-mediated changes are brought about by phosphorylation of the respective enzyme (by a protein kinase) when cyclic AMP is high, and by a dephosphorylation of the enzyme (by a phosphatase), when cyclic AMP is low.

(b) Alterations of enzyme synthesis

Pathway	Enzyme	Regulator	Alteration
Glycolysis (liver)	Glucokinase Phosphofructokinase Pyruvate kinase	Insulin	Induction
Gluconeogenesis	Phosphopyruvate carboxykinase Fructose 1,6 di-Pase Glucose 6-Pase	Glucagon; cortisol	Induction
	Phosphopyruvate carboxykinase Fructose 1,6 di-Pase Glucose 6-Pase	Insulin	Repression
Pentose phosphate cycle	Glucose 6-P dehydrogenase 6-Phosphogluconate dehydrogenase	Insulin	Induction
Fatty acid synthesis	Citrate cleavage enzyme Acetyl CoA carboxylase Fatty acid synthetase	Insulin	Induction
Triglyceride utilization (by adipose tissue, etc.)	Lipoprotein lipase	Insulin; heparin	Induction
Amino acid degradation	Transaminases	Cortisol; glucagon	Induction
	Arginase and other urea cycle enzymes	High protein in diet	Induction

In the case of regulators that *induce* synthesis of a particular enzyme, their lack generally *represses* synthesis, *and vice versa*. See also Figs. 9.31 and 9.32.

leads to the formation of the regulator; in many cases in which control is exerted through a change in enzyme activity, the regulator is cAMP. In cases where hormonal control is exerted through a change in enzyme synthesis, the hormone itself has been shown in Fig. 2.67 as regulator. It should not be assumed, however, that this is strictly true. In most cases, the detailed mechanism by which enzyme synthesis is altered by hormones is not yet known (see section 9.4.2). On the other hand, the hormones listed in Fig. 2.67 *are*

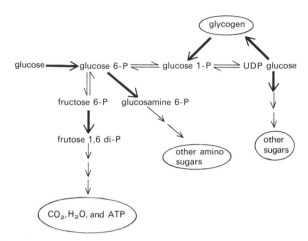

Fig. 2.68 Control points in carbohydrate metabolism

Bold arrows indicate enzymes that are subject to metabolic control.

the triggers that initiate the events leading to an altered rate of enzyme synthesis, and their description as 'regulators' is therefore not unwarranted.

2.7 Summary

Intermediary metabolism

Foodstuffs are required by the body

i. to provide energy. Energy is used for muscle contraction, for ion pumps, and for synthesis of cell constituents; the form in which energy released by the oxidation of foodstuffs is 'trapped' and transferred, is as the compound ATP;

ii. for the synthesis of cell constituents, which are necessary for growth and for replacement of components that are degraded by 'turnover'.

The initial pathways by which foodstuffs are metabolized for provision of energy and for synthesis of cell constituents are the same: degradation to low molecular weight 'building blocks' (such as sugars, fatty acids, and amino acids), followed by further degradation to '**common intermediates**' (such as pyruvate and acetyl CoA). Thereafter the pathways are different: metabolism for provision of energy involves oxidation of the common intermediates to CO_2 and H_2O, with concomitant synthesis of ATP; synthesis of cell constituents, (which occurs from 'building blocks' *or* from 'common intermediates'), is by pathways that resemble the degradative pathways in reverse.

Most of the degradative and synthetic pathways, which together are known as 'intermediary metabolism', occur in the cytoplasm of cells. Unless stated otherwise, the pathways occur in most cell types.

Degradation of foodstuffs and cell constituents to 'common intermediates'.

1. Carbohydrates

Carbohydrate that is present within cells (predominantly **glycogen** of liver and muscle) is degraded to sugar (in the form of glucose 1-P) by phosphorolysis; the enzyme system catalysing the reaction (**phosphorylase**) is complex. The further metabolism of glucose 1-P to pyruvate occurs by a sequence of reactions known as **glycolysis**. Formation of pyruvate requires oxygen; in its absence, **lactate** accumulates. The enzyme converting pyruvate to lactate is not the same in all tissues; the different forms are known as **isoenzymes**.

Carbohydrate that is present in food (predominantly glycogen and starch) is degraded (in the gut) to glucose by hydrolysis. Glucose and other sugars (predominantly galactose and fructose, derived from such dietary sources as **lactose** and **sucrose**), are converted in cells to glucose 6-P and then to pyruvate and lactate by the pathway of glycolysis. Some **ATP** is formed during the process. Further metabolism of pyruvate, to CO_2 and H_2O, occurs by the enzymes of the tricarboxylic acid cycle (present in the mitochondria of cells); in this process much **ATP** is formed. An alternative pathway for the degradation of glucose, that is present in the cytoplasm of most cells, is the **pentose phosphate cycle**; by this pathway, **ribose** and hence **deoxyribose** (required for the synthesis of RNA and DNA respectively) are formed.

2. Fats

Triglycerides and **phospholipids**, whether dietary or cellular, are degraded by hydrolysis to fatty acids, glycerol, and other products. In cells, the enzymes degrading phospholipids are present in organelles known as **lysosomes**. Fatty acids are further degraded by oxidation either:

(a) to the **ketone bodies** acetoacetate and β-hydroxybutyrate (in liver mitochondria) or

(b) to acetyl CoA and thence to CO_2 and H_2O by the enzymes of the tricarboxylic acid cycle (in mitochondria of all cells;

ATP is formed at all stages of the oxidative degradation. Ketone bodies are oxidized to CO_2 and H_2O (to form more **ATP**) in the mitochondria of muscle and other cells.

Cholesterol is only partially oxidized, as far as bile acids (chiefly cholic acid); it is not therefore a major source of ATP.

3. Proteins and amino acids

Proteins, whether dietary of cellular, are degraded by hydrolysis to amino acids; in cells, the degradative enzymes (proteases) are present in lysosomes. The first stage in the subsequent oxidation of amino acids is removal of the amino group by **transaminases**. The rest of the molecule is then oxidized to pyruvate and/or acetyl CoA, and thence to CO_2 and H_2O (forming ATP). The amino groups are excreted in the form of urea.

Synthesis of cellular 'building blocks'

The '**building blocks**' from which carbohydrates, fats, proteins, and nucleic acids are formed are sugars (in the form of nucleoside diphosphate derivatives); fatty acids (in the form of CoA derivatives) and glycero-P; amino acids (in the form of transfer RNA derivatives); and nucleotides (in the form of pyrimidine and purine nucleoside triphosphates). Sugars, fatty acids and glycerol, amino acids, and pyrimidine and purines are derived either from foodstuffs (carbohydrates, fats, proteins, and nucleic acids) or are synthesized (by ATP-requiring reactions) as follows.

1. Sugars

The synthesis of **glucose** from non-sugar precursors, such as lactate, glycerol, or certain amino acids, takes place in liver. The pathway (**gluconeogenesis**) is largely the reverse of glycolysis; at certain stages, the reactions of gluconeogenesis and glycolysis are different. The conversion of glucose to other sugars that are constituents of glycoproteins, glycolipids, and proteoglycans (galactose, mannose, fucose, xylose, uronic acids, and amino sugars) takes place in most cells. The conversion involves nucleotide derivatives (UDP- and GDP-sugars).

2. Fatty acids and cholesterol

The synthesis of fatty acids from non-fat precursors such as glucose (**lipogenesis**), occurs mainly in liver and adipose (fat) cells. Glucose is degraded by the pathway of glycolysis to pyruvate and thence to acetyl CoA. Acetyl CoA units are condensed together and reduced to form long-chain fatty acyl CoA derivatives; **malonyl CoA** is an intermediate.

Cholesterol is also formed by the condensation and reduction of acetyl CoA units; the mechanism is different to that of fatty acid synthesis.

3. Amino acids

Certain amino acids (known as non-essential amino acids) can be derived from glucose and a source of amino groups. Mostly, however, amino acids are derived from the diet or from the breakdown of cellular proteins.

Amino acids are activated by a two-stage process. First, an amino acyl-AMP derivative is formed; next, the amino acyl group is transferred to a molecule of transfer RNA (tRNA). Both stages are catalysed by the same enzyme (**amino acid activating enzyme**); some 20 different activating enzymes exist, each being specific for a particular amino acid. The tRNA molecules to which amino acids are bound are also specific for each of the amino acids.

4. Nucleotides

The synthesis of **pyrimidine** and **purine nucleotides** occurs in every cell capable of synthesizing RNA and DNA. The precursors of pyrimidine nucleotides are aspartate, CO_2, NH_3 (in the form of glutamine), ribose 5-P, and ATP. The precursors of purine nucleotides are glycine, serine, NH_3 (in various forms), ribose 5-P, and ATP.

Pyrimidine and purine deoxynucleotides are formed from the respective nucleotides by a reductase enzyme.

Synthesis and function of coenzymes

Many reactions of intermediary metabolism require the presence of a **coenzyme**. Such coenzymes are derived from the water-soluble, or B, group of **vitamins**.

Oxidation–reduction reactions involving the transfer of H atoms require the vitamins nicotinic acid (in the form of **nicotinamide–adenine dinucleotide, NAD^+**), and riboflavin (in the form of **flavin nucleotides**). The reduction of nucleotides to deoxynucleotides probably requires the vitamin B_{12}.

Oxidative decarboxylation reactions, such as the metabolism of pyruvate to acetyl CoA, require the vitamins thiamin (in the form of **thiamin PP**), **lipoic acid**, and pantothenic acid (in the form of **coenzyme A**); coenzyme A is also involved in the oxidation of long-chain fatty acids and in the tricarboxylic acid cycle. Transamination and other reactions involving the α-carbon atom of amino acids require the vitamin pyridoxine (in the form of **pyridoxal-P**).

Certain 1-carbon transfer reactions involving CO_2, require the vitamin **biotin**. Other 1-carbon reactions, such as the biosynthesis of purine nucleotides and the methylation of uridine deoxynucleotide (to form thymidine deoxynucleotide), require the vitamin folic acid (in the form of **tetrahydrofolate, THF**); THF acts to transfer a 1-carbon unit generated by the metabolism of serine (C_3) to glycine (C_2).

Synthesis of cell constituents

1. Carbohydrates

Glycogen is formed by transfer of glucose units from UDP glucose to a growing chain of 1:4 linked glucose units; the enzyme (**glycogen synthetase**) is complex.

1:6 branches are initiated by a separate enzyme. **Glycoproteins**, **glycolipids**, and **proteoglycans** are synthesized by the sequential addition of sugar units to previously formed protein or lipid (ceramide) chains; the sugars are transferred from UDP or GDP derivatives by various glycosyl transferases. Transfer occurs within an organelle known as the **Golgi apparatus**.

2. Fats

Triglycerides are synthesized by sequential addition of the CoA derivatives of long-chain fatty acids to glycero-P; diglycerides are intermediates. **Phospholipids** (such as phosphatidylcholine) are synthesized from diglycerides by addition of CDP derivatives (such as CDP choline).

3. Membrane biogenesis

The components of **membranes** (proteins, phospholipids, and other molecules) are synthesized from their respective 'building blocks' on cytoplasmic membranes (endoplasmic reticulum and Golgi apparatus). From there the components are somehow transferred and inserted into other membranes, such as nuclear, mitochondrial, or plasma membrane.

Insertion is accompanied by removal; removed components become transferred to lysosomes, where they are degraded (to amino acids, fatty acids, sugars and so forth). Membranes are thus in a continual state of turnover.

Control of intermediary metabolism

The reactions of intermediary metabolism are controlled by the amount and by the activity of **enzymes**. Changes in the rate of enzyme synthesis (**induction** and **repression**) affect **amount**; changes in the concentration of low molecular **activators** or **inhibitors** affect enzyme **activity**. Many pathways are self-regulating by **negative feed-back**; that is, an increased production of end-product leads to a decreased rate of the pathway. Several **hormones** control the rates of intermediary metabolism by affecting the amount and the activity of enzymes.

FURTHER READING

A general account of metabolic pathways

A.L.Lehninger (1975). *Biochemistry* (2nd edition). Worth Publishing Inc., New York.

Control of metabolism:

E.Hofman (1978). Phosphofructokinase—a favourite of enzymologists and of students of metabolic regulation. *Trends biochem. Sci.* **3**, 145.

S.D.Killilea *et al.* (1976). Modulation of protein function by phosphorylation: the role of protein phosphatase(s). *Trends biochem. Sci.* **1**, 30.

E.A.Newsholme and C.Start (1973). *Regulation in metabolism.* Wiley and Sons, London.

P.J.Roach and J.Larner (1976). Regulation of glycogen synthase — a relation of enzymic properties with biological function. *Trends biochem. Sci.* **1**, 110.

Structure and function of lysosomes:

A.C.Allison (1974). *Lysosomes* (Oxford Biology Reader no. 58). Oxford University Press.

R.T.Dean (1977). *Lysosomes* (Studies in Biology, No. 84). Edward Arnold, London.

Biochemical changes in disease:

F.Dickens, P.J.Randle and W.J.Whelan (eds) (1968). *Carbohydrate metabolism and its disorders.* Academic Press, London, New York.

R.H.S.Thompson and I.D.P.Wooton (eds) (1970). *Biochemical disorders in human disease* (3rd edition). J. and A.Churchill, London.

Some topics currently being studied:

F.Huijing (1975). Glycogen metabolism and glycogen storage diseases. *Physiol. Rev.* **55**, 609.

J.J.Volpe and P.R.Vagelos (1976). Mechanism and regulation of biosynthesis of saturated fatty acids. *Physiol. Rev.* **56**, 339.

J.Bremer (1977). Carnitine and its role in fatty acid metabolism. *Trends biochem. Sci.* **2**, 207.

S.J.Benkovic and C.M.Tatum (1977). Mechanisms of folate co-factors. *Trends biochem. Sci.* **2**, 161.

3

Mitochondria: bioenergetics

3.1 Introduction

Mitochondria have been described as the 'power house' of cells. It is here that the chemical energy derived by the **oxidation of foodstuffs** is released in the form of **ATP**. ATP is then used to drive the major energy-requiring processes of the body: muscle contraction, ion pumping, and biosynthetic reactions. That is, the chemical energy inherent in ATP is translated into mechanical energy (muscle), into osmotic pressure (ion pumping), or it is retained in the form of chemical energy and used to drive energetically unfavourable chemical reactions. The amount of energy that is required by a human being for each of the three categories is difficult to assess. Under normal conditions, some 60 per cent might be used for muscle contraction, about 30 per cent for ion pumps, and 10 per cent for biosynthesis. During violent exercise, when the intake of oxygen can increase more than ten-fold, the amount used for muscle contraction is correspondingly greater, and can reach 90 per cent.

In this chapter the manner in which ATP is generated is described. It is preceded by a discussion of elementary thermodynamics, since the concepts of free energy and entropy are important for an understanding of energy-transfer reactions. Note the comment regarding use of the phrases 'energy-requiring' and 'energy-supplying' in section 1.1.1.1. The conversion of ATP into mechanical energy is discussed in section 10.2. Conversion into osmotic energy is discussed in section 6.3.3; although osmotic energy is generated in every cell, it becomes of especial importance in cells of the kidney (section 7.2.2), the intestine (section 8.3.2), and the nervous system (section 12.3.1). Utilization of ATP for biosynthetic reactions is discussed in section 2.5 (carbohydrates and fats), sections 4.5 and 4.6 (DNA and RNA), and section 5.2 (proteins).

3.2 Mitochondrial structure

In contrast to lysosomes, mitochondria are not just membranous bags containing a number of enzymes. On the contrary, the intramitochondrial space is filled with invaginations of the inner mitochondrial membrane, giving a maximum of surface area within a confined space (Fig. 3.1). The invaginations, called cristae, are thus analogous to the microvilli found on the surface membrane of certain cells (Fig. 6.4), which maximize surface area. It will be noted that mitochondria are actually made up of two types of membrane, an inner membrane which constitutes approximately 70–80 per cent, and an outer membrane which makes up the remainder. The space inside the inner membrane, known as matrix, contains the enzymes of fatty acid oxidation and of the tricarboxylic

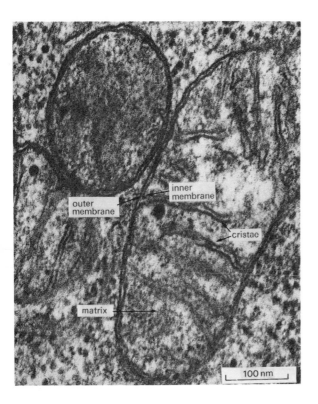

Fig. 3.1 Mitochondrial structure

Thin-section electron micrograph of (rat) kidney mitochondria. Two mitochondria are seen in longitudinal section, one in transverse section. Note that the inner membrane is continuous with cristae. Photograph by courtesy of the Electron Microscopy Unit, Chelsea College, London.

acid cycle. In addition, it contains DNA (mitochondrial DNA distinct from chromosomal DNA, accounting for some 0.1–0.2 per cent of cellular DNA) and RNA in the form of ribosomes (distinct from cytoplasmic ribosomes), transfer RNA (mitochondrial tRNA distinct from cytoplasmic tRNA), and some species of messenger RNA (distinct from cytoplasmic mRNA). In short, mitochondria have their own genes and the mechanism for translating them into proteins. The amount of DNA is, however, insufficient to code for more than a few species of mRNA, in addition to mitochondrial ribosomes and mitochondrial tRNA. Most mitochondrial proteins are therefore synthesized by the cytoplasmic machinery (Chapter 5) and are transported into mitochondria.

The **inner mitochondrial membrane** is one of the most important membranes of animal cells, for it is the site at which the oxidation of foodstuffs is completed. The oxidation of foodstuffs may be divided into two parts. In the first, pyruvate and long-chain fatty acids are converted to CO_2, with concomitant reduction of coenzymes (NAD^+ and FAD) to their reduced forms (NADH and $FADH_2$). In the second part, the reduced coenzymes are reoxidized by the transfer of hydrogen to molecular oxygen, with the formation of H_2O (Fig. 3.2). The first part occurs in the mitochondrial matrix; the second part occurs on the inner mitochondrial membrane.

The transfer of hydrogen takes place by way of iron-containing intermediates, the cytochromes. Because it is electrons that are actually transferred by the cytochromes, the transfer process is known as the **electron transport chain.** It is through the electron transport chain that the energy derived from the oxidation of foodstuffs is utilized to form ATP; the

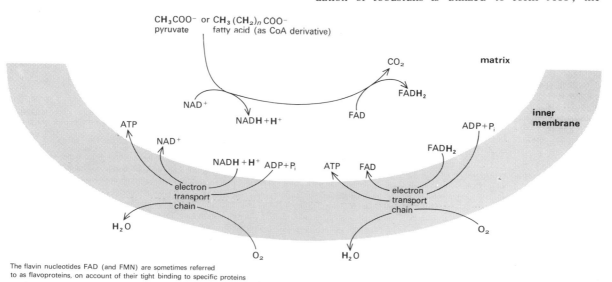

The flavin nucleotides FAD (and FMN) are sometimes referred to as flavoproteins, on account of their tight binding to specific proteins

Fig. 3.2 Mitochondrial oxidation of foodstuffs

formation of ATP by this process is therefore known as **oxidative phosphorylation.** Phosphorylation and electron transport are intimately linked, through enzymes embedded in the inner mitochondrial membrane.

The number of mitochondria per cell, and the number of cristae per mitochondrion, vary greatly, depending partly on the energy needs of a particular cell. Note that the number of mitochondria as determined by electron microscopy of thin sections may be an overestimate, due to the slicing of the long sausage-shaped structures at several points, which gives the appearance of separate structures.

The biogenesis of mitochondria is a complex process. Mitochondrial DNA and RNA are duplicated *in situ*, presumably by enzymes similar to those involved in nuclear DNA and RNA synthesis (sections 4.5.2 and 4.6.2). Mitochondrial proteins are, as mentioned above, synthesized both from within and from outside mitochondria. The mitochondrial protein-synthesizing enzymes differ from cytoplasmic ones in that they are sensitive to antibiotics such as chloramphenicol which inhibit bacterial protein synthesis but to which cytoplasmic animal protein synthesis is insensitive. This, and the presence of DNA and a self-replicating system within mitochondria, has led to the concept of mitochondria as being evolved from symbiotic bacteria.

Mitochondrial phospholipids are for the most part synthesized in the cytoplasm (section 2.5.2.2) and inserted from outside. One type of phospholipid that is specific to mitochondria, namely cardiolipin (Fig. 2.62) is synthesized inside mitochondria. Cardiolipin is present in bacterial membranes also, lending further support to the evolutionary origin of mitochondria referred to above.

3.3 Thermodynamics

If one gram of glucose on a spoon is ignited and allowed to burn to completion, approximately 17 kJ of energy are evolved as heart (3200 kJ/mol of glucose). In the body, in which this process is kept below 37 °C (section 1.2.1), up to 25 per cent of the energy released is harnessed as chemical energy. The energy is used to drive three processes crucial to man,

i.   muscle contraction, including the heartbeat,

ii. ion pumping (which is the basis of nerve conduction), and

iii. chemical synthesis of cellular constituents.

How is this transfer of energy achieved? The answer is by **separating** the **degradative process** into a number of **steps** (more than 20 in the case of glucose oxidation to CO_2 and H_2O) and by transferring energy at some

of the steps in small 'quanta' (approximately 30 kJ/'quantum').

The quanta are in the form of the chemical 'high-energy intermediate', adenosine triphosphate (ATP); the transfer of energy by oxidative degradation into a molecule of ATP occurs in the mitochondria. The utilization of ATP to drive the three types of energy requiring process occurs in other parts of the cell. Mitochondria are a crucial part of most living cells. Since ATP does not move out of one cell and into another, unlike glucose or fatty acids which may be synthesized in one type of cell and re-utilized in another, ATP is utilized within the cell in which it is synthesized. Without mitochondria no oxidative phosphorylation can take place and very little ATP can be synthesized. Ths small amount that can be formed by non-oxidative pathways, that is, by anaerobic glycolysis (Fig. 2.5), outside mitochondria is insufficient to keep the body alive. In any case, significant rates of anaerobic glycolysis occur in only a few cell types such as red cells (which have no mitochondria) and skeletal muscle (which is able to function for short periods without sufficient oxygen supply from the lungs). In order to appreciate why some reactions are able to regenerate ATP and others not, the concepts of free energy and entropy have to be understood.

3.3.1 Free energy and entropy

The oxidation of glucose to CO_2 and H_2O, then, releases some 3200 kJ of energy per mol. This amount is the same irrespective of whether the oxidation is carried out by burning the sugar in a spoon, or by metabolizing it in the body. In fact the energy yielded by any chemical reaction is the same, irrespective of what intermediate pathway is followed. What differs is the **form** in which that **energy** is released (Fig. 3.3). In the case of burning the glucose, all the energy is released as heat. In the case, of metabolizing it by means of enzymically catalysed reactions, approximately 75 per cent is released as heat and 25 per cent as chemical energy.

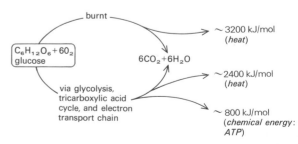

Fig. 3.3 Release of chemical energy and heat by oxidation of glucose

Although 25 per cent is a relatively high figure for energy harnessing (most industrial processes achieve much lower efficiencies), the maximum that is theoretically possible is much higher; it is nearly 70 per cent. The remaining 30 per cent can never be transformed into chemical energy, or any other form of utilizable energy for that matter. In other words, the total energy change of a reaction can either be realized as heat, or partly as heat and partly as chemical energy. Another way of expressing this is to say that chemical energy can be 'converted' entirely into heat, but heat can be 'converted' only partially into chemical energy. This statement is known as the second law of thermodynamics, with the substitution of the term **'free' energy** (that is **utilizable energy**) for chemical energy. (The first law of thermodynamics simply states that the total energy content of a closed system, such as the universe, is constant.) That amount of heat that is not convertible into free energy is called **entropy**, and the second law may be expressed as

$$\Delta H = \Delta G + T\Delta S,$$

where ΔH is the total energy change of a reaction, ΔG the free energy change, and ΔS the entropy change. Note that the entropy term is multiplied by T, the absolute temperature at which the reaction occurs. In other words, the higher the temperature, the less free energy is obtainable from a fixed energy change. At absolute zero ($-273\,°C$), all energy can theoretically be converted into free energy.

In the oxidation of glucose to CO_2 and H_2O, the entropy change is less than a third of the total energy change. In other reactions, such as the denaturation of proteins, the entropy change is high and the free energy change is low. The reason for this is as follows.

The entropy of a system is a measure of its **'disorderedness'**. The more disordered, that is the more random, the greater the entropy. This is best illustrated by reference to the three forms in which a compound such as water can exist. Below $0\,°C$, the molecules are rigidly arranged in a crystalline lattice, namely ice. There is a high degree of order, that is, a low entropy. As ice melts and becomes liquid, the molecules are free to move about and become randomized; the entropy increases. Nevertheless the water molecules still interact with each other to a certain extent (through hydrogen bonds) and it is only when water is heated above its boiling point that the molecules become completely randomized. Hence the 'heat of melting' (that is, the heat required to melt ice at $0\,°C$) and the 'heat of vaporization' (that is, the heat required to vaporize water at $100\,°C$) are each made up of a large entropy term, with little change in free energy.

It can now be seen why the unfolding of a protein molecule by denaturation (section 1.1.1) results in a large entropy change. For a native protein is considerably more ordered, by virtue of specific hydrogen bonds, hydrophobic bonds, ionic bonds, and covalent S–S bonds that determine its secondary and tertiary structure, than a denatured protein in which the groups involved in forming such bonds have become randomized. On the other hand, the degree of randomness in one molecule of glucose ($C_6H_{12}O_6$) and 6 molecules of oxygen is not so different from that of 6 molecules of CO_2 and 6 molecules of H_2O. Five C–C bonds and seven C–H bonds are broken, but six new C–O bonds and seven new O–H bonds are made. Hence the entropy change is relatively low while the free energy change (which is high whenever oxygen becomes bonded to carbon in place of carbon or hydrogen) is large.

Since it is free energy, and not entropy, that is realizable as chemical energy, only those reactions in which there is a large free energy change are capable of being coupled to energy requiring reactions such as the synthesis of ATP. It goes without saying that the change has to be exergonic (decrease in free energy), not endergonic (increase in free energy).

The free energy change of a reaction is a measure of the extent to which substrate is converted into product, once equilibrium has been established. A reaction with a free energy change of approximately $-5.7\,kJ$ results in approximately 90 per cent conversion of substrate to product; a change of $-11.4\,kJ$ in approximately 99 per cent, and a change of $-17.1\,kJ$ in approximately 99.9 per cent and so forth. A change of 0 means that at equilibrium there will be 50 per cent substrate and 50 per cent product. In other words, the equilibrium constant of a reaction is directly related to the free energy change. In fact

$$\Delta G^{\ominus} = -RT \log_e K$$

where K is the equilibrium constant, R is the 'gas constant', T the absolute temperature (assumed to be 298 K, or 25 °C, for the above calculations), and $\Delta G^{\ominus}$ is referred to as the 'standard' free energy change. Note that the standard free energy changes quoted in this book refer to reactants that are in *solution*. They are therefore not directly comparable to the total energy changes (ΔH) quoted for the oxidation of *solid* carbohydrate, fat, or protein ('caloric value' of foods, Fig. 8.1). For example $\Delta G^{\ominus}$ for the oxidation of glucose in solution is $-2880\,kJ/mol$, whereas ΔG° for the oxidation of solid glucose is probably around $-2200\,kJ/mol$.

The relation between $\Delta G^{\ominus}$ and K, then, refers to the situation at equilibrium. But many of the reactions in cells take place under conditions in which the substrates and products are *not* at equilibrium. What is the free energy change under those circumstances? For a reaction $A \rightarrow B$, it is given by the equation

$$\Delta G = \Delta G^{\ominus} + RT \log_e \frac{[B]}{[A]},$$

where [A] and [B] represent the concentration of A and B. In other words ΔG depends on the ratio between the concentration of substrate and the concentration of product. Only in the special case where the concentration of substrate and product is equal, that is at equilibrium, does the term $RT \log_e [B]/[A]$ become 0 (because $\log_e 1 = 0$) and $\Delta G = \Delta G^{\ominus}$. In other cases, ΔG will be either larger or smaller than $\Delta G^{\ominus}$, depending on the concentrations of substrate and product. If [B] is greater than [A], the term $RT \log_e [B]/[A]$ is positive (because $\log_e > 1$ is positive); if [B] is smaller than [A], the term is negative (because $\log_e < 1$ is negative). If the ratio [B]/[A] becomes so small that the term $RT \log_e [B]/[A]$ becomes negative, and bigger in amount than $\Delta G^{\ominus}$, then the sign of ΔG becomes different to that of $\Delta G^{\ominus}$. In other words, it is possible to make an energetically unfavourable (that is, endergonic) reaction favourable (that is, exergonic), by having sufficiently high concentrations of substrate. This is an illustration of Le Chatelier's principle, which states that a reaction can be made to go in either direction, depending on the substrate concentration (and other factors).

The degree to which energetically unfavourable reactions can be reversed depends on $\Delta G^{\ominus}$. If $\Delta G^{\ominus}$ is +17.1 kJ, for example, it requires a concentration of substrate 1000 times greater than that of product to drive the reaction forwards. By maintaining a high supply of substrate at the same time as effecting a very efficient removal of product, this may be feasible. In fact reactions having $\Delta G^{\ominus}$ of around +5 to +15 kJ are potentially reversible by this means. The synthesis of peptide, glycoside, or ester bond falls into this category. It is therefore theoretically possible to achieve synthesis by action of hydrolytic enzymes working in reverse. However, the intracellular concentration of amino acids, sugars, or fatty acids is never high enough to achieve such reversal, and synthesis is always coupled to the hydrolysis of a molecule of ATP. It is obvious that the synthesis of ATP itself ($\Delta G^{\ominus}$ of 30 kJ/mol) falls well outside this range and must clearly be coupled to some highly exergonic process. The major sequence to which ATP synthesis is coupled is the oxidation of the coenzymes NADH and FADH$_2$. The reaction sequence is confined to mitochondria and represents the most important aspect of mitochondrial function: the trapping of chemical energy resulting from reduction of molecular oxygen through a series of discrete energy-linked steps.

3.3.2 High-energy bonds: ATP

The free energy of hydrolysis of the most common bonds in biological molecules, namely ester, glycoside, or peptide, is approximately −8 to −16 kJ/mol. Some bonds, such as pyrophosphate and other acid anhydride bonds, however, have much greater standard free energies of hydrolysis (−30 to −60 kJ/mol). The **hydrolysis** of such **'high-energy' bonds** can be coupled to the **synthesis** of **low-energy bonds,** as seen in the many examples of Chapter 2, and again in Chapters 4 and 5; some mechanisms are summarized in Fig. 3.4. The way in which the high-energy bonds themselves are formed is discussed below (section 3.4). First, the chemical nature of high energy bonds will be considered.

At first sight, there appears to be little difference between a low-energy and a high-energy bond. Each is capable of being hydrolysed to yield two products, which can recondense (given sufficient input of energy) to form the bond again. On closer inspection, however, it is seen that all high-energy bonds are **acid anhydrides** (that is, the product formed by condensation of two acidic groups) or related compounds (Fig. 3.5). That enols (type 3) and thiols (type 4) are acids is seen by reference to common compounds such as phenol (carbolic acid) and hydrogen sulphide, each of which is an acid (stronger than carbonic acid). In other words, enols and thiols have a tendency to dissociate into enolate or thiolate anion and hydrogen ion. Alcohols and amines do not dissociate in this way, and hence their condensation products with an acid (ester and peptide) or with an aldehyde (glycoside) do not constitute anhydride bonds. That guanidino compounds (type 5), in their protonated form, are acids may be seen from the fact that on dissociation a proton is released:

$$\begin{array}{ccc} NH_2{}^+ & & NH \\ \| & \rightarrow & \| \quad\quad +H^+ \\ -C-NH_2 & & -C-NH_2 \end{array}$$

Exactly *why* an acid anhydride should have a high energy of hydrolysis is less clear. It has little to do with the fact that the hydroxyls of phosphate are ionizable; ATP has a high energy of hydrolysis whether it is hydrolysed at pH 6 (approx 3 hydroxyls as O$^-$; $\Delta G^{\ominus} = $ −28 kJ/mol) or pH 8 (all 4 hydroxyls as O$^-$; $\Delta G^{\ominus} = e$ −35 kJ/mol). It has, however, to do with the fact that in every case two positive changes are induced within the structures (partly through the presence of dipoles such as P$\rightleftharpoons$O and C$\rightleftharpoons$O), as shown in Fig. 3.6. There is a tendency for these partially positive atoms to 'pull away' from each other. They are prevented from doing so by the orbital electrons of the O (type 1,2 and 3) S, (type 4), or N (type 5) atoms holding them together. Only when the bond is broken by hydrolysis can their

Fig. 3.4   Utilization of ATP for biosynthetic (endergonic) reactions

Example	'Activating group'	Participation of ATP
(A) Formation of C–C bonds		
Long-chain fatty acids from acetyl CoA (Fig. 2.27)	CoA (also CO_2, Fig. 2.29)	Required to form cytoplasmic acetyl CoA (Fig. 2.28)
Cholesterol from acetyl CoA (Fig. 2.30)	CoA; pyrophosphate	Required to form isopentenyl-PP
Citrate from acetyl CoA and oxaloacetate (Fig. 3.16)	CoA	Not directly required (oxidation of pyruvate or long chain fatty acids generates acetyl CoA)
CO_2 fixation (e.g. in gluconeogenesis, Fig. 2.27, in fatty acid synthesis, Fig. 2.29, in purine synthesis, Fig. 2.38)	Biotin	Required to form biotin-CO_2 (Fig. 2.53)
N-acetyl neuraminic acid (C_9) from C_6 and C_3 (Fig. 2.25)	Phosphoenol group	Required if phosphoenolpyruvate is derived from pyruvate (Fig. 2.27), but not if derived from glucose (Fig. 2.5).
note. Fructose 1,6- diP from triose-P (Fig. 2.27)	None	—
and Reactions of pentose phosphate cycle (Fig. 2.14)	None	—
(B) Formation of C–N bonds		
Carbamoyl P (from CO_2 and NH_3 or glutamine, Figs. 2.23 and 3.17)	?	Required for overall reaction; intermediates not known
Carbamoyl aspartate (in pyrimidine synthesis, Fig. 2.33)		
Citrulline (in urea cycle, Fig. 3.17)	Phosphate	Required to form carbamoyl-P (see above)
IMP (several steps in purine synthesis, Fig. 2.38)		
AMP (from IMP, Fig. 2.39)	?	Required for overall reaction; intermediates not known
GMP (from IMP, Fig. 2.39)		
Arginine (in urea cycle, Fig. 3.17)		
(C) Formation of hydrolysable bonds		
O-glycoside (e.g. in carbohydrates, Fig. 2.25 and 2.56)	Pyrophosphate (as UDP or other nucleoside diphosphate derivative)	Required to form UDP or other nucleoside diphosphate derivative (Figs. 2.35 and 2.13)
N-glycoside (e.g. in nucleotides, Fig. 2.33 and 2.38)	Pyrophosphate (as PRPP)	Required to form PRPP (Fig. 2.33)
Phosphate ester (in phospholipids, Fig. 2.62) (in nucleic acids, Fig. 4.14 and 4.20)	Pyrophosphate (as CDP derivative) Pyrophosphate (as nucleoside triphosphate)	Required to form CDP derivative (Figs. 2.35 and 2.62) Required to form nucleoside triphosphates (Fig. 2.35)
Carboxylic ester (e.g. in triglycerides, Fig. 2.61, and phospholipids, Fig. 2.62)	CoA	Required to form CoA derivative of free fatty acids
Peptide (e.g. in proteins, Fig. 5.7)	tRNA and GTP	Required to form amino acyl tRNA (Fig. 2.31) and GTP (Fig. 2.35)
Sulphate ester (e.g. in proteoglycans Fig. 10.12)	Phosphosulphate	Required to form PAPS (Fig. 2.60)

positively charged groups pull away, and it is this that generates the extra free energy.

An alternative way of explaining the situation is in terms of 'resonance hybrids'. The more ways in which one can write the electronic configurations that contribute to a resonance hybrid, the greater is the stability of that hybrid. More configurations can be written for the hydrolysis products of ATP than for ATP itself (Fig. 3.6). This implies that a large decrease in free energy accompanies the hydrolysis of ATP. (Note: a resonance hybrid should not be confused with different isomeric forms of a compound having keto/enol or amine/imine forms (for example Fig. 1.17), each of which is a discrete compound.)

bond	formula	products of hydrolysis	examples
1 pyrophosphate	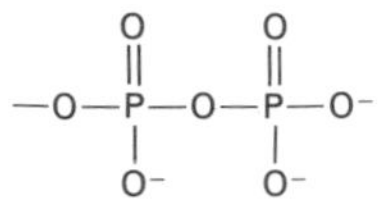 $-O-\overset{\displaystyle O}{\underset{\displaystyle O^-}{P}}-O-\overset{\displaystyle O}{\underset{\displaystyle O^-}{P}}-O^-$	phosphoric acid + phosphoric acid	ATP, ADP + other nucleotides, isopentenyl-PP
2 mixed anhydride	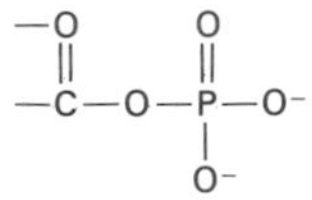 $-\overset{\displaystyle O}{C}-O-\overset{\displaystyle O}{\underset{\displaystyle O^-}{P}}-O^-$	carboxylic acid + phosphoric acid	1,3 di-P glyceric acid; aminoacyl AMP; carbamoyl-P
3 enol phosphate	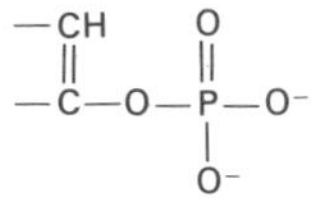 $\overset{\displaystyle -CH}{-C}-O-\overset{\displaystyle O}{\underset{\displaystyle O^-}{P}}-O^-$	enol + phosphoric acid	phosphoenol pyruvate
4 thioester	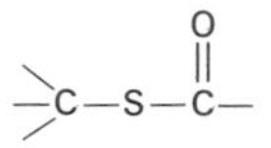 $>\!C-S-\overset{\displaystyle O}{C}-$	thiol + carboxylic acid	acyl CoA
5 guanidino phosphate	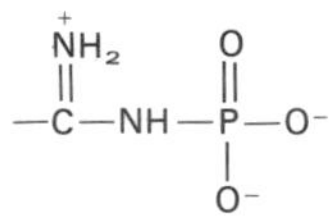 $-\overset{\displaystyle \overset{+}{N}H_2}{C}-NH-\overset{\displaystyle O}{\underset{\displaystyle O^-}{P}}-O^-$	guanidino derivative + phosphoric acid	P-creatine

All compounds are shown in the ionized forms that predominate between pH 7 and 8

Fig. 3.5 Some 'high energy bonds'

3.4 Electron transport: oxidative phosphorylation

When ATP is used to drive an endergonic process such as muscle contraction, active transport, or the biosynthesis of C–C, peptide, ester, or glycoside bonds, one of its terminal pyrophosphate bonds is split (Fig. 3.7). ATP is resynthesized as follows. If AMP is the product, a further molecule of ATP is first used to generate ADP. The enzyme responsible (adenylate kinase) is situated between the inner and the outer membrane of mitochondria in such a way that it acts cytoplasmically, since most of the reactions of type (2) such as amino acid activation (Fig. 2.35) and fatty acid activation (Fig. 3.14), occur outside mitochondria. The resynthesis of ATP from ADP occurs predominantly by oxidative phosphorylation, a process restricted to within mitochondria. ADP is transported back into mitochondria by a carrier system that specifically exchanges cytoplasmic ADP for mitochondrial ATP. Hence the carrier system is responsible for the release of ATP into the cytoplasm as well for the removal of ADP (Fig. 3.7). Inorganic phosphate diffuses into mitochondria by a separate carrier mechanism.

It will be appreciated from Fig. 3.7 that one of the most important reactions in animal cells is reaction 4(a), oxidative phosphorylation. Without a supply of ATP, no life is possible. It is not surprising that no hereditary disorders in which reaction 4(a) is missing have been described. On the contrary, an inhibition of reaction 4(a) by a poison such as carbon monoxide or cyanide, or a failure of the oxygen supply, which is required solely for the maintenance of reaction 4a, is lethal within a matter of minutes.

Oxidative phosphorylation is a complex process. It involves the stepwise **reduction** of molecular **oxygen** to **water** by coenzymes such as NADH and $FADH_2$ (Fig. 3.2), coupled to the **synthesis** of **ATP** from ADP and inorganic phosphate. The inner mitochondrial membrane is a key component of the reaction since (i) the enzymes catalysing the reduction of molecular oxygen are embedded within it and (ii) the energy derived from the reduction of oxygen is used to drive reaction 4(a) by means of an accumulation of **protons** (that is, H^+ ions) within, or on the outside of, the membrane. The membrane is rather impermeable to H^+ ions, so that they are prevented from diffusing across it. Hence a concentration gradient tends to build up. Just as energy is needed to establish a gradient, so a gradient can be used as a source of energy; that is, to drive other reactions. In this case, the reaction is:

$$ADP^{3-} + HPO_4^{2-} + H^+ \rightarrow ATP^{4-} + H_2O.$$

Wheather protons accumulate *outside* the membrane, or whether the production of protons is somehow coupled to the synthesis of ATP *within* the

Explanation A: charge repulsion

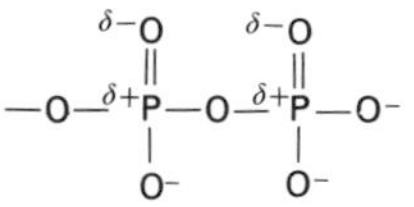

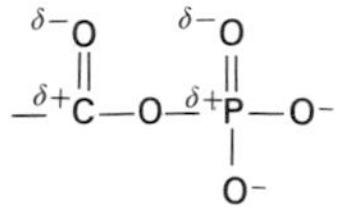

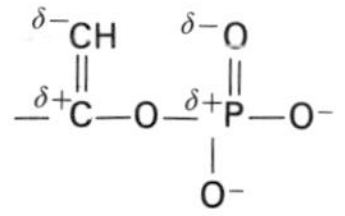

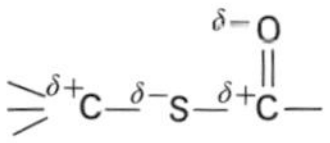

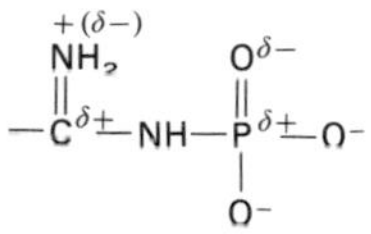

The main repulsive forces are between the atoms bearing partial positive charges (δ^+)

Explanation B: resonance stabilization

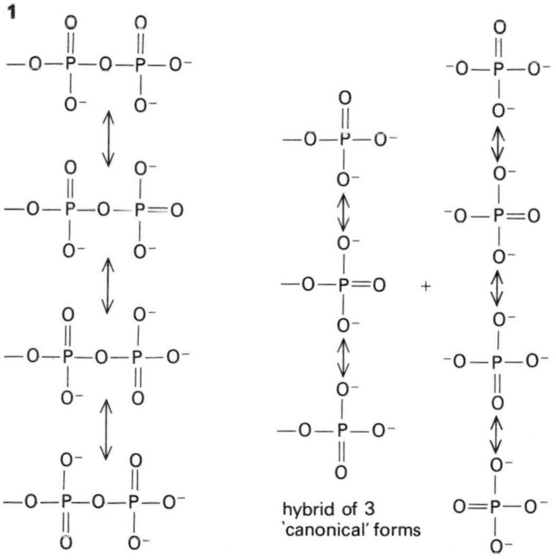

Explanation B *continued*

Fig. 3.6   Reasons for high energy of hydrolysis (continued on p. 90)

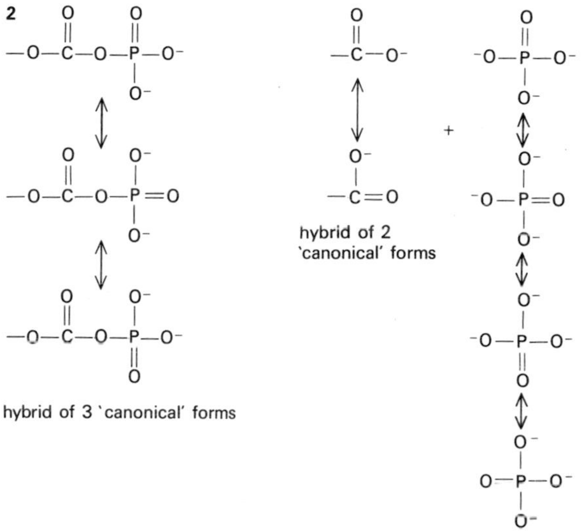

Explanation A underlies the fact that every high-energy bond is a type of acid anhydride. Explanation B is based on the fact that the products of hydrolysis can be written as a greater number of 'canonical' forms than the reactants: hence the products are more stable (i.e. of lower energy) than the reactants. Explanation A and explanation B are not mutually exclusive; they are merely different ways of looking at the same problem.

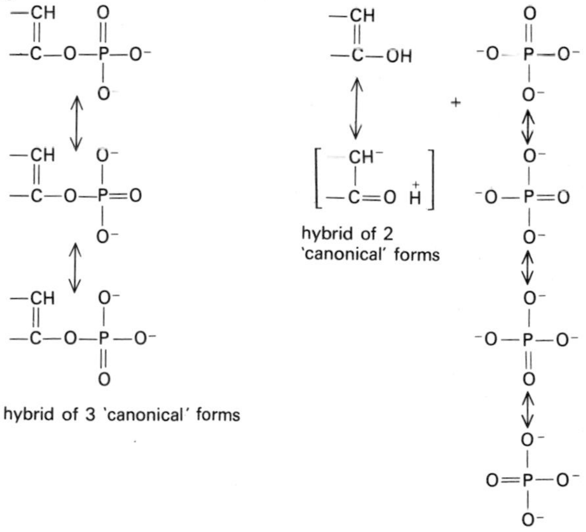

Explanation B *continued*

4 no obvious resonance stabilization in this case

before hydrolysis   *after hydrolysis*

5

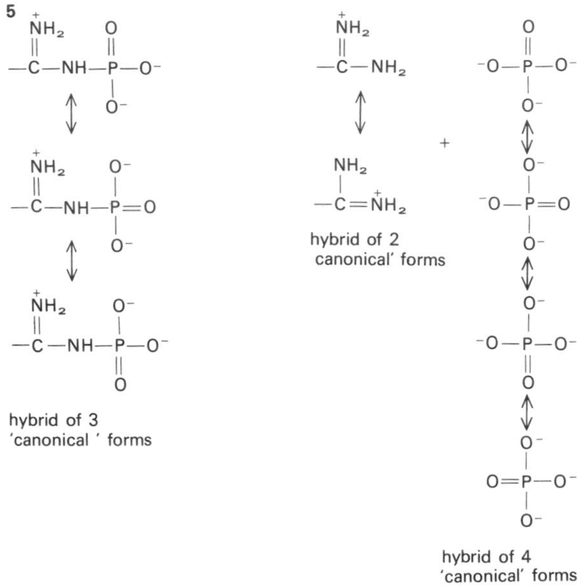

hybrid of 3
'canonical' forms

hybrid of 2
canonical' forms

hybrid of 4
'canonical' forms

Fig. 3.6 (contd.)

membrane, is not yet clear. Certainly oxidative phosphorylation is a process that well illustrates the importance of the link between structure and function in biological systems.

Molecular oxygen reaches the inner mitochondrial membrane as the dissolved gas (Fig. 3.8). An atom of oxygen is reduced to form water by electrons that are generated from the reduced coenzymes, via a quinone known as coenzyme Q, through a series of iron-containing proteins, the **cytochromes** (Figs 3.9 and 3.10). Cytochromes resemble haemoglobin in that they contain a porphyrin ring surrounding a central iron atom (Fig. 7.16). Cytochromes differ from haemoglobin in two respects. First, cytochromes are not in free solution, but are a part of the inner mitochondrial membrane (Fig. 3.11); cytochromes a and b are intrinsic membrane proteins, cytochrome c is a peripheral membrane protein (see Fig. 1.33). Secondly, the iron atom is continuously oxidized and reduced from Fe^{2+} and Fe^{3+} and back to Fe^{2+} again, whereas in haemoglobin, the iron remains as Fe^{2+} whether bound to oxygen or not. It is the oxido-reduction of the iron atom that is the basis of electron transport. The various cytochromes differ somewhat in structure of the porphyrin ring and in the amino acid composition of the attached protein.

The free energy changes of the various steps between NADH and molecular oxygen are illustrated in Fig. 3.12. By redox potential is meant the tendency for a compound to be reduced or oxidized. The **difference in redox potential** between the substrates and the

utilization;

1 Muscle contraction; active transport (Na^+ pump); some biosynthetic reactions

$$ATP \longrightarrow ADP + P_i \qquad\qquad \Delta G° \sim -30\,kJ$$

2 Other biosynthetic reactions

$$ATP \longrightarrow AMP + PP_i \qquad\qquad \Delta G° \sim -30\,kJ$$
$$PP_i \longrightarrow 2P_i \qquad\qquad \Delta G° \sim -30\,kJ \Big\} \sim -60\,kJ$$

resynthesis;

3 Adenylate kinase

$$AMP + ATP \rightleftharpoons 2ADP \qquad\qquad \Delta G° \sim 0\,kJ$$

4 a oxidative phosphorylation (mitochondria)

$$ADP + P_i \longrightarrow ATP \qquad\qquad \Delta G° \sim +30\,kJ$$

b anaerobic glycolysis (cytoplasm)

$$ADP + P_i \longrightarrow ATP \qquad\qquad \Delta G° \sim +30\,kJ$$

Fig. 3.7 Utilization and resynthesis of ATP

Note that reaction 4b is a relatively minor process in animals; it occurs to a significant extent only in cells such as skeletal muscle and red cells.

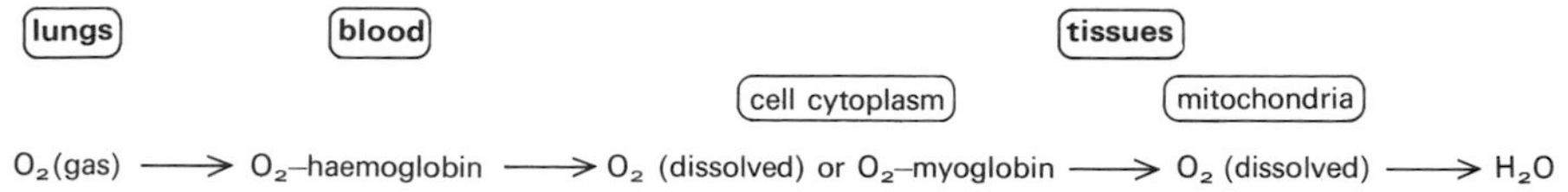

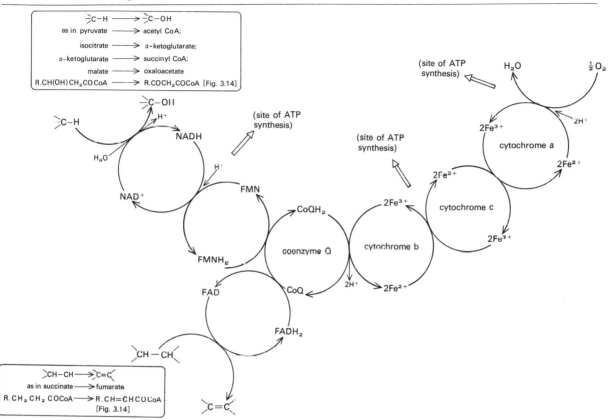

$$O_2 \text{(gas)} \longrightarrow O_2\text{–haemoglobin} \longrightarrow O_2 \text{ (dissolved) or } O_2\text{–myoglobin} \longrightarrow O_2 \text{ (dissolved)} \longrightarrow H_2O$$

Fig. 3.8   Transport of oxygen into cells

See also Fig. 7.25.

Fig. 3.9   The electron transport chain

A simplified version of the electron transport chain is shown; in fact several types of cytochrome b and cytochrome a, as well as other electron carriers, participate in the process.

For the structure of NAD^+, see Fig. 2.48; for FAD, see Fig. 2.49; for CoQ and cytochromes, see Fig. 3.10.

products of an oxido-reduction reaction is proportional to the **free energy change**. That is,

$$\Delta G^{\ominus} = -nF\Delta E^{\ominus},$$

where n is the number of electrons involved in the oxido-reduction, F is the Faraday constant and $\Delta E^{\ominus}$ is the difference in standard redox potential between substrates and products. Since it is by oxido-reduction reactions that the energy of food is converted into the energy of ATP, the concept of $\Delta E^{\ominus}$ is as important as that of $\Delta G^{\ominus}$ for an understanding of intermediary metabolism.

As seen from Fig. 3.12 the energy change of the overall reaction

$$NADH + H^+ + \tfrac{1}{2}O_2 \rightarrow NAD^+ + H_2O$$

is -220 kJ, divided into several stages. In three places the free energy change (for the transfer of two electrons) is greater than -30 kJ, and is therefore potentially capable of being coupled to the synthesis of ATP. This is exactly what happens, and in physiologically intact mitochondria, the 'coupling' ratio between phosphorylation and oxidation (P/O ratio) is 3, (that is, 3 moles of ATP per atom of oxygen). Under these conditions, some 40 per cent of the free energy released by the oxidation of NADH is converted into ATP. If the coupling reaction is somehow inhibited (by poisons such as dinitrophenol), the P/O ratio falls to 0. The value of the P/O ratio is often used as a criterion of the integrity of the mitochondrial inner membrane system in biochemical studies.

The total content of adenine nucleotides (AMP,

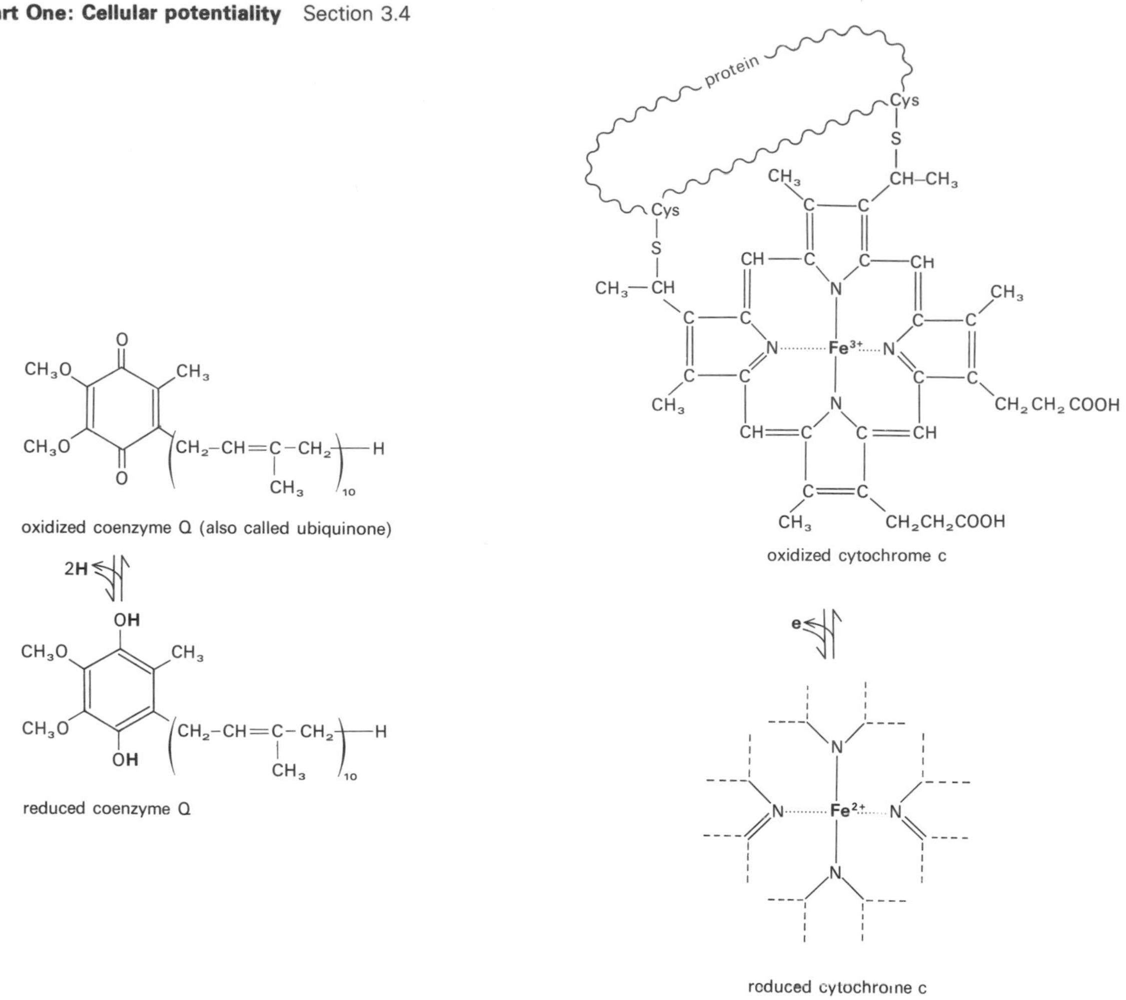

CH₃O ... (structure) oxidized coenzyme Q (also called ubiquinone)

CH_3O, CH_3, O, $CH_2-CH=C-CH_2$ — H, CH_3, 10

oxidized coenzyme Q (also called ubiquinone)

$2H$

OH, CH_3O, CH_3, CH_3O, $CH_2-CH=C-CH_2$ — H, OH, CH_3, 10

reduced coenzyme Q

protein

Cys, S, CH–CH₃, CH_3, C—C, CH—C N C—CH, CH_3, Cys, S, CH_3—CH, C—C N—Fe^{3+}—N C—C, CH_2CH_2COOH, CH_3, CH=C C=CH, CH=C C=CH, C—C, CH_3 CH_2CH_2COOH

oxidized cytochrome c

e

N, N—Fe^{2+}—N, N

reduced cytochrome c

Fig. 3.10   Structures of coenzyme Q and cytochromes

All cytochromes contain a haem group (4 pyrrole rings, known as porphyrin, surrounding a Fe atom) linked to protein; the substituents on the haem group vary among the different cytochromes.

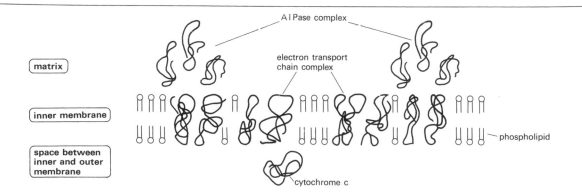

ATPase complex

matrix

electron transport chain complex

inner membrane

phospholipid

space between inner and outer membrane

cytochrome c

Fig. 3.11   Disposition of inner mitochondrial membrane components

The ATPase complex, responsible for the synthesis of ATP, is so-called because it also catalyses the reverse reaction:

$$ATP + H_2O \rightarrow ADP + P_i + H^+.$$

Cytochrome c and part of the ATPase complex are peripheral proteins; the rest of the ATPase complex, and the rest of the electron transport chain, are intrinsic proteins.

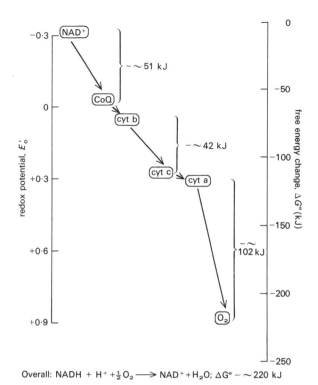

Overall: $NADH + H^+ + \frac{1}{2}O_2 \longrightarrow NAD^+ + H_2O$; $\Delta G^\circ - \sim 220$ kJ

Fig. 3.12 Redox potential and free energy change

Adapted with permission from A. L. Lehninger (1971) *Bioenergetics* 2nd edn). W. A. Benjamin Inc.

ADP, and ATP) in cells is relatively low. That is, ATP cannot be stored as an energy source, in the sense that glycogen and triglyceride are stored as energy sources. As soon as ATP is formed by oxidative phosphorylation, it is utilized for endergonic processes, regenerating ADP and P_i. In fact *demand* for ATP regulates its *supply*. The mechanism by which this is achieved is as follows. As the concentration of ATP within mitochondria falls, so the concentration of ADP rises. An increase in ADP concentration leads to a stimulation of the electron transport chain, and more oxygen and nutrient are utilized. A good example of this is seen when one walks or runs upstairs. The rate of breathing increases as a result of an increased demand for oxygen, brought about by an increase in the ADP level within the mitochondria of certain skeletal muscles, brought about by an increased rate of ATP-driven muscle contraction. In short, the basal metabolic rate (Fig. 8.1) is 'set' by demand for ATP.

Such a link between ATP demand and its supply can operate only in 'coupled' mitochondria. In 'uncoupled' mitochondria, no ATP is formed. The energy resulting from the oxidation of nutrients is then dissipated entirely as heat. The extent to which oxidation and

phosphorylation are coupled is referred to as **'respiratory control'**. In certain diseases, respiratory control falls markedly; the efficiency of energy transfer to ATP then becomes less than 40 per cent. Because of a reduced effectiveness of oxygen utilization, uptake of oxygen often increases. That is, the basal metabolic rate is increased. The increased basal metabolic rate resulting from thyrotoxicosis (increased levels of thyroxine in plasma (Fig. 9.33)) for example, may be due to lack of respiratory control.

In other situations, the respiratory control is deliberately lowered, in order to derive more heat, in place of ATP, from a given amount of nutrient and oxygen. This occurs in new-born babies. Their temperature control by other means (such as shivering) is not fully established; hence they derive extra heat by partial loss of respiratory control.

It must be emphasized that the P/O ratio is 3 only when NADH is the ultimate electron donor. When it is reduced flavoprotein, the ratio is 2 (Fig. 3.12). The dehydrogenation reactions by which substrates are linked to nicotinamide and flavin nucleotides are freely reversible (Fig. 2.48). That is, there is little change in the free energy. Hence the P/O ratio for the oxidation (that is, dehydrogenation) of alcohols and aldehydes (reaction a of Fig. 2.48) is 3, while that for the dehydrogenation of saturated C–C bonds (reaction b) is 2. This is why the dehydrogenation of saturated C–C bonds is coupled to the reduction of FAD and not to that of NAD^+. For if NAD^+ were the H acceptor, the reaction would run backwards:

$$Fumarate + NADH + H^+ \rightarrow Succinate + NAD^+.$$

In fact this is exactly what happens in the biosynthesis of fatty acids, in which NADPH drives the reduction of a β–γ unsaturated acyl derivative to the reduced form (Fig. 2.31).

Although the stages of electron transport at which ATP is synthesized are known (Fig. 3.9), the exact mechanism whereby this is achieved is not. The most widely held view at present is that the synthesis of ATP is driven by (Fig. 3.13) an accumulation of protons (H^+ ions). Exactly how the following three reactions of high energy change:

i. $NADH + \text{oxidized } CoQ \rightarrow NAD^+ + \text{reduced } CoQ$;

ii. $\text{reduced cyt b} + \text{oxidized cyt c} \rightarrow \text{oxidized cyt b} + \text{reduced cyt c}$;

iii. $\text{reduced cyt a} + \frac{1}{2}O_2 \rightarrow \text{oxidized cyt a} + H_2O$

are coupled to the production of protons, and at one side of the inner mitochondrial membrane only, is not known. What appears to be clear is that the passive flow of protons across the membrane is prevented. For it is known that if the inner mitochondrial membrane is allowed to become 'leaky', so that diffusion of protons *does* occur, the synthesis of ATP is prevented.

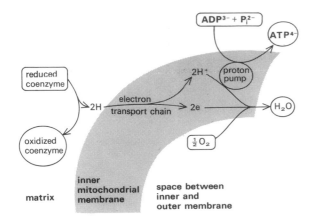

Fig. 3.13   Mechanism of oxidative phosphorylation

For every 2 H transferred from NADH, 3 ATP are formed (P/O ratio of 3). For every 2 H transferred from FADH$_2$, 2 ATP are formed (P/O ratio of 2) (See Fig. 3.8).

Note that O$_2$ and H$_2$O diffuse rapidly across membranes; the exact site of their reaction with the electron transport chain is not clear.

The source of mitochondrial NADH and FADH$_2$ by which oxidative phosphorylation is driven will now be considered.

3.5 Oxidative degradations

3.5.1 Fatty acid oxidation

The main tissues in which oxidation of fatty acids occurs are muscle and liver; the fatty acids are derived predominantly from triglyceride of dietary or cellular (fat depots) origin, as depicted in Fig. 7.7.

The sequence of reactions by which fatty acids are broken down is illustrated in Fig. 3.14. The first stage is an ATP-requiring reaction, namely the formation of the acyl derivative of CoA. In that sense the degradation of fatty acids is analogous to that of glucose (Fig. 2.5) in that in each case degradation is preceded by an activation reaction involving ATP. The formation of a fatty acyl CoA derivative is a reversible reaction sequence, in that the pyrophosphate bond of ATP, the carboxylic acid–AMP bond of the fatty acyl AMP, and the thioester bond of the CoA derivative are all 'high-energy' bonds (Fig. 3.5). In fact it is only the hydrolysis of inorganic pyrophosphate (Fig. 3.7) and the subsequent oxidation of fatty acyl CoA within the mitochondrion that drive fatty acids towards degradation.

As mentioned in section 2.3.2.1, CoA derivatives do not readily cross into or out of mitochondria. In this instance transport occurs by way of a carnitine derivative (Fig. 3.14). Within the mitochondria, fatty acyl CoA is regenerated. Two separate pools of CoA, a cytoplasmic and a mitochondrial one, therefore exist

in just the same way as separate pools of mitochondrial and cytoplasmic NAD$^+$/NADH exist.

The fatty acyl CoA derivatives are degraded by a sequence of reactions known as **β-oxidation.** The β-carbon atom becomes oxidized to the keto derivative by a series of dehydrogenations (Fig. 3.14). β-keto acids are relatively unstable, and the β–γ C–C bond is readily split by CoA. The products are acetyl CoA and the CoA derivative of a fatty acid two carbon atoms shorter than the original one. The process is then repeated over and over again until the entire fatty acid is degraded to acetyl CoA. In liver, the end-product is largely acetoacetyl CoA, which gives rise to ketone bodies (Fig. 8.17), not acetyl CoA.

In the case of odd-numbered fatty acids, degradation proceeds as before until propionyl CoA (C$_3$) is reached. This is metabolized to succinyl CoA by way of a biotin-requiring CO$_2$-fixation reaction, followed by a vitamin B$_{12}$-requiring isomerization. Since odd numbered fatty acids are rare, this is a relatively unimportant pathway in humans.

The site of fatty acid oxidation is in the mitochondrial matrix. The generation of NADH and FADH$_2$ therefore takes place at a site close to the electron transport chain, and hydrogen atoms are efficiently removed and oxidized. Five molecules of ATP are generated for every molecule of acetyl CoA that is formed, giving an overall P/O ratio for the oxidation of fatty acids as far as acetyl CoA, of 2.5.

Ketone bodies, secreted by liver for metabolism by muscle and other tissues, are degraded within mitochondria as follows (Fig. 3.15). β-Hydroxybutyrate is oxidized to acetoacetate, at the same time generating a molecule of NADH from NAD$^+$. Acetoacetate is converted to acetoacetyl CoA by reaction with succinyl CoA; in other words part of tricarboxylic acid cycle (Fig. 3.16) is 'bypassed' to enable acetoacetate to become activated to the CoA derivative. Acetoacetyl CoA then yields 2 molecules of acetyl CoA by reaction with CoA; this reaction is the same as the last stage of fatty acid oxidation.

3.5.2 Tricarboxylic acid cycle

The further oxidation of acetyl CoA formed from fatty acids by β-oxidation, from sugar by way of pyruvate (Fig. 2.5), or from ketone bodies, occurs by a cyclic mechanism (Fig. 3.16). Citric acid and two other tricarboxylic acids are intermediates, as a result of which the pathway has been termed citric acid cycle, **tricarboxylic acid cycle**, or, after its discoverer, Krebs cycle. It is the main source of reduced coenzymes for the electron transport chain, and occurs in mitochondria of every cell. As for fatty acid oxidation, the enzymes are located in the matrix region.

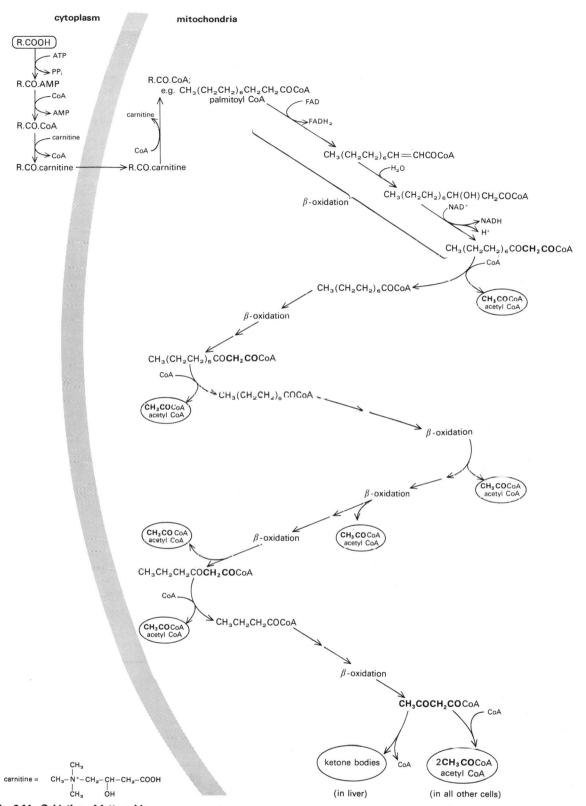

Fig. 3.14 Oxidation of fatty acids

Each $FADH_2$ generated during β oxidation yields 2 ATP when re-oxidized to FAD; each NADH yields 3 ATP. Each acetyl CoA oxidized to CO_2 and H_2O yields 12 ATP (Fig. 3.16). The complete oxidation of palmitoyl CoA therefore yields 136 ATP; since 1 ATP is required to form palmitoyl CoA from palmitic acid, the overall reaction

$$C_{15}H_{31}COOH + 23O_2 \rightarrow 16CO_2 + 16H_2O$$

is accompanied by the formation of 135 ATP.

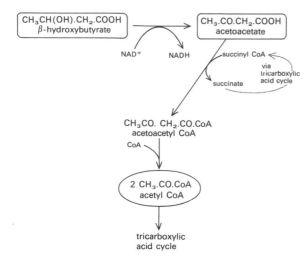

Fig. 3.15   Utilization of ketone bodies

ATP is derived by the re-oxidation of NADH (3 ATP), as well as by the oxidation of acetyl CoA through the tricarboxylic acid cycle (12 ATP; see Fig. 3.16).

No hereditary disorders of tricarboxylic acid enzymes are known, which is not surprising in view of the importance of the cycle. There are, however, some disorders known as mitochondrial myopathies, in which the entry of pyruvate into the tricarboxylic acid may be impaired. Such disorders are not lethal, since pyruvate is not the only precursor of acetyl CoA; long-chain fatty acids (and ketone bodies) provide an alternative source.

The key point to note about the tricarboxylic acid cycle is that the oxidation of CO_2 and H_2O is split into a number of separate reactions so that two atoms of hydrogen are oxidized at a time. The CO_2 that is formed diffuses out of mitochondria into the cytoplasma, and hence into the blood-stream. Unlike oxygen, most of it is not bound to a protein, but remains in solution until it is eliminated by the lungs (see Fig. 7.25).

Oxidation of isocitrate and of malate is of the type denoted as (a) in Fig. 2.48, and the hydrogen is transferred to NAD^+. Oxidation of α-ketoglutarate occurs simultaneously with decarboxylation; the reaction is analogous to the oxidative decarboxylation of pyruvate to acetyl CoA and like that reaction requires thiamin pyrophosphate, lipoic acid, and CoA (Fig. 2.51). The succinyl CoA that is formed from α-ketoglutarate is converted to succinate; the energy inherent in the thioester bond is transferred to form a pyrophosphate bond of ATP. In other words the P/O ratio of the overall conversion of α-ketoglutarate to succinate is 4. Oxidation of succinate to fumarate is of the type denoted as (b) in Fig. 2.48 and the hydrogen is transferred to flavoprotein. Thus for **every molecule of acetyl CoA** oxidized, **12 molecules of ATP** are

formed: 3 at the stage of isocitrate → α-ketoglutarate, 4 at the stage of α-ketoglutarate → succinate, 2 at the stage of succinate → fumarate, and 3 at the stage of malate → oxaloacetate.

For every molecule of pyruvate oxidized, 15 molecules of ATP are formed: (3 at the stage of pyruvate → acetyl CoA, and 12 by the oxidation of acetyl CoA). Assuming that the NADH produced in the cytoplasm when glucose is oxidized to pyruvate is transferred to mitochondria (by the indirect pathway indicated in Fig. 2.32) and oxidized to form another three molecules of ATP, it is seen that for every molecule of glucose oxidized to CO_2 and H_2O, $2 \times \times 18 = 36$ molecules of ATP are formed by oxidative phosphorylation, with another 2 formed by glycolysis (Fig. 2.28). Hence the efficiency of the oxidative process in terms of ATP (assuming $\Delta G^\ominus$ of -2880 kJ per glucose molecule oxidized, and $\Delta G^\ominus$ of 30 kJ per ATP molecule formed) is 40 per cent. For the oxidation of a fatty acid such as palmitate, the corresponding figures are $\Delta G^\ominus$ of -9780 kJ for the overall oxidation and $\Delta G^\ominus$ of 4050 kJ for 135 molecules of ATP (40 for palmitoyl CoA → 8 acetyl CoA, 96 for 8 acetyl CoA → 16 CO_2 and H_2O, *minus* 1 for the initial activation), giving an efficiency of 41 per cent. In short, the efficiency of oxidizing either of the main dietary sources of energy, sugar or fat, is about the same. It is somewhat less for oxidizing protein, as extra ATP is required for the synthesis of urea from ammonia (see below).

The intermediates of the tricarboxylic acid cycle act as coenzymes, in the sense that they are regenerated at every turn of the cycle. Unlike coenzymes, however, they enter into other metabolic pathways, such as gluconeogenesis (Fig. 2.27) and transamination (Fig. 2.18). Their concentration within the mitochondrial matrix has therefore to be maintained by synthesis from sugar or amino acid precursors (pyruvate, aspartate, glutamate, etc.). It is for this reason that a supply of carbohydrate is necessary for the degradation of fats (section 8.2.2).

3.6 Urea cycle

Ammonia, formed chiefly from dietary protein, is toxic in high amounts and is therefore metabolized to a non-toxic product before it is excreted by the kidneys. In humans and many other animals this takes the form of **urea.** The synthesis of urea from ammonia plus CO_2 is, like the oxidation of acetyl CoA, a cyclic process. Like the tricarboxylic acid cycle, it was discovered by Krebs and his associates and is sometimes referred to as the Krebs–Henseleit cycle. The reaction sequence, which occurs only in the liver, is depicted in Fig. 3.17. The enzymes are located in the matrix of mitochondria.

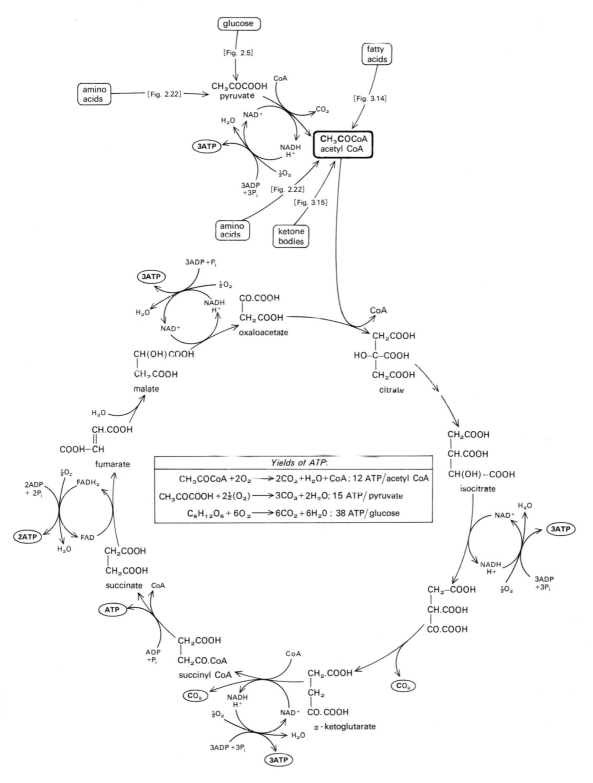

Fig. 3.16 The tricarboxylic acid cycle

The enzymes of the tricarboxylic acid cycle are named according to the substrates on which they act, e.g.
   pyruvate dehydrogenase (→acetyl CoA)
   isocitrate dehydrogenase (→α-ketoglutarate)
   α-ketoglutarate dehydrogenase (→succinyl CoA)
   succinate dehydrogenase (→fumarate)
   malate dehydrogenase (→oxaloacetate).

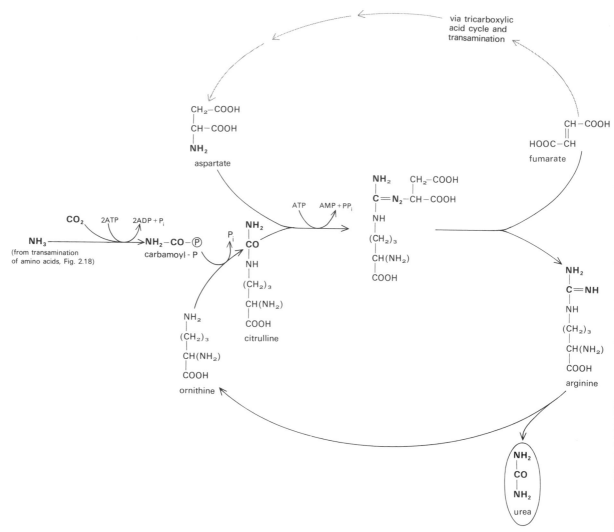

Fig. 3.17 The urea cycle

For every molecule of urea synthesized, 3 ATP are required. Note that the urea cycle operates to a significant extent only in liver cells.

The first reaction is the formation of carbamoyl phosphate from CO_2, NH_3 (see Fig. 2.18) and 2 molecules of ATP; (the reaction is similar to that involved in the biosynthesis of pyrimidine nucleotides, Fig. 2.37). Carbamoyl phosphate reacts with ornithine to form citrulline. A second amino group is introduced by condensation of citrulline with aspartate (see Fig. 2.18) to form arginine; argininosuccinate is an intermediate. Arginine is hydrolysed to form urea, which diffuses out of the cell and into the blood-stream. Note that the urea cycle does not effect a net synthesis of arginine, any more than the tricarboxylic acid cycle effects a net synthesis of, for example, oxaloacetate. Since ornithine is not synthesized to a significant ex-

tent from glucose by animal cells, arginine is an essential amino acid.

The overall synthesis of urea from 2 molecules of ammonia and 1 molecule of CO_2 is seen to require 3 molecules of ATP. This requirement for ATP makes the caloric value of protein less than an equivalent amount of carbohydrate or fat.

Despite the toxicity of ammonia, several hereditary disorders in which one or other enzyme of the urea cycle is missing have been described. Patients do of course suffer from the effects of excessive ammonia, especially after a protein meal. Ammonia toxicity affects the central nervous system in particular, and mental retardation is a common feature. By maintaining a diet low enough in protein, but sufficient to avoid a negative nitrogen balance (section 8.2.2), toxicity can usually be avoided. In any case one instance at least (citrullinaemia, due to a defect in argininosuccinate synthetase), the enzyme is not actually missing, but is merely less active than normal. The reason is due to a lowered affinity for its substrate (that is increased K_M). This is one of the few examples of an alteration in protein structure, due to an altered gene, underlying an enzyme defect; another example is G6PDH deficiency (section 2.2.1.4). Several such examples exist in the case of abnormal haemoglobins, and it is likely that as the sensitivity of enzyme assays in human tissue is increased, more examples will emerge.

3.7 Summary

Mitochondrial structure

Mitochondria are **membranous** structures found in all cells that are capable of **oxidizing foodstuffs**. The membranes contain an enzyme system that retains much of the energy released by oxidation of common intermediates in the form of **ATP**.

Thermodynamics

1. The **total energy** released by the oxidation of foodstuffs (such as the reaction $C_6H_{12}O_6 + 6O_2 \rightarrow 6CO_2 + 6H_2O$) can be measured as a heat change (ΔH). A part of this can be used to drive chemical reactions, such as the synthesis of ATP; this part of the total energy change is known as the change in 'utilizable', or **free energy** (ΔG). The remainder, which cannot be used to drive chemical reactions, is known as the **entropy** change ($T\Delta S$); it represents the increased 'disorder' of the products ($6CO_2 + 6H_2O$), compared with that of the reactants ($C_6H_{12}O_6 + 6O_2$).

2. The reason why **ATP** can be used to drive other reactions, such as the synthesis of carbohydrates, fats, proteins and nucleic acids, the movement of ions against a concentration gradient, or the contraction of muscle, is that hydrolysis of its terminal phosphate groups results in an exceptionally large change in free energy (ΔG).

Electron transport

1. During the oxidation of 'common intermediates', molecular oxygen (O_2) becomes reduced to water (H_2O); iron-containing proteins **(cytochromes)** are intermediates in the process, which is known as the **electron transport chain.** Cytochromes function in the transport of electrons by a change in their oxidation state between Fe^{3+} and Fe^{2+}.

2. The energy released by the reduction of oxygen can be utilized for the formation of ATP, by coupling ATP synthesis to the electron transport chain; the process is known as **oxidative phosphorylation.** The mechanism by which oxidative phosphorylation is achieved is through an accumulation of **protons** (H^+ ions). The extent to which electron transport is coupled to the synthesis of ATP, known as respiratory control, is decreased in certain situations.

Oxidative degradations

The oxidation of 'common intermediates' is coupled to the electron transport chain by the production of **reduced coenzymes** (NADH and FADH$_2$). These coenzymes are formed:

(a) by the formation of acetyl CoA from pyruvate (derived from carbohydrate and certain amino acids) and from long-chain fatty acids (derived from fats) and

(b) by the further degradation of acetyl CoA to CO_2 and H_2O; this occurs by a cyclical mechanism known as the **tricarboxylic acid cycle.**

FURTHER READING

General account of mitochondrial function:

J.B.Chappell and S.C.Reese (1972). *Mitochondria* (Oxford Biology Reader no.19). Oxford University Press.

W.M.Becker (1977). *Energy and the living cell.* J.B.Lippincott Co., Philadelphia.

Specific articles on oxidative phosphorylation:

A.P.Dawson and M.J.Selwyn (1974). Mitochondrial oxidative phosphorylation. In *Companion to biochemistry* (eds A.T.Bull, J.R.Lagnado, J.O.Thomas, and K.F.Tipton), p. 553. Longman, Harlow.

E.Racker (1976). *A new look at mechanisms in bioenergics.* Academic Press Inc., New York.

E.Racker (1976). Structure and function of ATP-driven ion pumps. *Trends biochem. Sci.* **1,** 244.

4
Nucleus: gene expression

4.1 Introduction

All cells of the human body are derived from a single cell, the fertilized ovum. During the many cell divisions that occur between fertilization and death, the same set of enzymes are synthesized over and over again. How is this constancy of composition maintained, and even passed to the next generation? The answer is that it is maintained not as the proteins themselves, which cannot self-replicate, but as a substance which codes for protein structure, namely **DNA**. Unlike protein, DNA can replicate itself and does so with complete exactitude at every cell division. As a result the DNA content of all cells of the human body is identical. What varies from cell to cell is the portion of DNA that is decoded into protein. In liver, for example, some 1 per cent of the DNA is translated into protein. In kidney, a different portion, also amounting to some 1 per cent of the total DNA, is translated into protein. Of course there is much overlap. Many of the enzymes and other proteins synthesized by liver are identical with those synthesized by kidneys; only some, perhaps less than half, are different (Fig. 4.1).

The part of the DNA molecule that codes for a protein molecule is a **gene**. Genes are not separate molecules, but are linked together in the form of a single linear molecule of DNA. Such molecules, together with attached proteins, are known as **chromatids.** Chromatids associate together in pairs to form **chromosomes** (Fig. 4.2). The number of chromosomes varies from species to species. In man there are 23 different chromosomes. The location of chromosomes is in the nucleus of cells. If the mitochondrion is the 'power house' of a cell, the nucleus is its 'records office'. All the information for synthesizing cellular proteins (apart from a few mitochondrial proteins) is contained in the nucleus. A cell that has no nucleus cannot replicate its genes, and therefore cannot divide into viable daughter cells.

The structure of chromosomes, as well as of other nuclear components, is considered in section 4.2. In section 4.3 the relation between chromosomal duplication and other cell events, known as the cell cycle, is discussed. It is a failure of the mechanism by which the cell cycle is controlled that leads to the unrestricted proliferation of neoplastic cells.

The duplication of chromosomes at every cell division accounts for the constancy of protein composition in liver, kidney, or other cell type (Fig. 4.1). The manner in which this constancy is transmitted from one generation to the next, in other words the biochemical basis of heredity, is considered in section 4.4. The importance of understanding this process is underlined by the number of diseases that have been described as being due to a hereditary defect in enzyme synthesis.

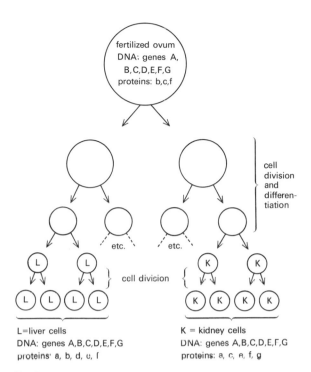

Fig. 4.1 **Relationship between constancy of DNA and variability of protein in cells**

The ability of DNA to self-replicate is a feature of its unique structure. This is described in section 4.5.1. In section 4.5.2 the enzyme system that catalyses DNA synthesis is considered.

The functional capacity of different cell types is a measure of which genes are expressed. How specific genes are selected for translation is not yet clear; what is established is that translation into protein is carried out by means of an RNA copy of the gene, the so-called messenger RNA (mRNA). The structure of mRNA and of other types of RNA, each of which is involved in protein synthesis, is discussed in section 4.6.1. Current views on the manner by which RNA is synthesized are presented in section 4.6.2.

In order to distinguish clearly the three steps leading to the synthesis of proteins, the following terminology is employed. The synthesis of DNA (on a DNA template) is known as **replication**; the synthesis of RNA (on a DNA template) is known as **transcription**; and the synthesis of protein (on an RNA template) is known as **translation**.

4.2 Nuclear structure

4.2.1 Chromosomes

Although the chromosomes are the most important components of nuclei, they are visible, by light or by

electron microscopy, only when the nuclear membrane is absent and the cell is undergoing cell division (mitosis). Chromosomes cannot otherwise be distinguished, largely because they are very diffuse structures, occupying most of the space within the nucleus. (The length of a chromosome, if its DNA were arranged in a straight line, would be approximately 10 000 times the diameter of the nucleus.) The DNA is wound around about itself to form a highly coiled structure and associated with it, by ionic and other non-covalent bonds, is a class of small (mol. wt 11 000–21 000) basic proteins, the **histones**. Histones are rich in the basic amino acids arginine and lysine, and it is presumably through these that the proteins are linked to the acidic phosphate groups of the DNA. Also present in chromosomes is a more heterogenous class of proteins called **non-histone** proteins, which are more acidic due to their relatively high content of aspartate and glutamate.

Exactly how non-histone proteins are linked to histones or to DNA is not known. Indeed the very function of histones and non-histones is not clear. All that is established is that the function of histones is a very fundamental one. For the amino acid composition of several histones is remarkably similar in all species

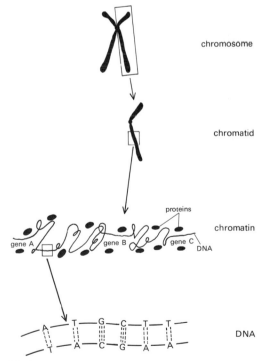

Fig. 4.2 **Structure of chromosomes**

Note: the structure of chromatin (i.e. the twisting of the double-stranded DNA and its relation to proteins) is not known.

101

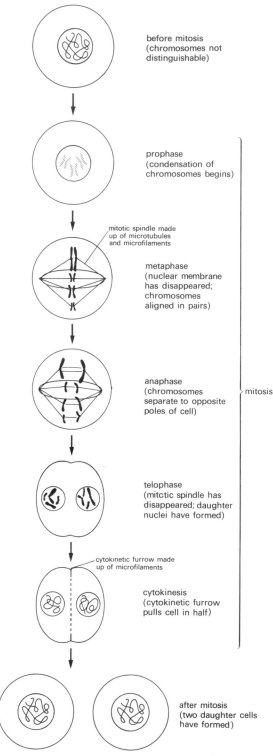

Fig. 4.3 Arrangement of chromosomes during mitosis

Four of the 23 chromosome pairs (in humans) are shown.

of animals and plants. Probably histones determine in some way the configuration of DNA in chromosomes, such that their biological activity, as expressed in the different stages of the cell cycle (see below), is controlled. Non-histones seem to be involved in some way in the transcription of specific genes into mRNA.

During cell division chromosomes can be distinguished by light microscopy (Fig. 4.3). In fact microscopic observation of chromosomes during mitosis has revealed the mechanism by which the correct number of chromosomes is apportioned into each of the two daughter cells. The reason why chromosomes are visible only during mitosis is presumably because their constituents are coiled so tightly around each other that they become extremely dense structures ('condensed').

If one examines the chromosomes in a mitotic cell carefully, one finds that there are two identical sets of 23 different chromosomes, half of which (that is 23 chromosomes) go into each daughter cell. All 23 chromosomes in females are made up of identical chromatids. In males one of the 23 chromosomes is made up of non-identical chromatids; together they are known as the XY chromosome; the corresponding chromosome in females is known as XX (Fig. 4.4). Because this is the only visible difference in chromosome pattern between males and females, the XY and the XX pairs have been termed the sex chromosomes.

It is probable that several enzymes controlling 'maleness' are on the Y chromatid. The X chromatid is unlikely to contain sex-specific enzymes, since it is present, and functional, in both males and females; in females one of the X chromatids of the XX chromosome is permanently inactivated, so that the 'dosage' of X-linked genes is the same in females as in males. In the case of the other 22 chromosomes, both sets of genes are expressed.

The genes specifying the different enzymes of a metabolic pathway are not clustered together on any particular chromosome. The genes specifying some of the enzymes of glycolysis, for example, are scattered on chromosomes 1, 4, 6, 19, and 23 (X); likewise the genes specifying the α- and β-chains of haemoglobin, or the light and heavy chains of immunoglobulins, are on separate chromosomes. Since a gene occupies <0.001 per cent of a chromatid, it is not surprising that hereditary disorders in which a specific enzyme is missing (see for example Fig. 2.25), are not discernible by visual examination of chromosomes. In some diseases, however, the whole pattern of chromosomes is altered. In several malignant tumours, for example, some of the chromatids are present in three or more copies. In certain types of Down's syndrome there is an extra chromatid on chromosome 21 (Fig. 4.4); the presence of three, instead of two, chromatids in a chromosome is referred to as trisomy.

A. Normal male

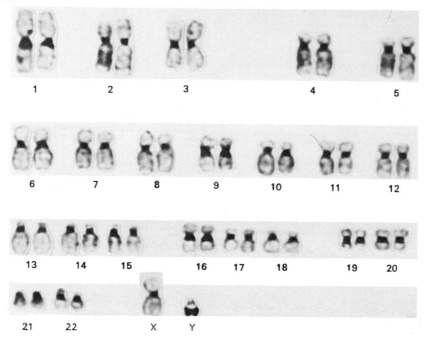

White blood cells just prior to mitosis are fixed, spread on a microscope slide and stained (Giemsa stain). The chromosomes may then be arranged in pairs, as shown.

B. Normal female

    Similar to A, except that Y is absent, and two X chromosomes are present.

C. Down's syndrome

    Similar to A or B except that an extra chromosome no. 21 is often present ('trisomy')

D. Various forms of cancer

    Similar to A or B, except that (i) a translocation of one part of a chromosome to another sometimes occurs, or that (ii) an additional chromosome (e.g.no. 7) is sometimes present ('trisomy')
    or that (iii) the whole complement of chromosomes is doubled or trebled (e.g. 46 instead of 23 chromosomes; polyploidy') or that (iv) a combination of (i)- (iii) occurs (e.g.47 chromosomes instead of 23)

Fig. 4.4 Normal and adnormal pattern of chromosomes

4.2.2 Nucleolus, nuclear membrane, and nucleoplasm

The nucleus consists of an inner structure, termed **nucleolus**, and an outer membrane surrounding the rest of the nucleoplasm (Fig. 4.5); sometimes more than one nucleolus is present. As mentioned above, chromosomes are not normally visible within the nucleus. During mitosis, when chromosomes become distinguishable, both nucleolus and nuclear membrane disappear from view.

Nucleoli contain most of the RNA that is in the nucleus and are the site of **ribosomal RNA** (rRNA) synthesis in cells; messenger RNA (mRNA) and transfer RNA (tRNA) are synthesized in the nucleoplasm.

Most of the nuclear DNA is outside the nucleolus. Exactly how those parts of the DNA that are destined for transcription into rRNA reach the nucleolus is not clear. In at least one instance a portion of a chromosome actually breaks off; this is during the formation of a female egg cell in frogs, in which the genes for synthesizing rRNA are found as discrete entities within nucleoli. Once formed, rRNA from nucleolus, and mRNA and tRNA from nucleoplasm, are somehow transported into the cytoplasm. In the case of mRNA, the RNA is translated into protein; in the case of rRNA or tRNA, the RNA is not translated but participates in a different way in protein synthesis.

Close inspection of the nuclear membrane reveals a number of pores (Fig. 4.5). It is presumably through

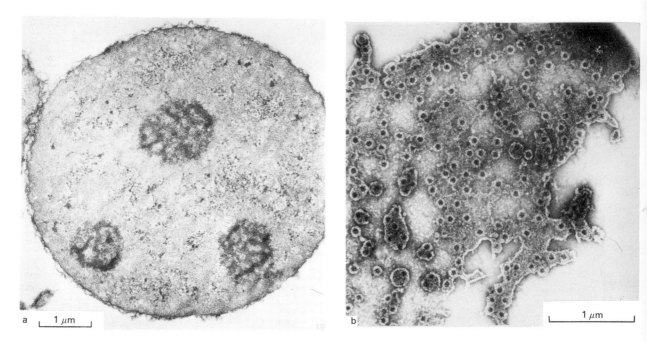

a ⌊ 1 μm ⌋

b ⌊ 1 μm ⌋

Fig. 4.5 Structure of nucleus

a Thin section electron micrograph of the nucleus of a (pig) liver cell. Note the presence of the dark-staining nucleoli (3 in this instance). The granular material is chromatin; since the cell is not in mitosis, distinct chromosomes are not seen.
b Thin-section electron micrograph of a part of the membrane of a (rat) liver nucleus (negatively stained). Note the many circular 'pores'.
  Photographs by courtesy of Dr. J. R. Harris.

these that RNA is transported to the cytoplasm. The pores appear to be valve-like in the sense that free diffusion between cytoplasm and nucleoplasm does not occur. The intranuclear concentration of Na^+, for example, exceeds that of K^+, whereas in the cytoplasm the opposite is the case. The composition of the nuclear membrane resembles that of the endoplasmic reticulum to a certain extent, and the suggestion has been made that the nuclear membrane is continuous with the endoplasmic reticulum. However the enzymic composition of the two membranes is different, and the presence of the pores argues against continuity. The nucleolus itself has no conventional membrane; how its contents, including the enzymes and other factors necessary for rRNA synthesis (section 4.6.2), are kept together is not clear. Chromosomes have no membrane either; the components are simply held together by ionic and other non-covalent bonds.

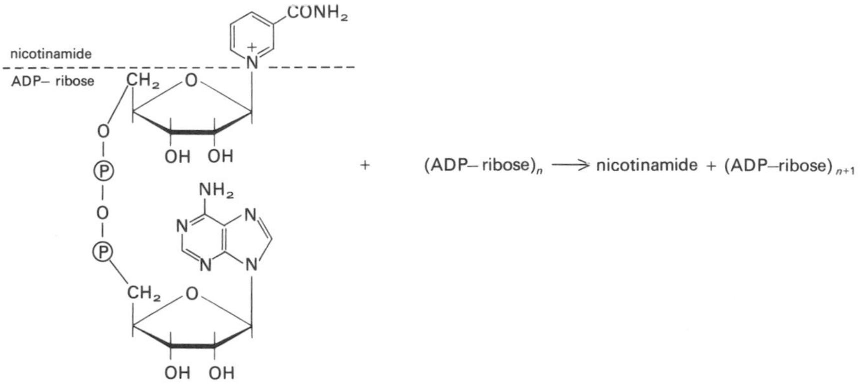

$$(ADP\text{--}ribose)_n \longrightarrow nicotinamide + (ADP\text{--}ribose)_{n+1}$$

Fig. 4.6 Formation of poly(ADP–ribose)

ADP–ribose is transferred by a similar reaction to proteins, such as histones, yielding (ADP–ribose)–protein complexes.

The nucleoplasm contains several enzyme systems, apart from those for synthesizing mRNA and tRNA. The most important, of course, is the system for synthesizing DNA (section 4.5.2). Other enzyme systems that may play a role in nuclear events are concerned with NAD^+. First, the enzymic pathway by which niacin is converted into NAD^+ (Fig. 2.47) and secondly, an enzyme that converts NAD^+ into poly(ADP–ribose) (Fig. 4.6), are present in the nucleus. NAD^+ thus has two functions: as H carrier in cytoplasmic and mitochondrial oxido-reductions, and as precursor of poly(ADP–ribose) in the nucleus. The fact that considerable amounts of NAD^+ become degraded by the latter process explains why the enzymes of NAD^+ synthesis are concentrated in the nucleus. The function of poly(ADP–ribose) is presumed to involve control of the transcription of DNA. How this is achieved is not clear.

Nuclear enzymes do not appear to be essential for keeping cells alive for limited periods of time. Cells lacking a nucleus, such as reticulocytes (Fig. 7.26), for example, are able to carry out all normal cellular functions except cell division.

4.3 Nuclear events

4.3.1 Cell cycle

Cell cycle is a convenient term for describing the events that occur between one **mitosis** and the next. The most important event is the replication of DNA and the other components of chromosomes, followed by their correct allocation into the two daughter cells (Fig. 4.3). DNA synthesis occurs at a discrete time during the cell cycle, the so called S or synthetic phase (Fig. 4.7). It is separated from mitosis (M) by two 'gap' periods called G_1 and G_2.

Nuclear events are synchronized with events that occur in the rest of the cell. For example **cytokinesis,** that is, the pinching in half of a cell during mitosis, is

related in space as well as in time to the movement of chromosomes: the cytokinetic furrow is formed exactly at right angles to the mitotic spindle, at a time just following the apportioning of the two sets of daughter chromosomes (Fig. 4.3). The mechanisms underlying mitosis and cytokinesis appear to be as follows.

The movement of chromosomes is dependent on structures known as **microtubules,** which constitute the **mitotic spindle,** while propagation of the **cytokinetic furrow** involves structures known as **microfilaments.** Microtubules are literally small tubes, diameter approximately 24 nm, consisting of a circular array of a polymerized protein called tubulin. Elongation of the mitotic spindle, by which chromosome movement occurs, is due to the assembly and insertion of successive microtubular subunits (Fig. 4.8). The subunits themselves are rigid; movement is due not to a stretching, but to a 'pushing apart' of the chromosomes by the growing mitotic spindle. How microtubules are attached to chromosomes and, even more important, how the correct pairing of chromosomes is achieved, is not at present known.

Microfilaments, which are fibre-like structures, narrower than microtubules, are made up of a protein called actin, that resembles, or is identical to, the actin of muscle fibres (section 10.2.3). Like muscle fibres, microfilaments are contractile, and the cytokinetic furrow appears to be formed by contraction of a ring of microfilaments pulling the surface membrane together (Fig. 4.8). The formation of two more or less spherical daughter cells from a spherical parent involves an increase in surface area of some 40 per cent. The increase is provided by an unfolding of previously accumulated convolutions (such as microvilli) of the surface membrane; membrane synthesis does not take place during mitosis. As in the establishment of the mitotic spindle, neither the attachment mechanism nor the spatial specificity of furrow formation is known. It is likely that some microfilaments are involved in spindle formation also, and conversely that microtubules may play a minor role in cytokinesis.

Other events during the cell cycle are less clearcut. The synthesis of RNA, protein, phospholipid, and carbohydrate occurs continuously throughout G_1, S, and G_2. During mitosis most biosynthetic activity is reduced to a minimum. Individual species of RNA, protein, and so forth are synthesized at specific times. Histones, for example, are synthesized together with DNA during S phase. The same is true of enzymes such as thymidine kinase that are involved in DNA synthesis. Other enzymes, such as lactate dehydrogenase, are synthesized continuously between G_1 and G_2. Many components of the cell surface are synthesized in G_1. Mitochondrial DNA and some mitochondrial enzymes are synthesized in S; other mitochondrial enzymes are synthesized continuously between G_1 and

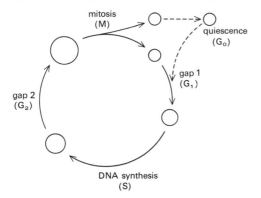

Fig. 4.7 The cell cycle

(a) chromosome
    movement

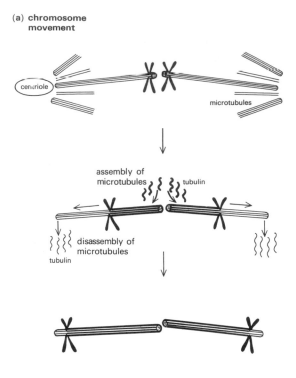

(b) constriction of cell

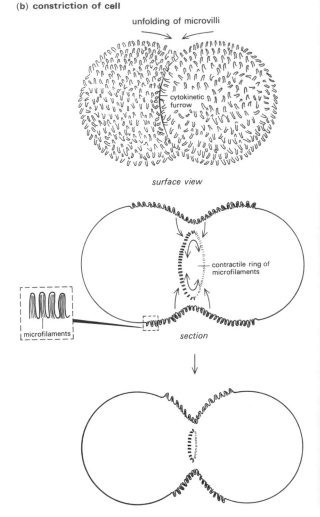

Fig. 4.8 Function of microtubules and microfilaments in mitosis and cytokinesis

(a) the mitotic spindle is first assembled (probably by growth from both ends). Movement of chromosome then occurs by simultaneous assembly (shown in bold) and disassembly of microtubules.
(b) During cytokinesis microvilli (which are also illustrated in Fig. 6.4) begin to unfold in such a way that surface membrane 'flows' towards the cytokinetic furrow. This releases microfilaments which re-assemble to form a contractile ring, which constricts the cell in half; the exact mechanism is not clear.

G_2. In each case an exact doubling in amount takes place between successive cell divisions.

In the case of proteins that are synthesized in S, immediately following duplication of the respective gene and its transcription into mRNA, the switch for initiating protein synthesis may be the doubling of the DNA; in other words the amount of DNA that is present determines the amount of protein that is synthesized. Such an effect is known as 'gene dosage'. In the case of proteins that are synthesized before, during, or after the synthesis of DNA, the amount of DNA present cannot determine the amount of protein formed. In this case protein synthesis is controlled by an effect other than 'gene dosage'.

When cells are dividing rapidly, as in the early stages of embryonic growth or during the proliferation of blood cell precursors in the bone marrow, the cell cycle time is short. The time between successive mitoses can be as little as 24 hours. As cell division is reduced, concomitantly with differentiation into specific cell types, the length of the cell cycle increases. This is due largely to an increase in the time of G_1, with the S, G_2, and M phases remaining approximately the same. In some cell types division virtually ceases altogether; for such cells the concept of G_0 (Fig. 4.7) has been introduced. Circulating lymphocytes, for example, do not divide at all, unless they are specifically stimulated to do so by presence of an antigen or other molecule (section 11.2.1.2). Progression from G_0 back into the cell cycle then occurs.

4.3.2 Cancer

One of the key differences between cancer cells and normal cells is the inability of cancer cells to stop dividing, that is, to enter G_0. At present the mechanism by which non-cancer cells become committed to G_0, or to an indefinitely long G_1 period, is not known.

The cell surface seems somehow to be involved. Elucidation of the exact mechanism would provide an important advance in our understanding of the aetiology of cancer. At the present time one can only speculate as follows.

The initial event is probably a **genetic** one. Many carcinogenic agents are mutagenic; that is, they increase the rate at which mutants appear. Radioactive emission following a nuclear explosion, or ultraviolet light, are both carcinogenic and mutagenic. So are many drugs. Certain viruses, the DNA of which may become incorporated into host chromosomes, appear to be carcinogenic. In some cancers an altered chromosome pattern is discernible (Fig. 4.4). Moreover cancer cells breed true; that is their defect is hereditable. That every type of malignancy involves a change in one or more genes, however, has not been proved. It is possible that in some instances gene *expression* (that is, the transcription of DNA into RNA) rather than gene *content* is permanently altered in a manner akin to that in cell development (Fig. 4.1).

The phenotypic outcome of the genetic change is not clear. If the cell surface is affected, it would explain two properties characteristic of cancer cells: an insensitivity to the environment, such that the normal signals for the cessation of growth are unheeded, and

an ability to metastasize, that is to invade and grow, in other tissues (Fig. 4.9). Failure to enter G_0 (or ability to remain in G_1 for long periods of time) means that cancer cells continue to divide unchecked and therefore have the enzymic pattern of constantly dividing cells like those of the early embryo. In other words the high content of enzymes of glycolysis and of DNA synthesis in some cancer cells is a consequence, and not a cause, of their unrestricted growth.

4.4 Heredity

The duplication of DNA and the subsequent allocation of identical sets of chromosomes into daughter cells explains how DNA is conserved in all cells of the body, and how the potentiality for cell differentiation is retained in every cell (Fig. 4.1). It does not explain how an offspring inherits some genes from one parent and some from another, and yet is able at the same time to inherit some genes from all four of his grandparents. The answer lies in the fact that in the reproductive cells of male and female, a special mechanism of cell division, known as meiosis, takes place.

4.4.1 Meiosis

When a sperm fertilizes an ovum, the resultant egg contains chromatids from each parent. Since the number of chromatids in the fertilized egg is the same as that in any other cell in the body, namely 23 pairs before 'S' phase, it is clear that in sperm or ovum the number of chromatids is only half the usual amount, that is 23 single copies. The way in which reduction from 46 to 23 chromatids occurs is simple: cell division without chromatid duplication. The process is called **meiosis**. It explains how the most important characteristic that an offspring inherits from either his father or his mother, namely his sex, is determined (Fig. 4.10). The same applies to other genes such as those specifying height, colour of eyes, of hair, and so forth. It does not explain how an individual can inherit genes from all four of his grandparents.

In fact the process of meiosis is somewhat more complicated than that shown in Fig. 4.10. It involves a special cell cycle, in which the chromatids become paired during replication, before the reductive division (Fig. 4.11). As a result of pairing during replication, occasional **'crossing over'** of genes occurs. That is, newly synthesized DNA contains segments of DNA from both the parental chromatids, instead of from only one (Fig. 4.11). This mechanism explains how a chromatid can come to carry genes from both parents; when such a chromatid participates in fertilization, the genes from both parents are passed to the offspring.

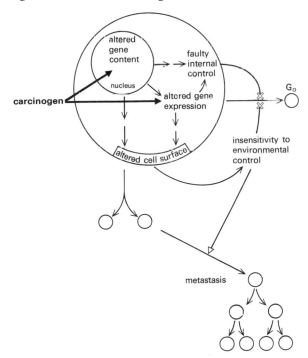

Fig. 4.9   Possible defects of cancer cells

The primary defect (altered gene content or altered gene expression) leads to a failure to enter G_0 (i.e. to cease cell division), possibly due to a defect at the cell surface. The result is continued cell division, which may lead to metastasis (invasion of other tissues).

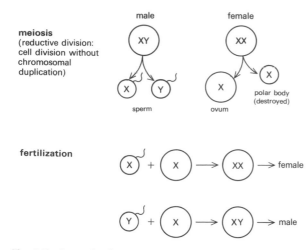

meiosis
(reductive division:
cell division without
chromosomal
duplication)

fertilization

Fig. 4.10   Determination of sex

Since this mechanism applies to either sperm or ovum, an individual can inherit genes from each of his four grandparents.

The distribution of chromatids into daughter cells during reductive division is entirely random. Just as there is a 1 in 2 chance that the sperm contains an X rather than a Y chromatid, so there is a 1 in 2 chance of the sperm or ovum containing the maternal rather than the paternal version of any of the other 22

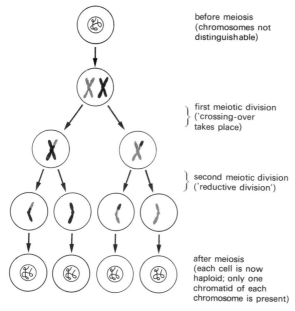

before meiosis
(chromosomes not
distinguishable)

} first meiotic division
('crossing-over
takes place)

} second meiotic division
('reductive division')

after meiosis
(each cell is now
haploid; only one
chromatid of each
chromosome is present)

Fig. 4.11   Arrangement of chromosomes during meiosis

Meiosis is a process that occurs only in the germ cells. The first meiotic division is in general similar to mitotic division (Fig. 4.3), except that 'crossing over' takes place. The overall process, which is quite complicated, is shown in outline only; one of the 23 chromosome pairs (in humans) is shown.

chromatids. In other words human characteristics that are specified by genes located on different chromosomes are inherited in a random manner, sometimes called Mendelian after its original discoverer (Fig. 4.12). Characteristics that are specified by genes on the same chromosome, on the other hand, are inherited in a linked manner, such that if one characteristic is inherited from one grandparent the other one is also. The latter statement is subject to the proviso that crossing over has not occurred. If it has, chromosomally linked characteristics will appear as unlinked.

4.4.2 Mutation

It must be emphasized that crossing-over is a relatively rare event (involving just a few chromatids), and therefore most genes on any one chromatid are inherited either from the maternal or the paternal grandparent. An even rarer event is **mutation** (about 1 out of 10^6 genes altered at every generation) (Fig. 4.13). The frequency of mutation is the same in meiosis as in mitosis. Mutation causes a change in the composition of DNA. A wrong base may become inserted, or an extra base inserted or removed, or a whole stretch of DNA may become deleted or inserted (gene duplication). Once it has occurred the change is generally stable, though in some instances DNA can be repaired (see below). The period of the cell cycle that is most sensitive to mutations is that of DNA synthesis, and most spontaneous mutations (that is, the very low level of mutation, calculated at approximately 1 in every 10^9–10^{10} base pairs replicated) occurs during DNA replication. But mutagens that alter the structure of DNA can act at any time during the cell cycle.

Mutation outside the germ cells is not passed to the next generation. Since most mutations are lethal, the cells in which a mutation has occurred simply die and little deleterious effect is seen. An exception is when a mutant clone is able to establish itself and grow. When this happens a tumour, which may be malignant, results.

Mutation within the germ cells can result in aberrant offspring. The most obvious example is that of one of the forms of Down's syndrome (Fig. 4.4). The afflicted offspring has an extra chromatid associated with one of the autosomal chromosomes (chromosomes other than the sex chromosomes). This in some way imbalances metabolism, including that of brain, resulting in subnormal behaviour and mental retardation. The incidence of Down's syndrome increases with the age at which the mother conceives: the chances of an afflicted child being born to a woman under 20 years old is 1 in 2500; to a woman over 45 years old, it is 1 in 45. Since it is known that all the ova which a woman will produce are laid down before birth (that is, well before

a autosomal genes

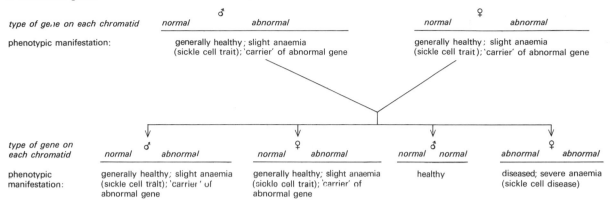

b X-linked genes

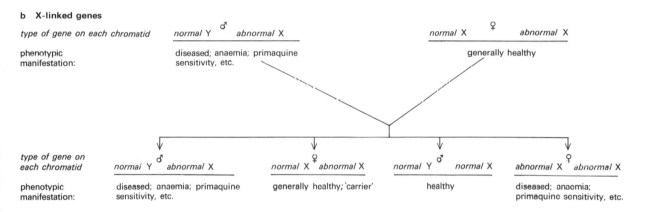

Fig. 4.12 Mendelian inheritance of genes

The anticipated distribution of genes among four offspring is shown in (a) and (b). Such an exact 2:1:1 distribution is observed only when a large number of families is analysed.

a The inheritance of the β-chain of haemoglobin and of its 'sickle cell' variant (Fig. 7.21), is shown. Note that the sex of the parent is immaterial to the transmission of autosomal genes.

b The inheritance of glucose 6-P dehydrogenase, and of its abnormal variant (G6PDH deficiency), is shown. All males carrying the abnormal gene are diseased; only females carrying the abnormal gene on both chromatids (homozygotes) are diseased.

Fig. 4.13 Frequency of inherited characteristics

Mechanism	Unit	Frequency
Meiosis	Entire chromosome	1 in 2 (i.e. as in determination of sex, Fig. 4.10)
Crossing over	Parts of chromosome	1 in 10 to 1 in 100 per chromosome
Mutation	Single gene[1]	1 in 10^5 to 1 in 10^8 per gene per generation

[1] Where the gene is responsible for a process like chromosome duplication, the result may lead to a change in an entire chromosome (as in Down's syndrome or certain forms of cancer, see Fig. 4.4).

the age of 20), it appears that the chance of a fault occurring during chromosomal duplication increases with time of exposure of the ovum to the environment. Such a time-dependence is typical of radiation damage, and it is possible that the mutation responsible for Down's syndrome results from the cumulative effects of atmospheric background radiation, or other environmental cause related to ageing. Certainly women exposed to high radiation as a result of fall-out from atomic bombs tend to give birth to genetically damaged offspring. The same is true of women exposed to drugs such as thalidomide, which are also mutagenic. In fact females prior to the age at which conception ceases are at great risk, with regard to their progeny, from exposure to any potential mutagen. The risk from exposure of males to mutagens is less, since fresh sperm is formed after every ejaculation, and the cumulative effect is therefore absent.

Not all mutations are deleterious. Many have no effect and some may even be beneficial. The fact that they are not detected does not mean that they do not occur. On the contrary, such mutations are the very basis of evolution, the selection of the fittest. In fact if one compares the structure of any particular enzyme or other protein in a large number of humans, one finds occasional mutants in which one or other amino acid is altered, without overt change in functional capacity. Some 25 per cent of all human enzymes appear to fall into this category. Such mutants, which are known as polymorphic enzymes, occur in about 10 per cent of the population. They are not isoenzymes in the sense that lactate dehydrogenase-1 and lactate dehydrogenase-5, or muscle phosphorylase and liver phosphorylase, are; in contrast to the lactate dehydrogenase or phosphorylase isoenzymes, which are coded for by two separate genes, polymorphic enzymes are variants (alleles) of the same gene. The evolutionary advantage of particular polymorphic variants is generally unclear. One can only surmise that if one mutant form predominates in the climatic conditions of Europe in the twentieth century, another may well do so in a different region today, or in the same region a thousand years hence.

In the case of haemoglobin, a specific advantage has been postulated with regard to the sickle cell gene (Fig. 7.21). This gene, which appears to provide some protection against malaria, occurs in some 20 per cent of the indigenous population in malaria-infested parts of West Africa. Among the descendants of West Africans now living in the USA, in which there is no malaria, the gene frequency is said to be approximately 9 per cent. In only some 200–300 years, therefore, there appears to have been a significant reduction in gene frequency. Although these figures are liable to considerable sources of error, they do illustrate in a dramatic form the principle of evolution by selection.

4.5 DNA

4.5.1 Structure

DNA is a polymer of **deoxyribose phosphate** units, linked through the 3:5-positions of the sugar. Attached to the 1-position of the sugar is one of the four nitrogenous bases, **cytosine** (C), **thymine** (T), **adenine** (A), or **guanine** (G) (Fig. 1.17). One of the first clues to the fact that DNA is made up of two chains wound round each other in a **double helix** was the discovery that the total number of pyrimidine molecules(C+T) is always equal to the number of purines (A+G), suggesting some kind of inter-chain purine–pyrimidine link. The nature of that link was indicated by the finding that the number of C residues is equal to the number of G residues and that the number of T residues is equal to that of A. A brilliant piece of X-ray crystallography confirmed the analytical results in showing the existence of a double helix, with pairs of bases held together by hydrogen bonds (Fig. 1.18).

The two chains of the double helix are separated by treatments that break hydrogen bonds (section 1.1.1): heating to around 60–70 °C, strong solutions of urea, and so forth. The fact that chains separated by heating are re-annealed by cooling has proved useful in biochemical studies of DNA–RNA interactions (Fig. 4.19).

Chromatids are made up of a single molecule (double strand) of DNA. In mitochondria the two ends of the molecule are attached to each other, making a circle; this is unlikely to be so in chromatids. The deoxyribose 3′:5′-phosphate chain is continuous from gene to gene. But not all regions of DNA specify genes. On the contrary, most of the DNA in a chromosome does not code for specific proteins; such regions probably contain sequences that regulate the transcription of genes. How much of the control sequences is transcribed into RNA is not clear. Although the organization of the primary structure of DNA is the same throughout a chromatid, the secondary and tertiary structure may vary between genetic and control regions. Such differences are probably stabilized by the relative attachment of histones and non-histone proteins.

Some genes are present not as a single copy, but are repeated several times; 30 per cent of DNA probably consists of sequence repeated 20 times or more. Reiteration of genes protects them against mutational change, since mutation affects regions shorter than a single gene. Hence even if one gene is destroyed, its reiterated copies continue to function. Repetitive genes have been found to code for structural components such as ribosomal RNA. Protection against mutation is not, of course, advantageous in the evolutionary sense, and most genes that code for specific enzymes are present as single copies.

The potential length of DNA in a chromosome exceeds the length of a chromosome by some 1000 to 10 000 fold. It is obvious that the DNA is coiled not merely as a double helix, but is supercoiled round itself (Fig. 1.19). The extent of coiling depends on the stage of the cell cycle. During mitosis it is very tight (condensed) indeed; during the rest of the cell cycle it is in a looser structure, such that the enzymes DNA and RNA polymerase have access to the parts of the DNA that are being copied. The mechanism of the above change from condensed to loose structure is not understood. It obviously affects also the histones and non-histones that are attached to DNA. In fact it is likely that structural alterations, such as phosphorylation of a serine residue in some of the histones attached to DNA, triggers the change.

4.5.2 Biosynthesis

In principle the biosynthesis of DNA is straightforward: the insertion of matching bases along each strand of a partially unwound stretch of chromosome by a non-specific **polymerase** enzyme (Fig. 4.14). Each newly synthesized strand remains paired to one of the existing strands, as a result of which the process is termed 'semi-conservative'. The polymerization step is potentially reversible, as in RNA, protein, and carbohydrate synthesis, and is pulled towards synthesis only by hydrolysis of inorganic pyrophosphate. In practice DNA synthesis is a good deal more complicated.

For one thing the polymerase enzyme works in one direction $(5' \rightarrow 3')$ only. Yet the two strands run in opposite directions. Since there is no room within the nucleus for the whole of the DNA of a chromosome to unwind at once, it is clear that DNA is replicated in short stretches at a time. The stretches appear to be as short as 100–200 nucleotide base pairs. In other words, there are many growing points (Fig. 4.15) along a chromosome during replication; there would in any case be insufficient time during the S phase of a cell cycle for a single polymerase enzyme to replicate an entire strand of DNA. Secondly, several different types of polymerase enzyme exist in cells. The exact role of each is by no means clear.

The mechanism by which DNA unwinds itself sufficiently, without breaking the covalently linked sugar phosphate backbone, to allow the mechanism depicted in Fig. 4.15 to take place, is also complex, and is not fully understood. It is possible that a sugar–phosphate bond is broken by an endonuclease enzyme, that is, an enzyme that breaks internal bonds of DNA, and that following replication the break is made good by a ligase enzyme. Hence at least three enzymes, nuclease, polymerase, and ligase may be involved in the synthesis of DNA. In addition the unwinding process may require yet another enzyme.

Another point that is not yet clear is the sequence in which the two strands are copied. Are they copied simultaneously by two enzymes, or is one strand copied before the other? All these are some of the details of DNA biosynthesis that have still to be resolved. Whatever the nature of the replicative process, it is clear that the enzymes are crucial to the

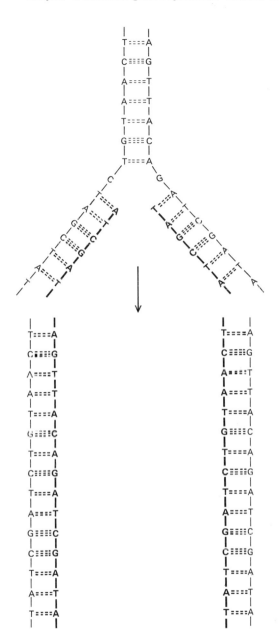

overall reaction

(1)

$$\text{dATP} + \text{dGTP} + \text{dTTP} + \text{dCTP} \xrightarrow[\text{polymerase}]{\text{DNA}} [\text{dAMP,dGMP,dTMP,dCMP}] + 4\text{PP}_i$$
$$\hspace{9cm}\text{polymer}$$

(2)

$$4\text{PP}_i \longrightarrow 8\text{P}_i$$

Fig. 4.14 Biosynthesis of DNA

Newly synthesized strands are shown in bold.

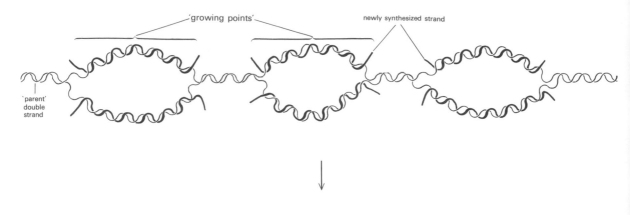

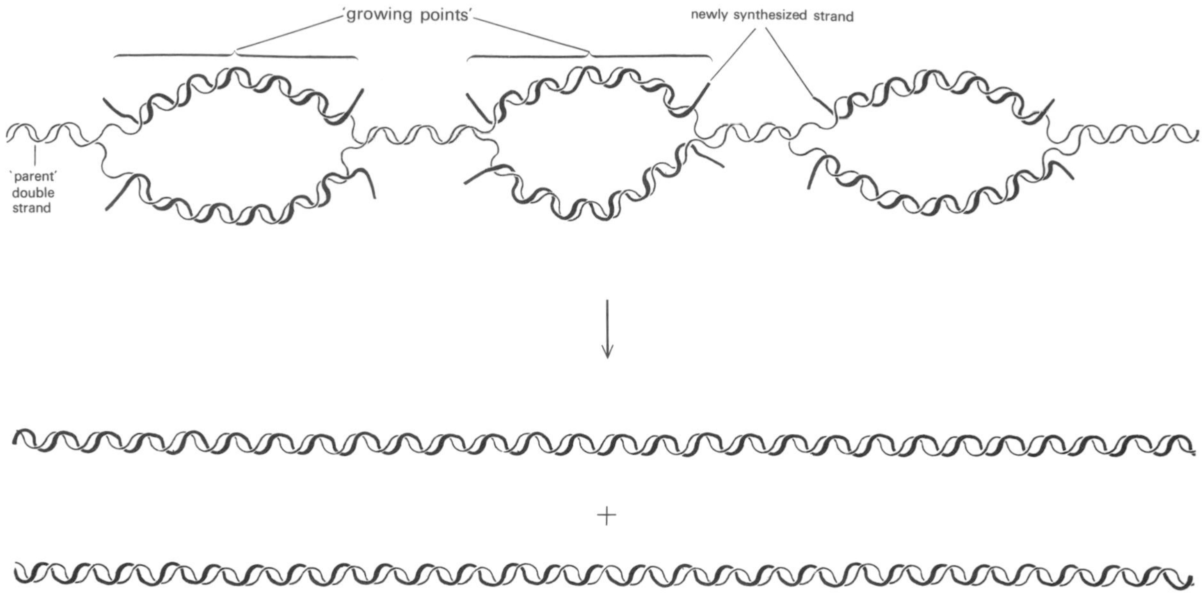

'growing points'

newly synthesized strand

'parent' double strand

Fig. 4.15 Mechanism of DNA replication

The newly-synthesized strand is shown in bold.

viability to cells; no hereditary disorders in which DNA synthesis is absent exist. There are however diseases in which enzymes *apparently* associated with DNA synthesis are non-functional.

A good example is xeroderma pigmentosum. Sufferers from this disease are exceptionally sensitive to the ultraviolet radiation present in bright sunlight, and tend to develop skin cancers after excessive exposure.

The missing enzyme is an endonuclease, like that mentioned above. Its function is to repair pieces of faulty DNA by excision of nucleotides. The way faulty stretches arise in UV light is by the cross-linking of adjacent thymine residues (Fig. 4.16). If the endonuclease is absent, the dimers are not removed and subsequent replication leads to DNA having a fault at this point. In other words a mutant gene results.

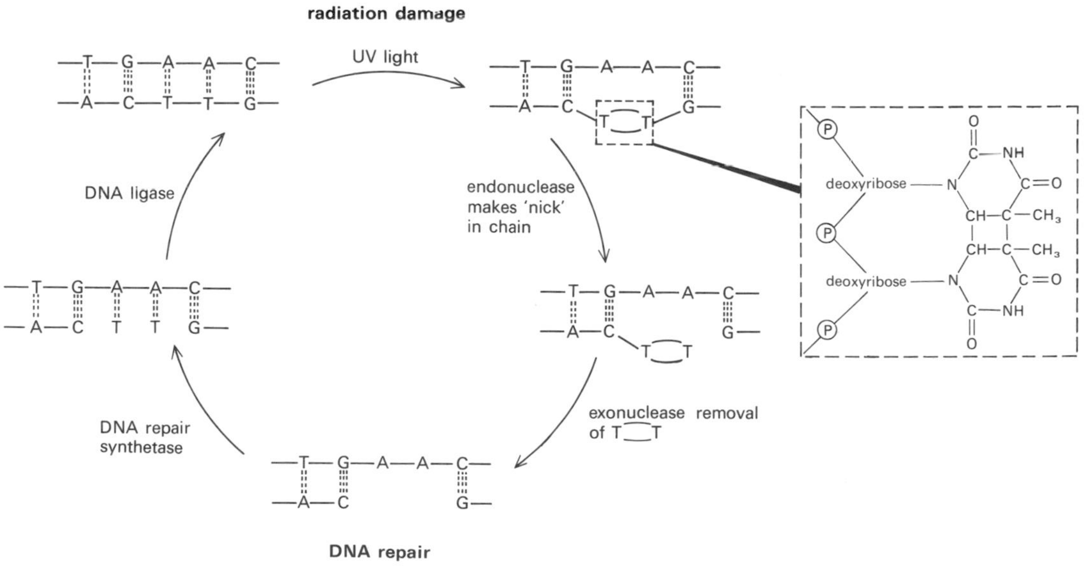

radiation damage

UV light

—T—G—A—A—C—
—A—C—T—T—G—

—T—G—A—A—C—
—A—C—T—T—G—

DNA ligase

endonuclease makes 'nick' in chain

deoxyribose

deoxyribose

—T—G—A—A—C—
—A—C—T—T—G—

—T—G—A—A—C—
—A—C—T—T—G—

exonuclease removal of T—T

DNA repair synthetase

—T—G—A—A—C—
—A—C—G—

DNA repair

Fig. 4.16 Formation of thymine dimers and their repair

Formation of thymine dimers and their repair is one example of a 'reversible' mutation. Another example is the insertion of a wrong base, followed by its subsequent excision.

Xeroderma pigmentosum is probably only one of several diseases in which the repair of faulty DNA is impaired. The underlying cause for the emergence of malignant tumours, for example, may be the failure to repair aberrant stretches of DNA. In all these situations DNA *synthesis* proceeds normally. If an endonuclease is required for DNA synthesis, it must be an enzyme different to the repair endonuclease.

The intracellular signals that trigger the start of a round of DNA synthesis, and that ensure that only one copy of each chromosome is made, have not been elucidated. Several possibilities exist. If the polymerase or other enzyme necessary for replication is very short-lived, for example, its synthesis just prior to the S stage may be the trigger. Alternatively some molecule that activates chromosomes so that unwinding of DNA begins may be the stimulus. In certain situations, such as in foetal development or in the liver during childhood, more than one copy of each chromosome is made. The result is known as polyploidy. Unlike the case in Down's syndrome or in certain cancers (Fig. 4.4), all chromosomes are replicated to the same extent; the number of copies is a power of 2, that is 4, 8, 16, 32, etc., copies of each of the 23 chromosomes. The reason for this overproduction has not been established.

4.6 RNA

4.6.1 Structure

The structure of RNA resembles that of DNA in that it is a 3:5 linked sugar–phosphate polymer having a nitrogenous base on the 1-position of the sugar. It differs from DNA in that the sugar is **ribose**, that the bases are **uracil** (U), **cytosine**, **adenine**, and **guanine**, and that the chains are single and not double (Fig. 1.17).

Cells contain three classes of RNA. **Ribosomal RNA** (rRNA) makes up approximately 80 per cent of total RNA. It is present, together with protein, as **ribosomes** which occur either free in the cytoplasm, or bound to the endoplasmic reticulum. The function of ribosomes is to catalyse protein synthesis; their structure is accordingly described in Chapter 5. The constituent RNA is of four sizes, a $5s$ component (mol. wt about 4×10^4), a $6s$ component (mol. wt about 6×10^4), an $18s$ component (mol. wt about 7×10^5), and a $28s$ component (mol. wt about 2×10^6). The s value is the sedimentation coefficient measured in an ultracentrifuge, and gives a good indication of the overall size of a macromolecule.

The next most abundant class (approximately 15 per cent) is **transfer RNA** (tRNA). tRNA is a relatively small molecule (mol. wt about 25 000), having an s

value of approximately 4. The function of tRNA is to **recognize** activated **amino acids** (Fig. 2.35) and to insert them into a growing polypeptide chain. There are as many as 50 or 60 distinct molecular species, at least one being specific for each of the 20 amino acids.

The ratios between the different bases, as well as physical evidence, is indicative of considerable hydrogen bonding between complementary bases (A–U and C–G) in tRNA. Yet the molecule is a single chain. It turns out that the chain is folded back on itself in several places, such that hydrogen bonding can occur (Fig. 1.18). The result is an L-shaped structure (Fig. 4.17). It should be noted that tRNA contains a number of modified bases in addition to the four common ones. 5'-Ribosyluridine, ψ uridine, in which the sugar is linked to uracil through the C-6 carbon atom instead of the N-1 atom, and N-methylated purines, are present in all types of tRNA so far examined. They also occur to a lesser extent in rRNA.

Transfer RNA molecules have three binding sites: one that attaches the amino acid, one that binds the codon (that is, the triplet of bases) specifying that amino acid in mRNA, and one that finds the amino acid activating enzyme. The codon binding site is called the anticodon, and contains three bases complementary to the codon. Thus for the tRNA shown in Fig. 4.17, the anticodon GUA on the tRNA binds to the codon UAC on mRNA (note that tRNA and mRNA 'run' in opposite directions at the anticodon–codon binding site). Obviously all the binding sites on tRNA have to be accessible, and the three-dimensional structure for tRNA shows how this is achieved (Fig. 4.17).

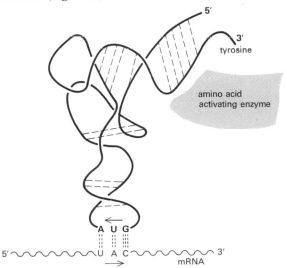

Fig. 4.17 Binding sites on t-RNA

An outline of the tertiary structure of Tyr-tRNA (in bold) is shown; H bonding is indicated by the dashed lines. The exact site for the amino acid activating enzyme is not known.

Messenger RNA (mRNA) makes up the remainder (approximately 5 per cent) of cellular RNA. Like rRNA and tRNA it is found mainly in the cytoplasm. Its function is to **translate** the **genetic code** of DNA into proteins. The size of an mRNA molecule is therefore the size of a gene, which works out at an average molecular weight of approximately 5×10^5.

Messenger RNA has sometimes been called dRNA or DNA-like RNA. Although all types of RNA, that is, rRNA, tRNA, and mRNA, are coded for by their respective genes, and therefore reflect the base composition of those genes on the DNA, the term dRNA has some validity. The reason is as follows. Most of the genes on a chromosome code for specific mRNA molecules. The number of genes coding for rRNA or

tRNA – even allowing for some reiterated genes – is very small in comparison. Hence the overall composition (in terms of the four bases) of the different mRNA molecules present in a cell together reflects the composition of the genetic part of DNA (or rather of one of its two strands) more closely than does the composition of tRNA or rRNA, despite the fact that the latter two species make up the bulk of cellular RNA (Fig. 4.18). But it must be stressed that the reflection is a relative one. Genes make up only a small part of DNA. Most of the DNA is not transcribed into RNA at all, and therefore the bulk of DNA does not correspond to any type of RNA molecule.

Hybridization experiments between DNA and RNA

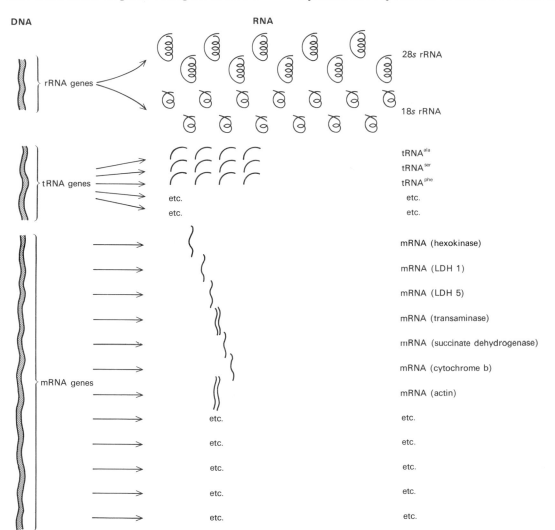

Fig. 4.18 Relative amounts of different types of RNA in cells

Cells contain *many* copies of rRNA and tRNA made from *few* copies of their respective genes, and *few* copies of mRNA made from each of *many* different genes. Note that the genes for rRNA and tRNA are on different chromosomes.

confirm the fact that mRNA resembles a greater part of the DNA than does tRNA or rRNA. That is, if single-stranded DNA is exposed to a mixture of cellular RNA molecules, radioactively labelled for ease of detection, the efficiency of binding is much greater for mRNA than for rRNA or tRNA, since the latter two species compete for a small number of binding sites. Such hybridization techniques have been used to assess the presence of specific mRNA molecules in cells. For example, in hereditary disorders in which a particular enzyme or other protein is missing, it might be advantageous from the therapeutic point of view to know whether the fault is in the gene (as generally supposed) or whether it lies in the translation of the particular mRNA. In the first case the mRNA would be absent, in the second case it would be present. Hybridization with excess unlabelled RNA from diseased cells, followed by hybridization with labelled RNA from healthy cells, allows one to distinguish the two possibilities (Fig. 4.19). Such experiments have shown that in some hereditary disorders, e.g. certain α-thalassaemias, it is the gene that is absent. In others, such as β-thalassaemias, the respective mRNA, and hence its gene, are present; the fault lies in the translation of mRNA, or in the stability of the protein that is formed.

In addition to the three main classes of RNA, cells contain minor amounts of mitochondrial and nuclear RNA. As mentioned in section 3.2, mitochondria contain their own rRNA and tRNA. The structures are similar to the respective cytoplasmic RNA species, except that mitochondrial rRNA is somewhat smaller (12s and 16s). Nuclear RNA, which represents some 5 per cent of total RNA, is made up largely of heterogenous (Hn) RNA. This contains much high molecular weight RNA (some as large as 100s). Most of this RNA never leaves the nucleus; it is degraded without ever apparently participating in protein synthesis. A small part of the Hn RNA is somehow selected for transport into the cytoplasm. It is degraded before, or during, transport and finishes up as mRNA in the cytoplasm. Ribosomal RNA and tRNA are also present in the nucleus as high molecular weight precursors, known as pre-rRNA and pre-tRNA respectively. Pre-rRNA, for example, is of size 45s; it is degraded to yield one molecule each of rRNA of size 5s, 7s, 18s, and 28s.

4.6.2 Biosynthesis

All RNA (with the exception of a small amount of mitochondrial RNA) is synthesized in the nucleus. The mechanism of RNA synthesis is broadly similar to that of DNA synthesis. A **polymerase** enzyme catalyses the successive condensation of nucleoside triphosphates along a DNA template (Fig. 4.20). Unlike DNA

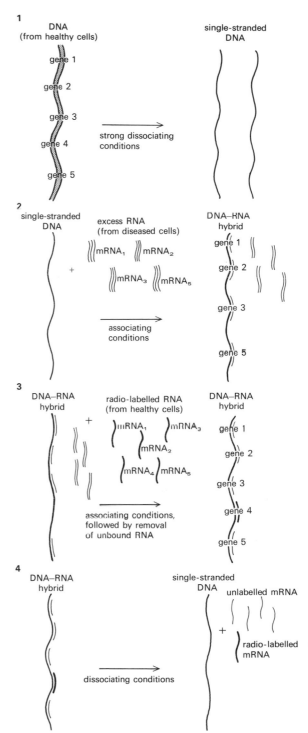

Fig. 4.19 Detection of specific mRNA molecules by hybridization techniques

The diagram illustrates how a biochemical defect in diseased cells can be attributed to a failure to synthesise a specific mRNA (e.g. due to loss of gene 4).

115

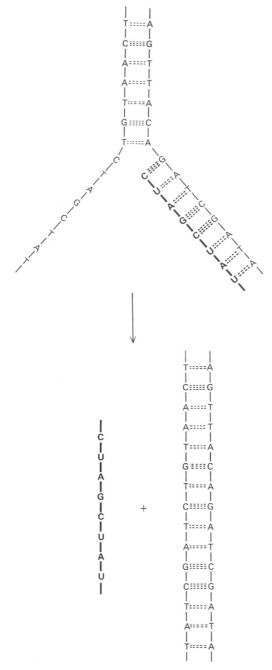

The newly synthesized strand is shown in bold

overall reaction

(1)

$$ATP + GTP + UTP + CTP \xrightarrow[\text{polymerase}]{\text{RNA}} [AMP, GMP, UMP, CMP] + 4PP_i$$
$$\text{polymer}$$

(2)

$$4PP_i \longrightarrow 8P_i$$

Fig. 4.20 Biosynthesis of RNA

synthesis, only a single strand is formed, and the molecule detaches itself from its template as polymerization proceeds. It seems likely that only one of the strands of DNA is used to make RNA. The other is probably not expressed. In certain DNA viruses, however, both strands are expressed in the following manner. One strand is transcribed first, to approximately half-way along the chromosome. The polymerase enzyme then jumps to the other strand and completes transcription on that strand.

RNA polymerase differs from DNA polymerase in several ways. First, separate RNA polymerases participate in the synthesis of Hn RNA, of pre-tRNA (both in the nucleoplasm), and of rRNA (in the nucleolus). Secondly, in bacteria, and perhaps in animal cells also, RNA polymerase requires specific proteins for the initiation and termination of transcription. It will be appreciated that a major difference between DNA and RNA synthesis is that in the latter case only certain regions of the DNA are copied. Moreover some regions are copied many times, as in rRNA and tRNA, and some only few times, as in mRNA (Fig. 4.18). None is copied only once (as in DNA biosynthesis). How initiators and terminator proteins act is not known.

The specification of particular genes for transcription appears to involve the non-histone proteins of the chromosome. For, just as the mRNA pattern of different cell types varies, so does the pattern of non-histone proteins. This is in contrast to the histones which, like DNA, are the same in all cell types. The exact relationship between non-histone proteins, initiation and termination factors, and DNA is not clear.

The majority of mRNA molecules appear in the cytoplasm with a stretch of AMP residues (up to 200 long) attached to the 3'-end of the molecule (Fig. 4.21). The addition of these adenine nucleotides takes place after transcription has occurred and is catalysed by an enzyme different from RNA polymerase. The other end of mRNA molecules is modified in a different way (Fig. 4.21). First, the last ribose residue is methylated at the 2-position. Secondly, the 5'-phosphate group has two further phosphate groups attached. Thirdly, the last phosphate group has attached to it an N-methylated guanosine residue, linked through the 5-position of its ribose. The function of the polyA regions at the 3'-end, and of the methylated 'caps' at the 5'-end, is not known. It is possible that the modifications influence the stability of mRNA.

4.7 Summary

Gene expression

1. The functions of the human body are specified by its **genes**. Genes are composed of **DNA**. The property

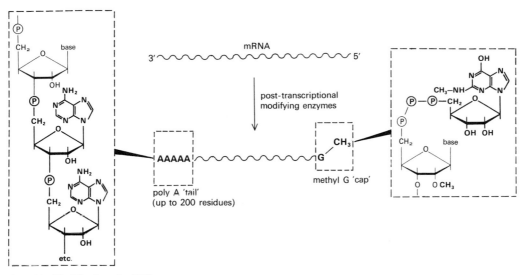

Fig. 4.21   Modification of mRNA

Modifications are shown in bold.

of a gene is

i. to duplicate itself exactly at every cell division (**replication**) and

ii. to cause a specific molecule of **RNA** (mRNA) to be synthesized (**transcription**), which then causes a specific **protein** to be synthesized (**translation**).

Hence the complement of proteins (including enzymes) in the human body is a reflection of its genes.

2. Because the genes are identical in every cell of the body, whereas the proteins are not, cell specialization involves a differential expression of genes (by transcriptional and translational control).

Nuclear structure

1. Genes, which are present in identical form in every cell, are contained within the **nucleus** of cells. The DNA that constitutes the genes has certain proteins (known as **histones** and **non-histone** proteins) bound to it; the complex is known as **chromatin**. During cell division (**mitosis**), when the nuclear membrane disappears, chromatin becomes distinguishable as a number (23 pairs) of structurally different **chromosomes**.

2. The nucleus also contains a region known as the **nucleolus**; this contains most of the nuclear RNA. The nucleus is separated from the cytoplasm by a membrane, which has pores through which RNA can pass.

Nuclear events

1. The events that occur in dividing cells are collectively known as the **cell cycle**. Following division,

synthesis of cell constituents (carbohydrate, fats, protein, nucleic acids, and so forth) starts, and continues until all components have doubled in amount; synthesis of DNA occurs only during a discrete part of the cell cycle (known as the S phase). During mitosis, the chromosomes become distributed, by means of a structure known as mitotic spindle, in such a manner that each of the two daughter cells receives a complete set of 23 chromosomes; the daughter cells separate from each other by a process known as **cytokinesis**. The formation of the mitotic spindle, and cytokinesis, involve cytoplasmic structures known as **microtubules** and **microfilaments**.

2. As cells of the body become differentiated (into muscle, nerve, liver, fat cells, and so on), cell division gradually decreases until organs and tissues of fixed size are formed. Occasionally malignant cells, in which cell division continues unchecked, arise; this leads to **cancer**. It is probable that cancer cells have altered genes.

Heredity

1. Certain cells of the body (germ cells) contain only half as much DNA as other cells; only a single set of chromosomes is present. The process by which cells, containing a double set of chromosomes, divide to form the germ cells, is known as **meiosis**. During sexual mating, a germ cell from a male individual (sperm) combines with a germ cell from a female individual (ovum), to form a fertilized cell that contains both sets of chromosomes.

2. During the life of an individual, the DNA may

become altered; such an event, which is very rare, is known as a **mutation**. Mutation may lead to the synthesis of inactive, or otherwise altered, proteins, or it may lead to a failure to synthesize certain proteins at all. A mutation that occurs in the germ cells may be passed onto offspring. A mutation that occurs in non-germ cells affects only the life of the individual; the development of cancer may result from such mutations.

DNA

1. DNA exists as a **double strand** of repeating 2-deoxyribose phosphate units, with a nitrogenous base (T, C, A, or G) attached to each deoxyribose unit. The two strands are held together by specific H bonding between a base in one strand and the adjacent base in the other strand: **T** of one strand is bonded to **A** of the other strand (and vice versa) and **C** of one strand is bonded to **G** of the other strand (and vice versa). The two strands are said to be complementary to each other. DNA occurs almost entirely in the nucleus of cells.

2. The biosynthesis of DNA (replication) occurs by separation of the two strands, followed by the formation of new strands complementary to each of the separated strands; that is, A is inserted opposite T (and vice versa) and C is inserted opposite G (and vice versa). The result is an exact duplication of the existing double strand. The enzyme responsible for the synthesis of DNA (**DNA polymerase**) is complex.

RNA

1. RNA exists as a single strand of repeating ribose phosphate units, with a nitrogenous base (U, C, A, or G) attached to each ribose unit. Three different types of RNA are present in cells: ribosomal RNA (**rRNA**), transfer RNA (**tRNA**), and messenger RNA (**mRNA**). Most of the RNA is present in the cytoplasm; rRNA is bound to proteins in the form of **ribosomes;** tRNA and mRNA are largely unbound. The three types of RNA participate in **protein synthesis** (See Chapter 5). Some RNA is present in the nucleus, where all types of RNA are synthesized.

2. Each species of RNA is synthesized from a complementary DNA molecule. That is, one of the strands of DNA is **transcribed** into a complementary strand of RNA (T, C, A, and G of the DNA being replaced respectively by A, G, U, and C of RNA), synthesized by the enzyme **RNA polymerase**. Only discrete regions of DNA (corresponding to different genes) are transcribed into RNA; the mechanism whereby certain regions of DNA are selected for transcription is not known.

FURTHER READING

Genetics

For a general introduction to genetics, including specific enzyme deficiencies:

C.A.Clarke (1978). *Human genetics and medicine* (2nd edition) (Studies in Biology No.20). Edward Arnold, London.

H.Harris (1975). *The principles of human biochemical genetics* (2nd edition), North-Holland, Amsterdam.

A.M.Srb, R.D.Owen, and R.S.Edgar (1970). *Facets of genetics.* (Readings from *Scientific American*) W.H. Freeman and Co., San Francisco.

Cell cycle

For a brief account:

C.A.Pasternak (1974). Biochemical aspects of the cell cycle In *Companion to biochemistry* (eds A.T.Bull, J.R.Lagnado, J.O.Thomas, and K.F.Tipton), p.399. Longman.

Cell differentiation:

N.MacLean (1977). *The differentiation of cells.* Edward Arnold, London.

C.A.Pasternak (1970). *Biochemistry of differentiation.* John Wiley, Chichester.

The biology of mitosis and meiosis:

B.John and K.R.Lewis (1972). *Somatic cell division* (Oxford Biology Reader No.26). Oxford University Press.

B.John and K.R.Lewis (1973). *Meiotic mechanism* (Oxford Biology Reader No.63). Oxford University Press.

A detailed discussion of chromatin:

G.Felsenfeld (1978). Chromatin. *Nature* (*Lond.*) **271,** 115.

M.MacLean and V.A.Hilder (1977). Mechanisms of chromatin activation and repression. *Int. Rev. Cytol.* **48,** 1.

A detailed discussion of microtubules:

R.E.Stephens and K.T.Edds (1976). Microtubules: structure, chemistry and function. *Physiol. Rev.* **56,** 709.

A biochemical discussion of mutation and cancer:

W.L.Nyhan (1977). Genetically determined molecular variation in Man. *Trends biochem. Sci.* **2,** 121.

R.B.Setlow (1978). Repair deficient human disorders and cancer. *Nature* (*Lond.*) **271,** 713.

B.Singer (1977). Nucleic acid alkylation, mutation and carcinogenesis: is there a relationship? *Trends biochem. Sci.* **2,** 180.

A general account of the structure and biosynthesis of nucleic acids:

Davidson's biochemistry of the nucleic acids (8th edition) (revised by R.L.P.Adams, R.H.Burdon, A.M.Campbell, and R.M.S.Smellie). Chapman and Hall, London.

A.Kornberg (1974). *DNA synthesis.* W.H.Freeman, San Francisco.

J.S.Krakow and S.A.Kumar (1977). Biosynthesis of RNA. In *Comprehensive biochemistry* (eds M.Florkin, A.Neuberger, and L.L.M.Van Deenen), Vol. 24. Elsevier, Amsterdam.

5

Cytoplasmic membranes: protein synthesis

5.1 Structural components

The cytoplasm of cells is filled with a network of membranous and non-membranous structures (Fig. 5.1). The membranes, known as **endoplasmic reticulum,** are of two types: smooth and rough. Rough endoplasmic reticulum resembles smooth endoplasmic reticulum except that particles known as ribosomes, are attached to it. Smooth and rough endoplasmic reticulum together account for most of the membranous structures in cells. Mitochondrial membranes, nuclear membranes, lysosomal membrane, Golgi membrane, and surface membrane make up the remainder.

The non-membranous structures comprise free ribosomes, microtubules, and microfilaments. The function of ribosomes, whether free or attached to endoplasmic reticulum, is to catalyse protein synthesis (section 5.2). The function of microtubules and microfilaments is to organize and to maintain the internal structure of cells: the movement of chromosomes and the establishment of the cytokinetic furrow (Fig. 4.8); the elaboration of the cell surface into microvilli (Fig. 6.4), and other protuberances; in fact the very determination of cell shape and cell movement.

5.1.1 Endoplasmic reticulum

The endoplasmic reticulum is similar to any other biological membrane. It is made up predominantly of proteins embedded in a phospholipid bilayer. There is rather little cholesterol, sphingomyelin, glycolipid, or glycoprotein.

The function of endoplasmic reticulum is as follows. First, several enzymes of 'cytoplasmic' pathways such as **cholesterol** and **phospholipid biosynthesis,** are embedded in the endoplasmic reticulum. Secondly, the **assembly** of **membranes** itself (section 2.5.3) probably takes place in the endoplasmic reticulum. The two functions are related, in that the second is a consequence of the first: phospholipids and cholesterol are the main lipid components of membranes. In other words, the enzymes that catalyse the synthesis of a lipid-soluble compound from a water-soluble precursor are situated at the boundary between aqueous and lipid enviroment (Fig. 5.2). As a result, synthesis and insertion are simultaneous and the assembly of membrane lipids becomes a spontaneous, thermodynamically favourable process (Fig. 5.3). The reverse mechanism operates in the degradation of membrane lipids (Fig. 5.2). In this instance, the enzymes causing the change of state (from hydrophobic to hydrophilic) are embedded in the lysosomal membrane, so that fusion between lysosomal membrane and endoplasmic reticulum (or other membrane) is a prerequisite for degradation (Fig. 2.24).

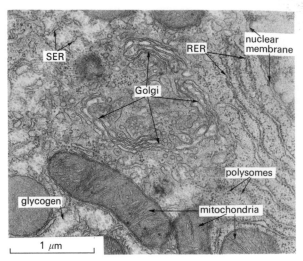

Fig. 5.1 Cytoplasmic membranes

Thin-section electron micrograph of (rat) liver. Rough endoplasmic reticulum (RER) containing bound ribosomes, smooth endoplasmic reticulum (SEM) devoid of ribosomes, Golgi membranes and polysomes can be distinguished. Nucleus, mitochondria and glycogen granules are also present. Photograph by courtesy of Dr. E. R. Weibel.

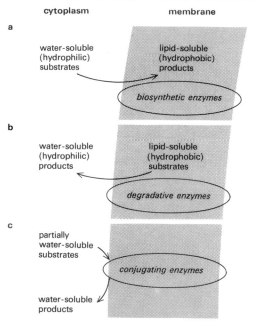

Fig. 5.2 Synthesis and degradation by membrane enzymes

See also Fig. 2.64.

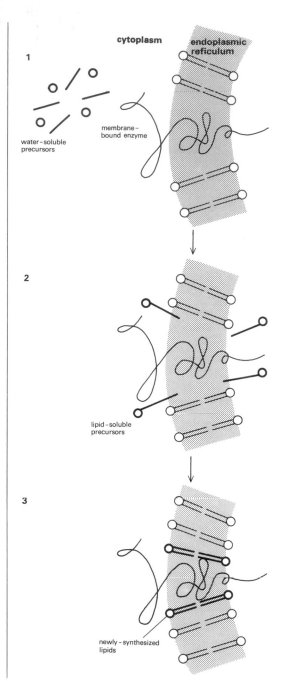

Fig. 5.3 Assembly of membrane lipids

Water-soluble precursors are compounds such as CDP choline, CDP diglyceride, etc. Lipid-soluble precursors are compounds such as lysophospholipids, diglycerides, etc.

The relationship between the **site** of enzymes and their **function** is a recurring theme in biochemistry. It is particularly important in the case of reactions, such as those mentioned above, involving a change of state.

The endoplasmic reticulum has another function. In liver, and to some extent in intestinal epithelium, smooth endoplasmic reticulum contains enzymes that catalyse a variety of reactions leading to **detoxication** of drugs and other compounds. In general the metabolic alteration makes the compounds more

Fig. 5.4   Conjugation of compounds on the endoplasmic reticulum

Compound +	Conjugate →	Product
Bilirubin; steroids; (foreign compounds containing —OH or —COOH groups)	UDP glucuronic acid	Bilirubin glucuronide (Fig. 7.27), etc.
Steroids, phenols, alcohols and other foreign compounds containing —OH groups	PAPS (Fig. 2.60) ('active' sulphate)	Steroid sulphate, etc.
Cholic acid; foreign compounds containing —COOH groups	Glycine + ATP	Glycocholic acid, etc.
Cholic acid	Taurine + ATP	Taurocholic acid

'Foreign compounds' includes many therapeutic drugs such as phenobarbitone, morphine, aspirin, sulphonamides, etc.

The endoplasmic reticulum also contains active *hydroxylating* enzymes that act on steroids, hormones, neurotransmitters, and various foreign compounds.

water soluble, and hence more readily excreted by the kidneys or the intestine. This involves hydroxylation of aromatic and aliphatic compounds, and conjugation of hydroxylated derivatives with glucuronic acid, sulphate, glycine, or taurine. Some examples are listed in Fig. 5.4. Because of the low specificity of the enzymes, many different compounds can act as substrates. Conjugation with glycine and taurine plays an important role in the excretion of cholic acid, the breakdown product of cholesterol, in the bile.

Since the reactions listed in Fig. 5.4 increase the water solubility of compounds, the spatial relationship between enzyme and substrate is clearly different from that shown by A or B in Fig. 5.2. Rather it is a third type, schematically shown as C in Fig. 5.2, in which the hydrophobic nature of the substrate, for example phenobarbitone, or cholic acid, makes it partition into the lipid environment near the catalytic site of the enzyme. Following metabolic alteration, the more water-soluble product diffuses into the cytoplasm. In the case of bile, diffusion from the endoplasmic reticulum into the bile canaliculi occurs.

5.1.2 Ribosomes

Ribosomes are knob-like granules, attached to endoplasmic reticulum (in other words, rough endoplasmic reticulum) or free in the cytoplasm. Their function is to participate in **protein synthesis**. The majority of the cytoplasmic ribosomes are not in fact entirely free: groups of ribsosomes are attached to a strand of mRNA; but because strands of mRNA are not revealed by electron microscopy, the ribosomes appear to be 'free'. Both types of ribosome are of similar size, (approximately 80s, approximately 22 nm in diameter) and structure.

Ribosomes are made up of two subunits: a larger, 60s, and a smaller, 40s, subunit. The two are linked by non-covalent bonds to form a doughnut-shaped structure (Fig. 5.5). Ribosomes contain only RNA and protein. No lipid is present; in other words they are granular not membranous. The RNA is known as ribosomal RNA (rRNA). Four different types of rRNA are present in ribosomes: three in the large

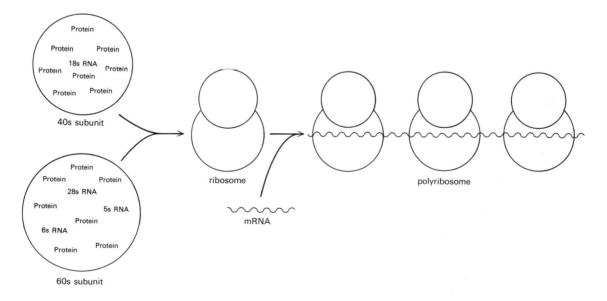

Fig. 5.5   Structure of ribosomes

The detailed configuration of rRNA, mRNA, and ribosomal proteins, and the nature of their interaction, is not known.

subunit and one in the small. The large subunit contains a molecule each of rRNA of size $28s$, $6s$, and $5s$. The small subunit contains a single molecule of rRNA of size $18s$. Each ribosomal subunit contains many different protein molecules: the large subunit about 40, and the small subunit about 30. Neither the structure, nor the function, of all these proteins is as yet clear, except that several of them, if not all, are part of the enzyme system responsible for peptide bond formation ('peptidyl transferase').

The exact mechanism by which ribosomes become assembled is not known. It is likely to be largely through self-assembly, in the way that proteins that are made up of more than one subunit aggregate spontaneously, because the final arrangement of subunits probably represents the state of lowest energy.

The components of ribosomes are synthesized in different parts of the cell: ribosomal RNA is synthesized in the nucleolus, whereas ribsomal proteins are synthesized in the cytoplasm. The proteins diffuse into the nucleus and bind to ribosomal RNA, as a result of which assembly of ribosomes takes place. Ribosomes then leave the nucleus and enter the cytoplasm. Exactly how ribosomes are attached to endoplasmic reticulum, or to mRNA, is unclear.

5.2 Protein synthesis

5.2.1 Genetic code

The sequence of amino acids in the protein determines its structure and hence its function. The mechanism by which sequence is specified is through translation of messenger RNA (mRNA). mRNA is itself specified by a stretch of DNA within the chromosome, that is, by a gene. The relationship between the bases of mRNA and the amino acids of a protein is known as the **genetic code.** The code is a **triplet** code, that is to say three different bases code for each amino acid (Fig. 5.6). The third base need not be accurately specified; as a result the code has 'wobble'. This is one reason why there is more than one tRNA coding for a particular amino acid. In addition to triplets coding for each amino acid, there are 'nonsense' triplets that result in the termination of a polypeptide chain (UAA, UAG, and UGA). The start of a polypeptide chain is specified by the triplet for methionine (AUG).

A glance at the genetic code shows how mutation of a single base in a gene leads to the insertion of a different amino acid in the protein corresponding to that gene. In sickle cell haemoglobin (HbS) for example, the glutamate normally at position 6 of the β-chain is replaced by valine. This results from a change of triplet from GAA to GUA, that is, from A to U. If substitution of a base by another base leads to one of the 'nonsense' triplets, no protein — or incomplete protein — is formed. The same is true if an extra base

Fig. 5.6   The genetic code

Amino acid	Base triplets
alanine	GCU, GCC, GCA, GCG
arginine	CGU, CGC, CGA, CGG, AGA, AGG
asparagine	AAU, AAC
aspartate	GAU, GAC
cysteine	UGU, UGC
glutamate	GAA, GAG
glutamine	CAA, CAG
glycine	GGU, GGC, GGA, GGG
histidine	CAU, CAC
isoleucine	AUU, AUC, AUA
leucine	UUA, UUG, CUU, CUC, CUA, CUG
lysine	AAA, AAG
methionine	AUG
phenylalanine	UUU, UUC
proline	CCU, CCC, CCA, CCG
serine	UCU, UCC, UCA, UCG, AGU, AGC
threonine	ACU, ACC, ACA, ACG
tryptophan	UGG
tyrosine	UAU, UAC
valine	GUU, GUC, GUA, GUG
start	AUG (methionine)
stop	UAA, UAG, UGA ('nonsense')

is inserted (or removed) in such a way that the whole 'frame' is shifted by one. If a whole region of DNA becomes deleted, it is again unlikely that an intact protein is formed. Incomplete or faulty proteins are generally more susceptible to intracellular degradation than normal proteins. Of course if a mutation occurs towards the end of a gene, it is possible for a protein to be formed with only a few incorrect, or missing, amino acids at one end. Such a protein may be stable and have a sufficient part of its structure intact to retain biological activity.

5.2.2 Peptide bond formation

The mechanism by which peptide bonds are formed involves the condensation of an amino acyl tRNA with the carboxyl group of a growing polypeptide chain. An ester bond (between the carboxyl of an amino acid and the $3'$-OH of the terminal ribose of the tRNA to which it is attached) is split and a peptide bond is formed instead (Fig. 5.7). There is little change in free energy. The overall process of protein synthesis involves three distinct stages: **initiation, elongation,** and **termination.**

All species of mRNA start with the codon for methionine (AUG); translation is from the $5'$-end to the $3'$-end, concomitant with growth of the polypeptide chain from the $-NH_2$ end to the $-COOH$ end. The process of initiation begins with the binding of Met-tRNA to the smaller ($40s$) ribosomal subunit at a site known as P (peptidyl). In the presence of various **protein factors** (termed initiation factors) that are present in the cytoplasm, messenger RNA binds to the complex, though hydrogen bonding between the triplet AUG on mRNA and the respective anticodon

(CAU read in the opposite direction) on Met-tRNA; **GTP** is also required. The resulting complex then attaches to the larger ($60s$) ribosomal subunit; during the process GTP is hydrolysed to GDP and P_i. The complex of Met-tRNA, mRNA, and ribosomal subunits, which is known as an initiation complex, is now ready to accept a second molecule of charged tRNA, and thus to commence elongation (Fig. 5.7).

The second tRNA that binds to the initiation complex is specified by the triplet adjacent to the methionine triplet (AUG). If this is UUU, for example, phenylalanine-tRNA becomes bound. The site on the ribosome at which the Phe-tRNA binds is known as the A (aminoacyl) site. Binding involves the participation of a cytoplasmic protein factor (termed elongation factor), as well as the hydrolysis of another molecule of GTP. The —NH$_2$ group of the phenylalanine is now so close to the —COOH group of the methionine that peptide bond formation readily occurs. That is, the Met–tRNA bond is split, and a Met–Phe peptide bond is formed instead. An enzyme, peptidyltransferase, which is one of the proteins of the $60s$ subunit, catalyses the transfer. The uncharged tRNA$_{Met}$ now dissociates from the P site and the Met Phe-tRNA moves from the A site to the P site. A molecule of GTP becomes hydrolysed in the process. At the same time, the mRNA moves relative to the ribosomal complex, such that the next triplet is exposed near the vacated A site (Fig. 5.7).

Such a 'shuttle' mechanism, involving two separate binding sites, operates not only for growing polypeptide chains, but for growing fatty acid chains (Fig. 2.31) and for growing proteoglycan chains also.

The process of elongation is repeated over and over again, until one of the termination triplets (UAA, UAG, or UGA) appears at an empty A site. No tRNA having the complementary anticodon to UAA, UAG, or UGA exists. The A site is filled instead by a cytoplasmic protein factor (termed release factor). The presence of the factor somehow causes peptidyltransferase to hydrolyse the bond between the last amino acid on the polypeptide chain and its tRNA; that is to transfer the polypeptide chain, not to an amino group of another amino acid, but to water (Fig. 5.7). The completed protein diffuses away, and the residual uncharged tRNA leaves the ribosomal complex. The complex dissociates into its component parts, which can participate in a new round of protein synthesis.

Few proteins have methionine as NH$_2$-terminal amino acid. Following the release of intact protein, degradation by a protease removes the NH$_2$-terminal methionine; other amino acids, in the form of a peptide, are often removed at the same time. In fact, many proteins are synthesized as larger precursor molecules which are degraded to the final product, in a manner analogous to the way that RNA precursor molecules are 'processed' by degradation into shorter forms.

Note that although the process of peptide bond formation is not energetically unfavourable from amino acyl tRNA, two extra high energy bonds (of GTP) are split for every peptide bond synthesized. As a result, peptide bond formation is driven towards protein synthesis.

Another point to note is that at no time is an amino acid of a growing polypeptide chain itself bound to a ribosomal subunit. Binding occurs between mRNA and the smaller ribosomal subunit, between mRNA and a tRNA molecule (through hydrogen bonding between codon on mRNA and anticodon on tRNA), and between tRNA at its 3′-end (to which amino acid is linked) and the large ribosomal subunit (Fig. 5.7). The ribosomal components to which binding occurs are mainly the ribosomal proteins. The function of rRNA may be no more than to arrange the ribosomal proteins in correct disposition relative to mRNA and tRNA.

The specificity of protein synthesis thus depends on two processes. First, the recognition by amino acyl activating enzyme of its correct tRNA; once an amino acid is attached to tRNA, it is committed to insertion into a growing polypeptide chain, whenever its anticodon is recognized by an mRNA triplet having a codon-exposed ribosome. The second is that of the recognition process itself. It depends solely on hydrogen bonding between three adjacent bases on mRNA and the complementary three bases on tRNA. In other words a particular tRNA is selected from a pool of 20 or more tRNA molecules, on the basis of forming suitable bonds with mRNA. The process is analogous to the selection of a deoxyribonucleoside triphosphate, or of a nucleoside triphosphate, to form hydrogen bonds with a strand of DNA during DNA, or RNA, synthesis, respectively.

It is a remarkable feature of hydrogen bond formation that mistakes in protein synthesis are as rare as in DNA or RNA synthesis (about 1 in every 10^9–10^{10} molecules synthesized). In each case, however, correct base pairing is only marginally more favourable thermodynamically than incorrect base pairing, or not forming base pairs at all.

Nevertheless the process occurs at a surprisingly high rate. Protein synthesis, for example, occurs at approximately 2–6 peptide bonds formed per second, and DNA and RNA synthesis is even faster. Just as DNA synthesis occurs at several points along a gene, so protein synthesis occurs at several points along a mRNA molecule. In other words a strand of mRNA will have several ribosomes attached to it simultaneously. The resulting structure is known as a polyribosome (Fig. 5.5).

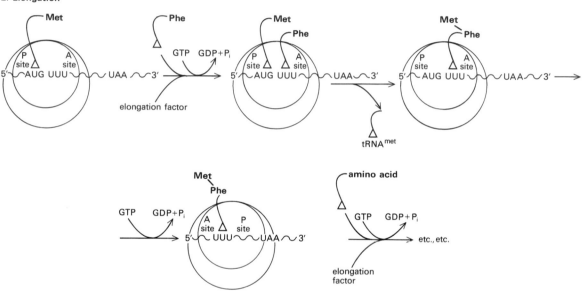

1. Initation

2. Elongation

peptide bond formation causes a polypeptide to 'grow' from the $-NH_2$ terminal towards the $-COOH$ terminal as follows:

Fig. 5.7   Mechanism of peptide bond formation

Note that the exact positioning between the two ribosomal sub-units and mRNA is not known.

3. Termination

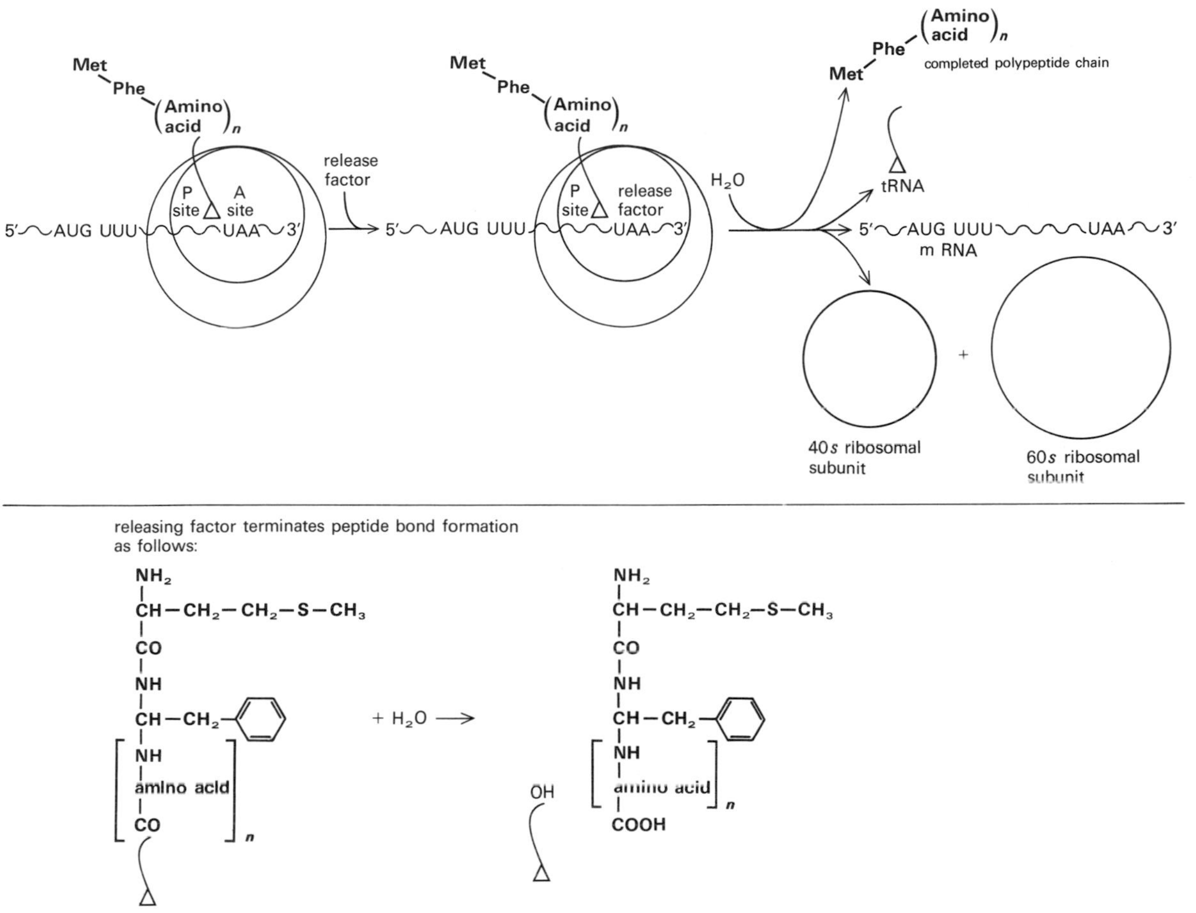

releasing factor terminates peptide bond formation as follows:

Fig. 5.7 (cont.)

It is not yet clear what proportion of protein is synthesized by free ribosomes, and what proportion by ribosomes attached to the endoplasmic reticulum (rough endoplasmic reticulum). In general, intracellular proteins are synthesized on free ribosomes, and proteins secreted for export by rough endoplasmic reticulum; in liver, in which most of the plasma proteins are synthesized, approximately 75 per cent of protein synthesis is on the rough endoplasmic reticulum.

A suggestion as to how secreted proteins become committed to synthesis on rough endoplasmic reticulum is the following. The mRNA specifying such proteins begins with a stretch that codes for a 'secretory peptide'. This peptide, which may contain some 10–20 largely hydrophobic amino acids, is common to all secreted proteins. As soon as it has been translated, it not only binds to the endoplasmic reticulum, but actually 'buries' into it, making a tunnel for the rest of

the protein to enter. As a result, the rest of the protein is synthesized inside the endoplasmic reticulum and so becomes committed to movement towards the Golgi system and eventual secretion by exocytosis. During the process, the 'secretory peptide' is removed. Secreted proteins such as insulin and collagen, which exist in a precursor form (pro-insulin and pro-collagen) within the cells from which they are secreted, therefore undergo two consecutive stages of proteolytic modification. First, removal of the 'secretory peptide' at the start of translation, and second, removal of further peptides after translation has been completed.

Various antibiotics exert their action by inhibition of protein synthesis. Streptomycin and chloramphenicol, for example, bind specifically to the ribosomal proteins of bacteria, and thus prevent bacterial growth. Ribosomal proteins of animal cells are different, as a result of which animals are insensitive to the action of streptomycin or chloramphenicol. Some toxicity is nevertheless observed, which is probably due to the small amount of mitochondrial protein synthesis that takes place in animal cells; for mitochondrial ribosomes resemble bacterial ribosomes more than they do the cytoplasmic ribosomes of animal cells. Unlike protein synthesis on cytoplasmic ribosomes, protein synthesis on mitochondrial ribosomes is inhibited by streptomycin and chloramphenicol.

Other substances inhibit cytoplasmic protein synthesis. An antibiotic known as puromycin, for example, interferes with the elongation step of protein synthesis and causes release of incompleted polypeptides. It has been tested as a potential anti-cancer drug (since cellular growth depends on protein synthesis), but as might be expected its considerable toxicity limits its usefulness. Another inhibitor of the elongation step is diphtheria toxin. Until the development of vaccines, the disease caused by *Corynebacterium diphtheria* was lethal due to the inhibition of protein synthesis.

Protein synthesis is as vital a part of cellular metabolism as is the synthesis of RNA or DNA. No hereditary defects in which protein synthesis cannot occur are known. There are of course many diseases, both hereditary and enviromental, in which synthesis of a *specific* protein is impaired. And nutritional deficiencies, such as lack of amino acids, oxygen, or metabolic fuel, result in a decreased *rate* of protein synthesis in general; so does a decrease in intracellular K^+. Conversely, diseases of the endocrine system, especially those involving malignant growths, often lead to an excessive production of specific proteins. Under normal conditions, however, the synthesis of proteins is controlled in a precise manner. The nature of the controlling mechanisms is described below.

5.3 Transcriptional and translational control

The amount of a particular protein that is present in a cell is dependent on two factors (Fig. 5.8): first, the rate at which it is synthesized from its mRNA, and secondly, the rate at which it is degraded by protease activity or otherwise eliminated from a cell. The rates of synthesis and degradation are normally equal, and define the turnover of a protein. If the rates are high, the protein has a high rate of turnover; if

they are low, the protein has a low rate of turnover. Half-lives of proteins vary from a few hours to several days or weeks.

The amount of mRNA that is present is likewise controlled, first by the rate of synthesis from its respective gene, and secondly by the rate of its degradation by RNAase. Again, synthesis and degradation are normally matched, and the rate at which they occur is a measure of the turnover of that particular species of mRNA. The amount of DNA, or rather the number of copies of a particular gene that is present in a cell, is also under some control; in general, however, the number of reiterated genes per chromosome, and the number of chromosome copies per cell (that is, diploid, tetraploid, etc.) does not vary within a tissue or even within an organism.

Of all these factors, the rate of protein synthesis (translation) and the rate of mRNA synthesis (transcription) are usually the most important. In other words, a change in the amount of protein present is most often achieved by increase or decrease in the rate of its synthesis, rather than by alteration in the rate of its degradation. The same is thought to be true for a change in the amount of mRNA in a cell. The mechanism by which some proteins, and some species of mRNA, are degraded faster than others, is not clear; the enzymes that degrade proteins and RNA are rather non-specific. Presumably the tertiary structure of those proteins and RNA molecules that are especially sensitive to degradation is different to the structure of the more stable molecules.

An important group of effectors of **transcriptional** and **translational control** are **hormones.** Some, such as the steroid hormones, increase transcription of DNA by binding directly to chromatin (Fig. 9.30). Others, such as growth hormone or insulin, increase translation of mRNA by mechanisms that are not yet clear. In each case, increased synthesis of specific proteins results. Non-hormonal effectors also act at both sites. The initiation of histone synthesis through activation of the relevant genes, by as-yet unidentified components during the S phase of the cell cycle, is an example of transcriptional control. The increased translation of haemoglobin mRNA brought about by haem in reticulocytes is an example of translational control.

In general, transcriptional control appears to be exercised in initiating cell-specific protein synthesis during differentiation. Translational control modulates the rate at which that synthesis occurs. Where the protein that is being synthesized is an enzyme, a change in amount is reflected in a change in activity. Another way of altering the activity of enzymes is by inhibition or activation of existing protein (Fig. 2.67). The activation of adenyl cyclase by hormones, the activation of glycogen phosphorylase by cAMP, or the

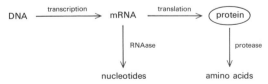

Fig. 5.8   Control of protein synthesis

The four stages at which the amount of protein present in cells can be controlled, are indicated.

inhibition of glycolysis by ATP are examples. Alteration of the activity of existing enzymes is known as 'short-term' control. Alterations in the amount of enzyme, by transcriptional or translational control, is known as 'long-term' control.

5.4 Summary

Structural components

1. The cytoplasm of cells contains a network of membranes known as the **endoplasmic reticulum.** Parts of the endoplasmic reticulum have **ribosomes** attached; these are referred to as **rough endoplasmic reticulum** (in contrast to non-ribosomal regions, known as **smooth endoplasmic reticulum**).

2. One function of smooth endoplasmic reticulum is the synthesis of phospholipids, cholesterol and other **membrane components;** another function is the detoxification of drugs and other substances.

3. The major function of rough endoplasmic reticulum is the synthesis of **proteins;** proteins are also synthesized on ribosomes that are not attached to the endoplasmic reticulum.

Protein synthesis

1. The sequence of amino acids in a protein is specified by the sequence of bases in one of the strands of the DNA that makes up that particular gene; the sequence of bases in DNA is transcribed into a complementary molecule of mRNA, which is translated into a molecule of protein. Translation of mRNA into protein is achieved as follows. Each amino acid is specified by a different combination of three consecutive bases in mRNA; the relation between an amino acid and its corresponding three bases (referred to as a triplet) is known as the **genetic code.** In addition to specifying the 20 amino acids, there are triplets specifying 'start' and 'stop' of protein synthesis.

2. The mechanism whereby amino acids are joined together to form proteins is as follows. Amino acids are activated and transferred to their respective **tRNA** molecule (see section 2.3.3). Each tRNA contains within its structure a triplet of bases that is complementary to the triplet of bases in mRNA, coding for that particular amino acid. This allows binding between successive **tRNA** molecules and **mRNA,** such that amino acids become aligned in the correct sequence. **Peptide bond formation** is catalysed by enzymes contained in **ribosomes,** to which mRNA is bound.

Transcriptional and translational control

The types of protein that are synthesized in cells are regulated at two points: by selection of specific regions of DNA (that is, genes) for **transcription** into mRNA, and by selection of specific molecules of mRNA for **translation** into protein. Neither mechanism of control is understood; each is modulated by certain **hormones.** Together, transcriptional and translational control regulate the **type** and **amount** of proteins (including enzymes) that are synthesized in cells.

FURTHER READING

Ribosomes

H.Bielka (1978). The eukaryotic ribosome. *Trends biochem. Sci.* **3,** 156.

Protein synthesis

For tRNA, mRNA, and the genetic code:

B.F.C.Clark (1977). *The genetic code* (Studies in Biology no.83). Edward Arnold, London.

W.E.Cohn and E.Volkin (eds) (1977). mRNA: the relation of structure to function. *Progr. nucleic Acid Res. mol. Biol.* **19.**

S.K.Mitra (1978). Recognition between codon and anticodon *Trends biochem. Sci.* **3,** 153.

A.Rich and S.H.Kim (1978). The three-dimensional structure of transfer RNA. *Scient. Am.* **328,** 52.

For other aspects of protein synthesis:

R.Mazumder and W.Sza (1977). Protein biosynthesis. In *Comprehensive biochemistry* (eds M.Forkin, A.Neuberger, and L.L.M.Van Deenen). Vol.24. Elsevier, Amsterdam.

V.M.Pain and M.J.Clemens (1978). Protein synthesis in mammalian systems. In *Comprehensive biochemistry* (eds M.Florkin, E.H.Stoz, A.Neuberger, and L.L.M.Van Deenen), Vol.19B. Elsevier, Amsterdam.

M.Szekely (1978). *Nucleic acids and protein synthesis.* Macmillan, Basingstoke.

6

Cell surface: transport

6.1 Introduction

The cytoplasm of cells is separated from the external environment by the plasma membrane. Like other membranes, plasma membrane is essentially a phospholipid bilayer containing integral proteins embedded in, and peripheral proteins bound to it (Fig. 1.33). Certain cells have in addition an outer layer or 'surface coat', which is loosely attached to the plasma membrane. Plasma membrane and surface coat together make up what is commonly referred to as the cell surface.

The shape of most cells is not that of a smooth sphere. Many cells are elongated or flattened, and their surface is elaborated into a variety of protuberances such as microvilli, cilia, and other structures. As cells differentiate, their shape often changes: precursors of skeletal muscle cells, for example, fuse together in long rows; precursors of nerve cells extend long, thin processes which eventually become axons. The precursors of cells that enter the circulation, such as erythroblasts and other 'stem' cells, retain their spherical shape during maturation but their size, which is initially quite large, decreases as the cells mature. In short, much of the specificity of cellular function resides at the cell surface. In section 6.2, the composition and structural diversity of the cell surface is discussed.

The cell surface has three main functions:

i.   to **insulate** the **interior** of the cells from its environment,

ii. to **transport** substances into and out of cells, and

iii. to **transmit** hormonal and other **signals** to the cell cytoplasm.

Insulation is necessary to retain the intracellular building blocks (section 2.3) and other metabolic intermediates at sufficiently high concentration for biosynthetic and other pathways to operate at adequate rates. The same is true of intracellular organelles such as mitochondria. Optimal activity of the tricarboxylic acid cycle is possible only because the intermediates of the cycle, such as citrate, α-ketoglutarate, or NAD^+, are kept concentrated within mitochondrial matrix; diffusion into the cytoplasm is minimized. At the same time specific transport processes take up ADP and release ATP. Insulation of the cytoplasm also insures that fluctuations of pH and other changes that occur in extracellular fluid are prevented from affecting cellular activity.

Transport is necessary to bring nutrients such as glucose, amino acids, or fatty acids into the cell in order to maintain energy production and biosynthetic activity. Waste products, such as H_2O, CO_2, and urea, have to be excreted, though this is generally achieved by simple diffusion.

Hormones control the metabolic activity of cells. Lipid-soluble hormones such as steroids enter cells by diffusion across the cell surface. Polypeptide hormones, on the other hand, act by binding to specific receptors at the cell surface, as a result of which a signal is transmitted to the interior of the cell.

In a sense the functions of insulation and transport are the opposite of each other, and the structure of membranes represents a compromise between them. In general the lipid bilayer provides the insulation, while the membrane proteins catalyse transport. The mechanisms by which the latter is achieved are discussed in section 6.3.1. Excretion of end-products is discussed in section 6.3.2.

In addition to transporting nutrients, the surface membrane transports ions. Whereas the extracellular fluid is high in Na^+ and low in K^+, the reverse is true of intracellular fluid. Such ionic asymmetries are achieved not by, for example, exclusion of Na^+ from cells, but by active extrusion of intracellular Na^+ by a specific enzymic pump. The nature of such pumps is discussed in section 6.3.3.

The transmission of signals from non-diffusible hormones is achieved by activation of specific membrane proteins. The proteins may be enzymes which are asymmetrically orientated within the membrane, such that binding of hormone at the outside leads to a reaction at the inside; or they may be transport proteins, such that binding from the outside increases the rate at which particular substances are transported into the cell; or they may be ion pumps, such that binding of hormone at the outside of the cell leads to transient abolition of the ionic asymmetry. Examples of each type of signal are discussed in section 6.3.4.

The above mechanisms for transporting substances across the cell surface involve changes in specific membrane components. The surface membrane as a whole remains intact; the substances that are transported are low molecular weight compounds or ions. But large molecular weight compounds such as proteins, viruses, and even bacteria are also transported into cells; and proteins and viruses that are synthesized within cells are secreted into the blood-stream. How is this achieved? The answer is that these compounds are transported by movement of whole portions of surface membrane. The elasticity of the cell surface is such that the engulfment of particulate matter into a vesicle, which then 'buds off' at the exterior or interior side of the cell, is possible; the surface is made good by fusion of the plasma membrane. Such processes of entry (absorption) and exit (secretion) are discussed in section 6.4.

6.2 Structure

6.2.1 Constituents

As in all membranes, the main constituents of the plasma membrane are **proteins** and **phospholipids**. The

Fig. 6.1 Composition of plasma membrane (approximate percentage by weight)

Constituent	Plasma membrane	Endoplasmic reticulum
Protein	68	80
Phospholipid	19 (PC, 43%; PE, 20%; PI, 7%; PS, 4%; Sph, 23%)	18 (PC, 59%; PE, 22%; PI, 8%; PS, 4%; Sph, 4%)
Cholesterol	7	1
Carbohydrate	6 (~90% glycoprotein; ~10% glycolipid)	1 (~99% glycoprotein; ~1% glycolipid)

Abbreviations: PC, phosphatidyl choline; PE, phosphatidyl ethanolamine; PI, phosphatidyl inositol; PS, phosphatidyl serine; and Sph, sphingomyelin.
Composition of endoplasmic reticulum is given for comparison. (Data are of (rat) liver and combined from a number of sources.)

plasma membrane also contains more cholesterol and sphingomyelin than other membranes. In addition, plasma membrane has the highest content of glycoproteins and glycolipids. In fact the presence of cholesterol, sphingomyelin, glycoprotein, and glycolipid is often taken as diagnostic of plasma membrane. A comparison between typical plasma membrane and endoplasmic reticulum is illustrated in Fig. 6.1.

The components of the plasma membrane are **asymmetrically distributed** between the outer and the inner half. Glycoproteins and glycolipids are almost entirely outward facing, with the sugar residues making the maximum number of hydrogen bonds with extracellular water. Of the phospholipids, phosphatidylcholine and sphingomyelin are largely in the outer leaflet; phosphatidyl ethanolamine and phosphatidyl serine

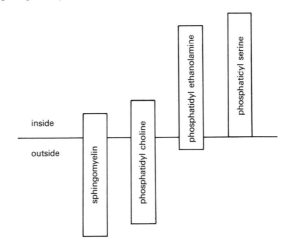

Fig. 6.2 Asymmetry of phospholipids in membranes

The distribution of phospholipids in plasma membrane (human red cells) is shown. The distribution in plasma membrane of other cells is probably similar; the distribution in endoplasmic reticulum is probably the reverse (i.e. phosphatidyl ethanolamine and phosphatidyl serine facing mainly 'outwards', etc.). Taken from Zwaal et al. (1973) *Biochim. Biophys. Acta* **300,** 159, with permission.

largely in the inner leaflet. Fig. 6.2 shows the arrangement for red cell plasma membrane, which is a particularly clear-cut example, since the red cell has no internal membranes and does not undergo any kind of membrane flow (section 6.4) by which components may become redistributed. In membranes from other organelles, phospholipids are also asymmetrically distributed. In endoplasmic reticulum, for example, the distribution is the opposite of that of the plasma membrane, which fits well with the fact that when vesicles of endoplasmic reticulum fuse with the plasma membrane during secretion, the inner and outer halves of the bilayer become reversed.

The proteins of the plasma membrane, as of all membranes, are also asymmetrically distributed. Extrinsic proteins are, by definition, situated on one or other side of the bilayer. Intrinsic proteins, which penetrate the bilayer, are also orientated to one or other side. Indeed the very function of such proteins (section 6.3) requires them to be positioned in such a way that one type of binding site is accessible to the extracellular fluid, and another to the cytoplasm (Fig. 6.14). The same is true of mitochondrial (Fig. 3.11) and other membrane proteins (Fig. 5.2) also.

Cholesterol, which is 'dissolved' in the central hydrophobic region of the plasma membrane, may be asymmetrically distributed in so far as there may be more at the inner or outer leaflet. This may be particularly marked at the base and tip of microvilli, where curvature is high.

The **cell coat,** sometimes referred to as glycocalix, is a rather ill-defined structure that is loosely bound to the exterior of cells. It is rich in **proteins, glycoproteins** and **proteoglycans,** some of which are sulphated; the cell coat contains little, if any, phospholipid. The cell coat has several functions. On epithelial cells, such as those of the respiratory or gastrointestinal tract, it forms a **protective layer** of mucus over the cell surface. This prevents mechanical and chemical damage, especially during the passage of food and faeces in the respective portions of the alimentary tract. On all cells that are part of a tissue, but especially in connective tissue, the cell coat has an **adhesive role,** cementing cells together. In fact the substance known as intracellular

matrix is in a sense merely a continuation of the cell coat. As might be expected from their different functions, the exact composition of mucus and intracellular matrix is different (Fig. 6.3). In each case material that is progressively further away from the plasma membrane may have a slightly different composition, modified by extracellular enzymes. It will be appreciated that it is therefore difficult to draw clear-cut distinctions between membrane-associated material and truly extracellular material.

6.2.2 Microvilli

In several cells the plasma membrane is convoluted into a variety of protuberances. The most characteristic of these are **microvilli** (Fig. 6.4). Microvilli are of fairly uniform diameter (approximately $0.1 \mu m$). Their length varies, especially in circulating cells (Fig. 6.4), type b), up to some $2 \mu m$. The overall composition of microvilli is the same as that of the rest of the plasma membrane. The type of protein or glycoprotein, on the other hand, may be different. Microvilli of epithelial cells (Fig. 6.4, type a), for example, contain specific enzymes for degrading disaccharides and peptides, whereas Na^+/K^+-ATPase is concentrated in other parts of the plasma membrane. In view of the regionalization of surface function in such cells (Fig. 8.12), this is hardly surprising. The shape of microvilli is maintained by a bundle of microfilaments which fills the cytoplasmic space inside them. They are not rigid, but have a waving motion, rather like cilia.

Microvilli of epithelial cells are quite tightly held together, and the entire brush border can be removed intact from segments of intestine. Microvilli of circulating cells are more loosely arranged. They unfold not only during cytokinesis (section 4.3.1), but also during hypotonic shock: one of the reasons why red cells, which are devoid of microvilli, lyse under conditions in which white cells, which have microvilli, do not, is that the unfolding of microvilli allows the latter cells to increase their surface area in response to an increase in volume; that is, to swell. Red cells are capable of some limited swelling by changing from a biconcave disc to a sphere.

Fig. 6.3 Composition of surface coat and extracellular material

	Surface coat	Mucus	Intercellular 'cement'
Predominant components:	Glycoproteins ~15% carbohydrate ~85% protein	Glycoproteins ~50% carbohydrate ~50% protein	Proteoglycans ~92% carbohydrate ~8% protein

None of these materials contains any significant amount of phospholipid or cholesterol (<5 per cent of dry weight).

The values quoted are taken from a variety of human and animal tissues; they should be taken as no more than indicative of the type of constituents found. Note that mucus has an exceptionally high water content (96 per cent).

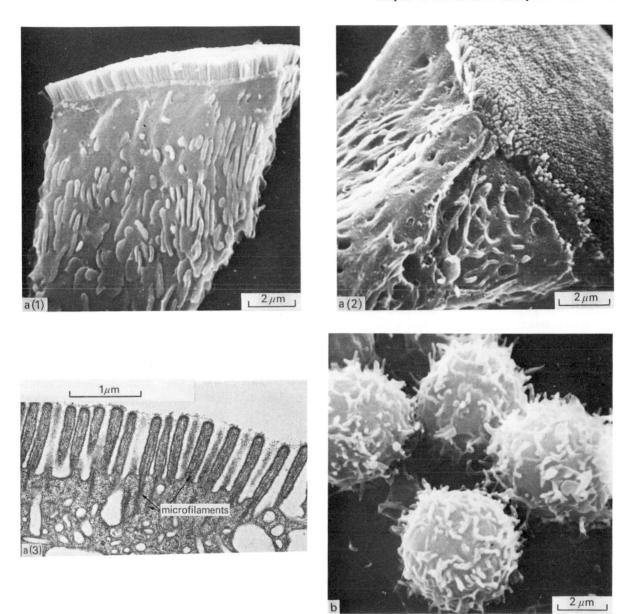

Fig. 6.4 Microvilli

a (1) Scanning electron micrograph of a (rat) intestinal epithelial cell. The cell is considerably longer than shown; only the upper third of the cell is seen. Note the regular array of closely-packed microvilli at the top (luminal) surface — the so-called brush border. At the sides of the cell some microvilli, blebs and other protuberances have arisen during preparation of the specimen. (2) A 'top' view of the same cell. (3) Thin-section electron micrograph of the same cell, showing bundles of microfilaments within each microvillus.
b Scanning electron micrograph of (human) lymphocytes. Note the numerous, randomly spaced, microvilli on the surface. Photographs by courtesy of Dr. S. Knutton.

The function of microvilli is thus related to the fact that they accommodate more surface membrane than would otherwise be possible in a cell of fixed size. In microvilli of epithelial cells, the increase is used to maximize the area across which the absorption of small molecules and ions takes place. In microvilli of dividing cells, the increase is used to carry out cytokinesis, to enable cells to flatten and spread during interphase, and to withstand osmotic shock, as described above. In malignant cells, microvilli may have another function, that of invading non-cancerous tissue.

6.2.3 Cilia and sperm tails

Epithelial cells of the respiratory tract have long (up to 10 μm), relatively thick processes, (0.5 μm in diameter) known as cilia, into which their plasma membrane is elaborated. Like microvilli, cilia are part of the plasma membrane. Unlike microvilli, cilia contain a characteristic arrangement of microtubules (Fig. 6.5); microfilaments may also be present, as well as a protein called dynein, which is in some respects similar to myosin. The function of cilia is to maintain the flow of mucus past the epithelial lining of the tract, by their beating motion. Exactly how the microtubules and other components are involved in this is not clear.

The sperm tail has a similar arrangement of microtubules inside it. In this case, the function of the elongation is to propel the sperm itself.

6.2.4 Other protuberances

Several other elaborations of the plasma membrane are seen in certain cells. Ruffles, which appear on cells observed by time-lapse cinematography, like the bubbles of a boiling viscous liquid, are probably portions of plasma membrane in the process of exocytosis and endocytosis (section 6.4). Protuberances known as pseudopodia, which are distinct from microvilli or cilia, are seen on certain cells during their developmental stages. Undifferentiated nerve cells, for example, elaborate such structures during the process of establishing contact with each other or with muscle cells. In this case pseudopodia are the precursors of axons. Like cilia they contain microtubules. Unlike

cilia they are rigidly embedded in extracellular matrix; in some axons layers of plasma membrane from other cells surround the axonal membrane in a spiral fashion until a thick sheath, the so-called myelin sheath, is formed (Fig. 12.3). Axons, which are rather thicker than cilia, can grow to a length of several millimetres or more.

6.2.5 Junctions

Many cells are connected by **junctions.** There are three main types, the so-called communicating (or 'gap') junctions, occluding (or 'tight') junctions, and desmosomes (Fig. 6.6); ('gap' and 'tight' refer to the spaces between adjacent cells). It can be seen that the junctions vary in structure. What they all have in common is a rigidity of **intercellular binding,** achieved through specific protein complexes. Occluding junctions and desmosomes prevent the flow of extracellular fluid into tissues. Communicating junctions bypass some of the permeability properties of the plasma membrane, and allow direct contact between cells. Thus Na^+ and K^+ exchange freely between cells, as a result of which cells joined by communicating junctions are electrically coupled. Dyes such as fluorescein, and phosphorylated compounds such as nucleotides, pass from cell to cell. The junctions are not just holes or non-specific pores, such as those formed during lysis of cells: high molecular weight compounds (proteins, nucleic acids, and so forth), which are lost from cells during lysis, do not pass between cells joined by communciating junctions.

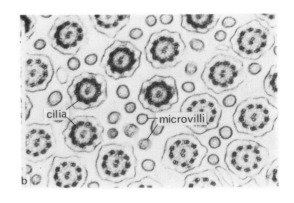

Fig. 6.5 Cilia

a Scanning electron micrograph of (human) nasal mucosa. The cilia emanating from the cell in the centre should be compared in size with the microvilli covering the surface of the surrounding cells. Photograph by courtesy of Mr. R. Moss.
b Thin section electron micrograph of cilia (rat tracheal mucosa) in transverse section. Note the characteristic circular arrangement of 9 microtubules (which are actually doublets) surrounding 2 microtubules (also doublets), inside each cilium. Microvilli, also in transverse section, can clearly be distinguished. Photograph by courtesy of Dr. P. K. Jeffery

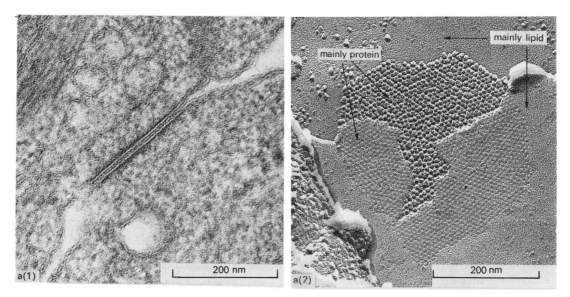

Fig. 6.6 Intercellular junctions

a Communicating or 'gap' junction. (1) Thin-section electron micrograph of (rat) gastric mucosa. The two plasma membranes are seen to lie close together, yet separated by a 'gap'. (2) Freeze-fracture replica of the same material taken at right-angles to that shown in (1) i.e. looking 'through' the two membranes. Note the clustering of particles at the site of the junction (roughly circular in this side-view). The particles, which are intrinsic membrane proteins, generate hydrophilic channels that span the gap between adjacent cells, and allow water-soluble substances to pass freely through. Photographs by courtesy of Dr. C. Peracchia and the Rockefeller University Press.

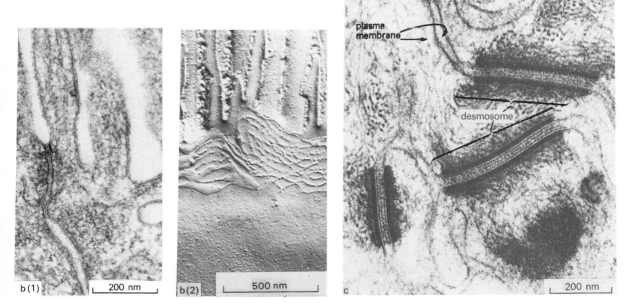

b Occluding or 'tight' junction. (1) Thin section electron micrograph of (rat) intestinal mucosa. The two plasma membranes just below the brush border make such close contact, that compounds on the outside are prevented from penetrating into the tissue (i.e. are 'occluded'). (2) Freeze-fracture replica of the same material taken at right angles to that shown in (1), i.e. looking 'through' the two membranes. The junctions are seen to be characterized by a 'belt' of particles just below the brush-border. Photographs by courtesy of Dr. S. Knutton.
c Desmosomes. Thin-section electron micrograph of (human) lower lip. The plasma membranes separating adjacent cells can be distinguished. The dark-staining material at the desmosomes is responsible for imparting rigidity to the junctions. Photograph by courtesy of Dr. A. Haywood.

A diagrammatic representation of the 3 types of junction is shown in Fig. 8.19.

The reason for the increased permeability at communicating junctions is not clear. A clustering of intramembraneous particles into regular patterns accompanies junction formation, with protein complexes spanning the plasma membrane of both cells. Exactly how such structures are related to an increase in permeability is not clear, though presumably hydrophilic channels (see for example Fig. 1.35) are involved. The permeability 'gates' are closed by an increase in intracellular Ca^{2+}. The cell–cell communication achieved by such junctions plays a role in the transmittion of electrical impulses in heart muscle and in the development and maintenance of co-ordinated tissue functions in general.

6.3 Regulation of cellular function

6.3.1 Nutrient uptake

Low molecular weight compounds such as sugars and amino acids enter cells by one of three processes: passive diffusion, facilitated diffusion, or active transport (Fig. 6.7).

In **passive diffusion,** nutrient molecules simply move down a concentration gradient at a slow rate. No binding between the nutrient and any component of the plasma membrane takes place and there is no specificity of uptake. The process varies relatively little with temperature. All molecules, whether free nu-

trients, phosphorylated derivatives, or other compounds have a slow but measurable rate of passive diffusion across the plasma membrane. The rate is the same for entry as for exit. Apparently inert substances such as anaesthetic gases have rather high rates of passive diffusion. This is due to their solubility in the hydrophobic regions of the plasma membrane. Since solubility implies some sort of interaction with the solvent, such processes should not be considered as truly passive.

In **facilitated diffusion,** molecules also move down a concentration gradient. In this instance binding between the molecule and a protein component of the plasma membrane occurs. As a result diffusion is greatly accelerated (that is, facilitated). In fact the protein acts just like an enzyme, showing (i) saturation of uptake at high nutrient concentration and (ii) specificity towards one or other molecule: a K_M value for uptake by its respective carrier protein can be calculated in the same way as can a K_M value for metabolism of a molecule by an enzyme. As with enzymes, carrier proteins distinguish between L- and D-amino acids, between L- and D-sugars, and between other closely related groups. As with chemical reactions, the uptake process is dependent on temperature, showing an approximate doubling in rate for a 10° C rise in temperature.

Molecules such as D-sugars and L-amino acids, for which specific carrier proteins exist, enter cells predominantly by facilitated diffusion at low concentrations.

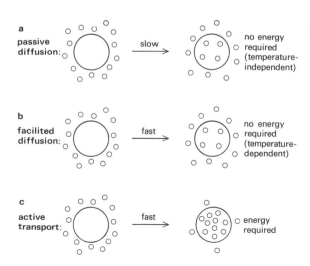

Fig. 6.7 Types of nutrient uptake

Examples of mechanism b are the entry of glucose and amino acids into muscle and red cells; examples of mechanism c are the entry of glucose and amino acids into the epithelial cells of small intestine and renal tubule (see Fig. 6.9).

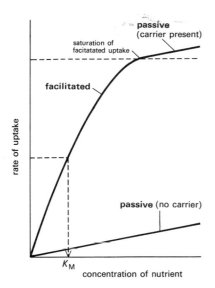

Fig. 6.8 **Rates of facilitated and passive diffusion**

Above the K_M, when carrier mediated uptake is saturated, entry by passive diffusion becomes appreciable (Fig. 6.8). All cells including red cells have carriers for facilitated uptake of D-glucose and related sugars, for the L-amino acids, for choline, for nucleosides, and for many other molecules. The carriers generally operate in the direction of entry only, so that loss from cells is by the relatively slow process of passive diffusion. In addition, most of the molecules mentioned above (with the exception of amino acids) become rapidly phosphorylated once they enter the cytoplasm. Since passive diffusion of phosphorylated compounds is slower even than that of unphosphorylated compounds, a very efficient uptake and retention of nutrients is achieved by the combination of facilitated diffusion and intracellular phosphorylation.

In **active transport,** nutrient molecules move against a concentration gradient (Fig. 6.7). The process is made possible by a coupling between metabolic energy and transport. The way in which energy, in the form of ATP, is coupled to transport is not clear. Phosphorylation of nutrient by known kinase enzymes is not involved: for sugars such as 6-deoxyglucose or amino acids, which are not phosphorylated, are taken up by active transport in certain types of cell. The most important of these are the epithelial cells of the small intestine which are responsible for the absorption of the bulk of dietary sugars and amino acids, and the epithelial cells of the renal tubules which are responsible for the reabsorption of sugars and amino acids into plasma. The mechanism of coupling is more likely to involve phsophorylation of a membrane protein – perhaps a protein similar to that catalysing facilitated diffusion. Certainly all forms of active transport are linked to facilitated, and not to passive, diffusion. In the case of active transport of cations, which occurs in most types of cell (see section 6.3.3), phosphorylated membrane protein intermediates have been isolated.

An alternative mechanism by which sugars and amino acids may be actively transported is by coupling transport to an ionic gradient. There is evidence that the accumulation of sugars and certain amino acids in small intestine and renal tubule is linked to the Na^+ gradient (high outside and low inside) across the cell membrane (Fig. 6.9). Metabolic energy is indirectly involved, in that maintenance of the Na^+ gradient is dependent on the intracellular supply of ATP.

In addition to active transport, sugars and amino acids are absorbed by quite a different mechanism. The degradation of carbohydrate and protein in the alimentary tract is incomplete, and intestinal juice contains a considerable quantity of disaccharides (that is, maltose, sucrose, and lactose) and peptides (section 8.3.2). These are degraded by disaccharidases and peptidases located in the brush border of intestinal epithelial cells. The spatial arrangement of the membrane enzymes is such that the products of degradation appear in the cytoplasm (Fig. 6.10). Thus a net accumulation of nutrients is achieved by the asymmetric disposition of a degradative enzyme.

A similar mechanism is responsible for the uptake of fatty acids. Dietary fat is present in the circulation largely as triglyceride. An enzyme, 'clearing factor lipase', situated at the surface of adipose and other cells, hydrolyses triglyceride in such a way that free fatty acids enter cells, leaving glycerol in the bloodstream. The fatty acids pass directly through the lipid bilayer into the cell. Glycerol is eventually taken up by the liver and metabolized by glycolysis or gluconeogenesis. Circulating free fatty acids, produced by lipolysis of triglyceride synthesized and stored in adipose tissue, are taken up by liver cells. Entry is predominantly by passive diffusion through the lipid bilayer. It should be noted that the plasma membrane of liver is particularly permeable not only to fatty acids but also to water-soluble compounds, such as sugars and amino acids, moving in either direction.

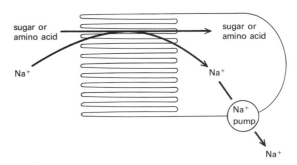

Fig. 6.9 **Na^+-linked transport of sugars and amino acids**

135

a disaccharides
e.g. lactose

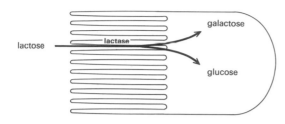

b peptides
e.g. leucyl-alanine

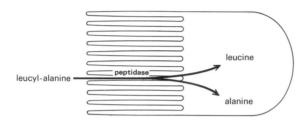

Fig. 6.10 Absorption of disaccharides and peptides

6.3.2 Excretion of end-products

The main end-products of cellular metabolism are water, carbon dioxide, and urea. All are transported out of cells by passive diffusion. The amount of water that leaves a cell is controlled by the relative amount of dissolved material (chiefly protein and low molecular weight compounds and ions) in cells and plasma; in other words by the osmotic pressure. Assuming that the concentration of dissolved molecules in plasma remains constant, an increased production of water in cells will lead to a net movement from cells to plasma (Fig. 7.5). Water is lost from plasma by passage through three types of tissue, each of which excretes water to the environment: kidney, sweat glands of the skin, and lung. The volumes normally excreted by an adult might be 1500 ml (kidney), 900 ml (sweat), and 400 ml (lungs).

The situation is actually more complicated than this, in that plasma is not in direct contact with cells; between blood-vessels (capillaries) and cells is a space, filled with what is known as **interstitial fluid.** Interstitial fluid contains virtually no protein; hence there is a tendency for water to move from interstitial fluid into cells. Equally there is a tendency for water from interstitial fluid to enter the capillaries. What prevents interstitial fluid from 'drying up' is (i) the outward pressure generated by the arterial blood-flow, and (ii) the operation of the Na$^+$ pump, which acts to keep water out

of cells (section 6.3.3). Interstitial fluid and plasma together make up what is referred to as **extracellular fluid.** (See Fig. 7.5).

Under normal conditions, the amount of water excreted is equal to the amount produced by metabolism; that is, by oxidation of carbohydrate, fat, and protein, and by absorption from the gut. If excessive sweating occurs, due to a fever, violent exercise, or a hot climate, excess water is lost from the body unless an equivalent amount is replaced by drinking. Conversely, if loss of water through the kidney is diminished, as in renal disease, excessive water is retained, giving rise to odoema.

Superimposed on net flow is a rapid exchange of water between the inside and outside of cells. The rate of passive diffusion of water, through the lipid bilayer as well as through hydrophilic channels created by intramembranous proteins, is faster than that of any other compound.

The loss of CO_2 from cells, partly as dissolved CO_2 and partly bicarbonate HCO_3^- ions, is controlled in a manner similar to that of water. In other words movement across the plasma membrane is passive, and is controlled largely by the acid–base balance of plasma (section 7.2.2.2).

Urea, which is formed only in the liver, is transported across the plasma membrane by passive diffusion. Loss from the blood-stream is predominantly through the kidney. Under normal conditions, renal excretion balances production in the liver.

6.3.3 Ionic asymmetry

6.3.3.1 Na$^+$ and K$^+$

It is a feature of all types of cells that the concentrations of Na$^+$ and K$^+$ inside them are the opposite of those in plasma (Fig. 1.20). That is, intracellular K$^+$ is high, whereas extracellular K$^+$ is low; intracellular Na$^+$ is low, whereas extracellular Na$^+$ is high. The establishment and maintenance of such asymmetry clearly requires energy. Since the plasma membrane is actually rather permeable to Na$^+$ and even more so to K$^+$ (moving, like water, in either direction by passive diffusion), ionic asymmetry is achieved by an ATP-driven pump that drives Na$^+$ outwards and K$^+$ inwards, each moving against a concentration gradient (Fig. 6.11). In fact the movement of K$^+$ is probably passive. Its entry into cells is due to the tendency to maintain electrical neutrality. That is, as Na$^+$ ions are **actively** transported out of cells, K$^+$ ions enter **passively.** For this reason the pump is generally referred to as a **Na$^+$ pump.** An additional driving force is the presence of impermeable, negatively charged, proteins in cells. These attract positive ions. Since any Na$^+$ that enters cells is immediately pumped out again, it is

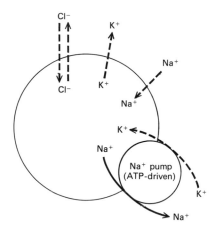

Fig. 6.11 Ion movements in cells

Dotted arrows indicate passive movement; the solid arrow indicates active movement. Movement is probably through aqueous channels, as indicated in Fig. 1.35. The ATP-driven Na^+ pump is a membrane protein complex having ATPase activity (Fig. 6.12).

K^+ that remains in cells. What the pump does is (a) to provide the energy for this thermodynamically unstable situation and (b) to ensure that it is K^+, and not any other cation, that enters cells. For K^+ is specifically required for intracellular processes such as glycolysis and protein synthesis; other cations are inhibitory.

The operation of the Na^+ pump achieves not only the maintenance of ionic asymmetry between intracellular and extracellular fluid, but also prevents osmotic entry of water into cells; this is more fully explained in section 12.3.1. In certain cells, the Na^+ pump serves additional functions. In the kidney it causes a net reabsorption of Na^+ (with accompanying passive movement of Cl^- and water) from glomerular filtrate into the tubule (section 7.2.2.1). In the intestine, the pump causes a similar movement of Na^+, Cl^-, and water to take place from intestinal lumen to the blood-

stream (section 8.3.2). In nerve and muscle cells, the pump 'repairs' the loss of Na^+/K^+ asymmetry that occurs every time a cell is stimulated (section 12.3.1).

The Na^+ pump has the properties of a **Na^+- and K^+-activated ATPase**; the manner in which ATP is probably involved is shown in Fig. 6.12. The important point is that a concerted movement of Na^+ and K^+, in opposite directions, occurs. The pump represents one of the few examples of a protein that distinguishes between Na^+ and K^+. The stoicheiometry is that for every molecule of ATP hydrolysed, 3 Na^+ ions leave, and 2 K^+ ions enter; maintenance of electrical neutrality presumably involves the movement of an anion.

6.3.3.2 Ca^{2+}

Another cation pump is concerned with the movement of Ca^{2+} against a concentration gradient. A membrane protein having this property has been isolated from the sarcoplasmic (endoplasmic) reticulum of muscle, in which Ca^{2+} movements are linked to the cycle of muscle contraction and relaxation (section 10.2.3). The protein has properties similar to the Na^+ pump of the plasma membrane. Since the internal concentration of free Ca^{2+} in cells, non-muscle as well as muscle, is maintained at a fraction of the concentration in the external environment (less than 10^{-6} M in cells; approximately 10^{-3} M in plasma), it is likely that some form of Ca^{2+} pump is present in the plasma membrane, and that its function is to pump Ca^{2+} out of cells. An alternative mechanism for excluding Ca^{2+}, that is present in certain cells, is a linked Na^+–Ca^{2+} exchange: as Na^+ enters, Ca^{2+} is removed; the Na^+ is itself subsequently removed by the Na^+ pump.

In addition to these mechanisms for transporting Ca^{2+} out of cells across the plasma membrane, there

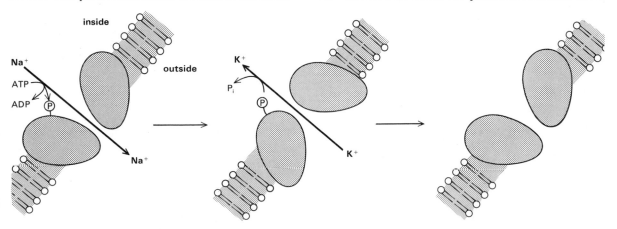

Fig. 6.12 Mechanism of Na^+/K^+-ATPase (Na^+ pump)

The mechanism by which selective movement of Na^+ and K^+ is achieved (indicated by a 'tilting' of membrane proteins) is not known.

are other mechanisms that maintain a low concentration of cytoplasmic Ca^{2+}. The Ca^{2+} pump of the endoplasmic reticulum of muscle, which causes a sequestration of Ca^{2+} within the spaces of the sarcoplasmic network, has already been mentioned. Mitochondria have special uptake processes for Ca^{2+} that are directly linked to respiration. Nuclei also sequester Ca^{2+} to a certain extent. Moreover physical binding of Ca^{2+} to anionic sites on proteins, phospholipids, and acidic carbohydrates is itself an important factor in limiting the concentration of free Ca^{2+} in the cytoplasm. The reason for a carefully controlled intracellular Ca^{2+} concentration is that many enzymes are sensitive to Ca^{2+}. Protein kinases, for example, which control the direction of metabolic pathways such as glycogen breakdown (Fig. 2.4) or resynthesis (Fig. 2.56) are activated or inhibited by low concentrations of Ca^{2+}. It is therefore to be expected that cytoplasmic Ca^{2+} is under some sort of hormonal control. This is indeed the case and several hormones act on cells by allowing a transient influx of extracellular Ca^{2+} to take place (Fig. 9.28).

6.3.3.3 H^+

Active transport of H^+ occurs in two important situations. Cells of the gastric mucosa secrete HCl into the lumen of the stomach. The extracellular pH falls to a value as low as 1, implying transport against a concentration gradient of the order of 10^6 (extracellular pH of $1 = 10^{-1}$ M H^+; intracellular pH of $7 = 10^{-7}$ M H^+). Chloride ions passively accompany H^+ in order to maintain electrical neutrality. The secretion of HCl uses ATP, generated by the oxidative metabolism of mucosal cells. The mechanism is not clear.

The other situation, which occurs in the mitochondria of all cells, is the local accumulation of H^+ ions resulting from the operation of the electron transport chain (section 3.4). In this instance the accumulated H^+ ions drive ATP synthesis, in contrast to the gastric secretion of H^+, in which ATP drives H^+ accumulation. Clearly the coupling between H^+ and ATP is in a sense reversible (Fig. 6.13).

6.3.4 Receptors

6.3.4.1 Hormones

The transport processes discussed in the preceding section control the entry of nutrients and ions into cells. This not only allows metabolic activity of the cytoplasm to take place, it also regulates it. The rate of glucose degradation in muscle, for example, is proportional to the intracellular concentration of glucose. But not all metabolic pathways are controlled by substrate concentration alone. The rate of glycogen breakdown in muscle and liver, and the rate of triglyceride break-

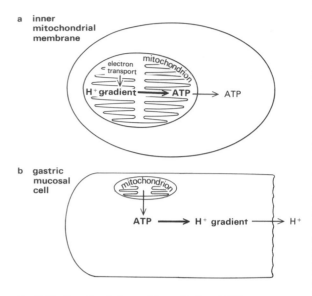

Fig. 6.13   Coupling between H gradient and ATP

The extent to which H^+ *gradient* is more of a localized *accumulation* of H^+ ions, is not clear.

down in fat cells, for example, are controlled by the activity of phosphorylase and lipase respectively. These enzymes are themselves controlled by the activity of other enzymes, the ultimate control being exercised by the concentration of compounds such as cyclic AMP or Ca^{2+}.

The metabolic activity of cells like muscle or liver is dependent on several conditions. The dietary supply of nutrients and the extent of muscular movement are two of the most important. The manner by which cells 'sense' extracellular changes is through the action of **hormones.** Hormones, then, control processes such as the entry of glucose into muscle and the raising or lowering of intracellular cyclic AMP and Ca^{2+} in liver and muscle. In each case the hormone, which is present in the extracellular fluid only, interacts with a **protein receptor** present on the plasma membrane. The receptors, which are specific for different hormones, transmit a signal into the interior of the cell. Binding between hormone and receptor can be dissociated from transmission of the resulting signal by lowering the temperature: that is, binding occurs at $0\,°C$, whereas membrane-mediated transmission is temperature-dependent. Three different types of receptor have been recognized (Fig. 6.14).

The first is the protein responsible for the facilitated uptake of glucose into muscle. It is maximally active only in the presence of the hormone insulin. In other words insulin acts as a cofactor. In the absence of insulin, entry of glucose is largely by passive diffusion (lower curve of Fig. 6.8); in the presence of insulin uptake is facilitated (upper curve of Fig. 6.8). The

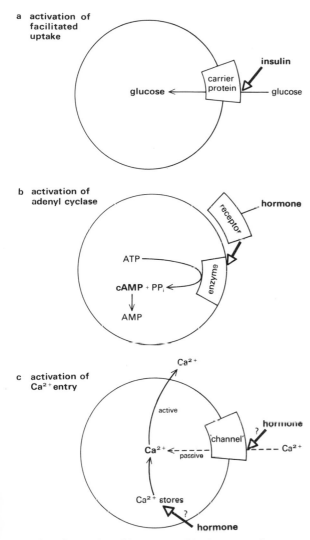

a activation of facilitated uptake

b activation of adenyl cyclase

c activation of Ca²⁺ entry

Fig. 6.14 Interaction of hormones with plasma membrane receptors

Note that in **c**, other mechanisms for increasing intracellular Ca²⁺ concentration may operate.

mechanism by which insulin increases the activity of the transporting protein, that is, decreases the activation energy of the entry process, is not known.

The transport protein itself has not yet been isolated or purified. It would appear to have two binding sites, one for glucose and one for insulin. Alternatively, there could be two separate proteins, one binding glucose and the other binding insulin, that somehow interact. The glucose-binding site has an affinity for other D-sugars that are transported, such as galactose. The entry of amino acids into muscle is by facilitated diffusion also. Like the uptake of glucose, the process is stimulated by insulin. The relation between the two classes of transporting protein is not clear. Insulin-

stimulated transport of sugars and amino acids takes place in fat cells also.

The second type of plasma membrane receptor is associated with the enzyme **adenyl cyclase.** The enzyme catalyses the intracellular conversion of ATP to cyclic AMP (Fig. 6.14). The metabolic consequences of an increased level of cyclic AMP in cells are manifold (Fig. 9.26; p. 211). In each case, phosphorylation of specific proteins, by stimulation of a protein kinase, is involved (e.g. Fig. 2.4). The level of cyclic AMP remains elevated for a limited period only. The original concentration is restored by action of the enzyme phosphodiesterase, which hydrolyses cyclic AMP to AMP. Adenyl cyclase is activated by several hormones. The enzyme is orientated in the membrane in such a manner that its catalytic site is on the inside; the hormone-receptor, which is a protein distinct from the enzyme, is orientated such that its hormone-binding site is on the outside. In other words ATP and hormone bind on opposite sides of the membrane, to two separate proteins that somehow interact with each other. Different hormones stimulate adenyl cyclase in different types of cell; that is, the hormone-receptor is different. The hormone-receptor is similar to the β-adrenergic site on other cells (section 12.3.2) in that β 'agonists' also stimulate adenyl cyclase. The exact mechanism by which the activation is achieved in either case is not clear.

The third type of membrane receptor controls the entry of Ca²⁺ into cells. As with cyclic AMP, an increase in intracellular Ca²⁺ has many metabolic consequences (Fig. 9.28). The entry of Ca²⁺ is stimulated by several cell-specific hormones. The hormone-binding site is similar to the α-adrenergic site on other cells, which also initiates entry of Ca²⁺ (section 12.3.2). The relation between hormone binding and Ca²⁺ entry is not clear for the reason that the nature of Ca²⁺ entry itself has not been elucidated. It may involve transient turnover of phosphatidyl inositol in the membrane. Increased entry is due either to an inhibition of the active exit mechanism, by Ca²⁺-ATPase or by Na⁺–Ca²⁺ exchange, or to a stimulation of the passive uptake mechanism. In the latter case some kind of opening of a Ca²⁺ channel or 'gate' would be involved. As is the case with cyclic AMP, the stimulus is transient: low levels of cytoplasmic Ca²⁺ are rapidly restored by active transport out of the cell, as well as by binding of Ca²⁺ to intracellular organelles.

6.3.4.2 Other compounds

Many compounds other than hormones bind to the plasma membrane. As with hormones, binding generally results in a change in the metabolic activity of the target cell; again the membrane-mediated stimulus is

Fig. 6.15 Non-hormonal interactions with plasma membrane receptors

Compound	Receptor	Cell type	Response
Foreign compounds (antigens)	Immunoglobulin M (antibody)	B lymphocyte	Transformation to plasma cell and secretion of antibody (IgG) (see section 11.2.1)
Foreign compounds (antigens)	Immunoglobulin E (antibody)	Mast cell	Histamine release (see section 11.2.2)
Foreign cells; virally modified cells	Histocompatibility antigens and other receptors	T lymphocyte	Destruction of foreign and virally modified cells (see section 11.3.1)
Mitogens	Glycoproteins and other receptors	Several	Cell-division in previously quiescent cell
Bacterial toxins; viruses	Glycoproteins, glycolipids and other receptors	Several	Cell damage by toxins and viruses; viral infection
Other cells	Glycoproteins	Several	Cell adhesion
Serotonin; other factors	?	Platelet	Release of ADP and more serotonin (see section 7.5)

temperature dependent, whereas the binding is not. Some examples are listed in Fig. 6.15.

The specificity of binding is greatest in the case of lymphocytes called B lymphocytes, which contain an immunoglobulin on their surface. The diposition of the protein in the membrane is not clear. Each circulating B lymphocyte contains just one out of some hundred thousand or more different types of immunoglobulin receptor, each of which is capable of binding to a specific type of antigen (Fig. 11.9). Binding results in the stimulation of the lymphocyte to undergo transformation and a few rounds of cell division. The antigen-immunoglobulin receptor complexes on the lymphocyte surface aggregate and flow to one end of the cell, forming a 'cap' of complexes. The 'cap' is then ingested by phagocytosis (section 6.4.2). The intracellular trigger for transformation may be an increase in Ca^{2+} concentration, following an increased Ca^{2+} uptake, brought about by one of the mechanisms referred to above. When cell division is complete, a plasma cell emerges. This continues to replicate itself under the appropriate stimulus and to secrete into the circulation that immunoglobulin species that was originally on the surface of the parent lymphocyte.

Immunoglobulin E (IgE) is a receptor present on the surface of mast cells. When the appropriate antigen binds to the receptor, the cell is stimulated to secrete histamine (Fig. 11.13). The trigger is an increased rate of Ca^{2+} entry.

The histocompatibility antigens are a class of membrane receptor of unknown physiological function. One property of histocompatibility antigens is the ability to participate in the rejection of foreign tissue grafts, but this can hardly be their physiological role. They appear to be involved in the general 'immune responsiveness' of the organism and may have a role to play in differentiation (section 11.3.1.2). What has been established is their structure: they are membrane-embedded glycoproteins, with the carbohydrate residues on the outer surface of the plasma membrane.

Several compounds, when bound to the surface of non-dividing cells, initiate a resumption of cell cycle activity and cell division. Such compounds have been termed mitogens. They include proteins known as lectins, which are derived from plants and certain animals, that bind specifically to the sugar residues of plasma membrane glycoproteins. Cells that are affected are lymphocytes and different types of cells in tissues. Physical manipulation of cells, such as partial hepatectomy or the wounding of skin cells, has a mitogenic action. The mechanism may, like that of immune stimulation of lymphocytes, involve a transient uptake of Ca^{2+} from the blood-stream. Compounds that are secreted by cells and that have an anti-mitogenic action have been termed 'chalones'; they are said to be specific inhibitors of cell proliferation, and to be missing or inactive in certain types of cancer cells. As yet a proper definition of chalones and their mode of action has not appeared.

Several types of plasma membrane constituent happen to be receptors for viruses and other toxic agents. Myxo- and paramyxoviruses, which include measles, mumps, influenza, and other respiratory tract infections, bind to N-acetyl neuraminic acid-containing glycoproteins and glycolipids. The attachment involves a specific viral protein that is embedded in the lipid envelope of each of these classes of virus. Binding to the cell surface is followed by entry of the virus by phagocytosis or by cell fusion, as described below (section 6.4). The fate of virus particles in cells depends on the efficiency of lysosomal degradation. Where lysosomes are plentiful and active, as in polymorphonuclear leucocytes (neutrophils) and macrophages, destruction of the virus ensues. Where they are not, replication of the virus particle and release from the cell, leading to infection of other cells, results. Toxins released by bacteria, such as cholera and tetanus toxin, bind to glycolipids in the plasma membrane of certain cells. Cell inactivation and destruction, by mechanisms not yet elucidated, ensues.

Finally, cells have receptors for recognizing each

other. This is important during embryonic development, when cell movement and the establishment of cell-specific colonies (that is, organs and tissues) takes place. In other words, recognition is followed by adhesion and the initiation of cell contacts such as the junctions referred to in section 6.2.5. A breakdown of these activities is a characteristic of malignant cells that invade and grow in organs unrelated to their site of origin. Glycoproteins and glycolipids have been implicated in cell–cell recognition and adhesion. One mechanism by which this may occur is by cross-linking of the carbohydrate residues at the surface of one cell to those at the surface of another. As yet this hypothesis is unproved.

6.4 Membrane movement

The regulation of cellular function by transport of low molecular weight compounds and ions into and out of cells is achieved by passage *through* the plasma membrane. Proteins and larger particles such as viruses and microbes also pass into and out of cells in certain conditions. In this instance passage *with* the plasma membrane takes place. The processes, known as **endocytosis** and **exocytosis** respectively, involve the addition or removal of a piece of membrane to or from the existing plasma membrane. These processes, and other types of absorption and secretion, are depicted in Fig. 6.16; types B and D involve fusion between two membranes.

6.4.1 Membrane fusion

Normally two membranes lying adjacent to each other do not fuse. This is not surprising, since the surface of cells is negatively charged (due to the polar groups of phospholipids and carboxyl groups on proteins, glycoproteins, and glycolipids), and two cells lying close to each other therefore tend to repel rather than to attract each other. Examples of the stable arrangement of adjacent membranes are the myelin sheath of nerve axons (Fig. 12.3) or the tight junctions between certain cells (Fig. 6.6). For fusion of adjacent membranes to occur, the repulsive forces that keep membranes apart have to be overcome. In other words, the membranes have to be destabilized (Fig. 6.17). The resulting structures can then undergo some kind of mixing of their components, which is followed by restabilization of the new membrane configuration.

The molecular nature of these changes is not known. Ca^{2+} is likely to play a major role. Certainly membranes depleted of Ca^{2+} are less stable than normal; readdition of Ca^{2+} restores stability (Fig. 6.17). Exactly what triggers such a transient change in membrane-bound Ca^{2+} is not clear. Another factor

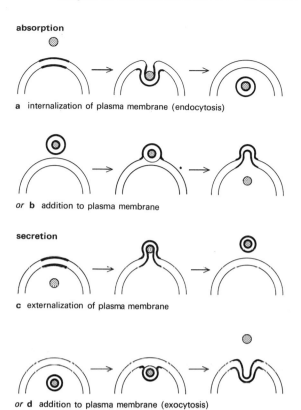

absorption

a internalization of plasma membrane (endocytosis)

or **b** addition to plasma membrane

secretion

c externalization of plasma membrane

or **d** addition to plasma membrane (exocytosis)

Fig. 6.16 Types of membrane movement involving the cell surface

that is involved in fusion is the lipid fluidity of the two membranes. Artificial membrane vesicles (called liposomes) composed of very fluid phospholipids fuse spontaneously, not only with each other, but with intact cells also.

Membrane fusion occurs in other situations besides endocytosis and exocytosis (Fig. 6.18). During fertilization, sperm plasma membrane fuses with the plasma membrane of the ovum. During differentiation of skeletal muscle, rows of myoblast cells fuse together to form myotubules. Certain enveloped viruses cause infected cells to fuse during viral release; viruses can also bring about cell–cell fusion in the test tube, and this has become a useful tool for preparing somatic cell hybrids for genetic analysis.

The fact that membrane fusion occurs between plasma membranes of different cell types, and between intracellular organelles and plasma membrane, shows

Fig. 6.17 Mechanism of membrane fusion

I. Destabilization of membranes
(? removal of Ca^{2+})

II. Mixing of membrane components

III. Restabilization of new membrane configuration
(? restoration of Ca^{2+})

Fig. 6.18   Situations involving membrane fusion

1. Endocytosis
   (including subsequent lysosomal fusion)

2. Exocytosis
   (including secretion of plasma proteins, hormones, neurotransmitters, etc., expulsion of 'polar body' (maturation of ovum);
   expulsion of nucleus (maturation of red blood cells)

3. Fertilization

4. Myotubule formation

5. Entry and release of certain viruses

that an exact similarity of composition of the two fusing membranes is not important. What is important is that each membrane is sufficiently fluid to allow rapid mixing to occur once the destabilization event has taken place.

6.4.2 Absorption and secretion

The types of membrane movement that involve the cell surface are illustrated in Fig. 6.16. Endocytosis has been divided into **phagocytosis** (uptake of insoluble particles and cells) and **pinocytosis** (uptake of water droplets containing dissolved substances). The distinction is not important, since the mechanism is virtually the same. More important is the distinction between pinocytosis and uptake of water-soluble substances by the transport processes discussed in section 6.3.1. The latter processes result in the translocation of substances directly into the cell cytoplasm. In pinocytosis, as in other forms of endocytosis, lysosomal digestion of the internalized ('pinocytotic' or 'phagocytic') vesicle has to take place before the dissolved molecules can mix with the cytoplasmic contents. The most common mechanism of absorption of low molecular weight substances is that of direct transport.

An important type of endocytosis is the phagocytic uptake of foreign particles by macrophages and neutrophils. Bacteria, viruses, and other infectious agents are removed from the blood-stream by this mechanism. Generally an immune response precedes uptake. That is, antibodies against the foreign agent (antigen) are first secreted into the blood-stream (section 11.2.1.2). These then bind to the antigen and it is an antigen–antibody complex that is taken up by endocytosis. The endocytic or phagocytic vesicle is then degraded, as in pinocytosis, by lysosomal fusion. The lysosomes of macrophages are particularly rich in degradative enzymes, including peroxidases, with which to break down the components that have been absorbed.

Macrophages ingest intact host cells. Phagocytosis of red cells, for example, is the mechanism whereby their life-span is terminated (Fig. 7.26). Macrophages also aid in the destruction of virally infected (Fig. 11.14) and other cells.

The endocytic process is non-specific and cannot distinguish between an intact virus and virus inactivated by combination with antibody. Uptake of intact virus, which occurs also by the mechanism of direct membrane fusion (Fig. 6.16), results in infection. Generally this is limited to certain types of cell, partly as a result of an initial binding between virus and the cell surface (Fig. 6.15). Influenza, 'common cold', and similar viruses, for example, invade cells of the respiratory tract, while poliomyelitis and meningitis viruses invade the nervous system.

The most important type of exocytosis is the **secretion** of **hormones** and **neurotransmitters** (Fig. 6.18). Hormones such as insulin, thyroxine, or the hormones of the anterior and posterior pituitary, and neurotransmitters such as the catecholamines and acetylcholine are synthesized within vesicles inside the secretory cell. When the cell is stimulated the vesicles move to, and fuse with, the plasma membrane, releasing their contents into the blood-stream (e.g. Figs 9.17, 9.20, 11.13, and 12.8).

Non-hormonal proteins are secreted by the same mechanism. The secretion of albumin by liver, milk proteins by mammary gland, and digestive enzymes by stomach and pancreas, are examples. Note that fusion of an intracellular vesicle with the plasma membrane, without secretion of anything into the blood-stream, is one postulated mechanism for the assembly of new plasma membrane (section 2.5.3).

Secretion of membrane-coated vesicles, such as the plasma lipoproteins or bile by liver, or milk fat droplets (distinct from milk fat proteins) by mammary gland, occurs by the mechanism shown in Fig. 6.16. That is, the compounds to be secreted are synthesized within the cytoplasm, then move to the plasma membrane and become enfolded within it in such a way that a vesicle is formed. This is then pinched off. In the case of plasma lipoproteins and other fat vesicles, a single phospholipid leaflet encloses the contents. This 'single' membrane has polar residues exposed on the outside, with the apolar fatty acid chains facing inwards and binding to the rest of the lipid constituents such as cholesterol and triglyceride (Fig. 7.6). Newly synthesized particles of enveloped viruses are secreted into the blood-stream by a similar mechanism of plasma membrane 'budding'. In this instance the membrane is a bilayer, that comes to surround the nucleocapsid and other viral proteins.

It will be noted that in the situations described in Fig. 6.16, pieces of membrane are either added to, or removed from, the plasma membrane. A non-membrane-coated particle arriving at one side of the plasma membrane is released at the other side with a membrane coat. Conversely a membrane-coated particle is released without its membrane coat. Such movements constitute the 'membrane flow' that is responsible for the high rate of phospholipid turnover (section 2.2.4). Situations in which a membrane-coated particle crosses the membrane intact, or picks up an extra piece of plasma membrane, are rare. An enveloped virus that enters the cell by phagocytosis rather than by membrane fusion is an example.

From what has been said it is clear that the movement of particles and vesicles across the cell surface involves more than just membrane fusion. A means of propelling the particle or vesicle towards, or away from, the inside of the plasma membrane has to exist. Although the mechanism of this movement is not established, it is known that the intracellular network of microtubules and microfilaments plays a part. The enveloping of the plasma membrane around the particle to form a vesicle involves another mechanism that is as yet unclarified.

6.5 Summary

Nature of cell surface

The surface or plasma membrane controls the passage of substances into and out of cells. In general, the phospholipid bilayer serves to **insulate** a cell from its surroundings, by preventing rapid diffusion of non-specific water-soluble molecules; the **proteins** of the surface membrane, on the other hand, serve to catalyse the **transport** of specific water-soluble molecules and ions. In addition, the proteins of the surface membrane serve to **transmit signals** initiated by the binding of **hormones** and other effectors to the cell surface.

Structure

1. The plasma membrane, like other membranes, consists primarily of proteins and phospholipids; it is also particularly rich in **cholesterol, glycoproteins** and **glycolipids.** Surrounding the plasma membrane of cells is an outer **coat** consisting of protein and **proteoglycans;** its functions are to provide an additional protective layer and to cement cells together.

2. The cell surface of many cells is elaborated into protuberances such as **microvilli,** which serve to increase the surface area of cells.

3. Between adjacent cells in tissues there are various types of **junctions.** Occluding junctions and desmosomes serve to maintain intercellular rigidity and structure. Communicating junctions serve to transmit molecules that do not easily cross the plasma membrane from cell to cell, and thus to co-ordinate the function of tissues and organs.

Regulation of cellular function

1. Uptake of nutrients and other molecules occurs by **passive diffusion,** by **facilitated diffusion** and by **active transport.** Diffusion of molecules occurs when the concentration outside the cell exceeds that inside; the presence of membrane proteins having affinity for specific molecules, such as amino acids and sugars, facilitates (that is, accelerates) diffusion. Movement against a concentration gradient requires the expenditure of ATP (active transport).

2. Active transport of Na^+ out of cells achieves an **ionic asymmetry** of cations: inside cells, K^+ is high and Na^+ is low (outside cells K^+ is low and Na^+ is high). A Na^+/K^+-specific ATP-ase known as the **Na^+ pump** is responsible for this asymmetry. The active transport of sugars and amino acids into certain types of cells is mediated by coupling to the Na^+ pump. The Na^+ pump also regulates the amount of water in cells.

3. High molecular weight substances that are present in the blood-stream, such as protein hormones, control metabolism within cells by the transmission of signals across the plasma membrane. This is achieved by the binding of hormones to specific **protein receptors** on the surface membrane. This results in a membrane modification, such that (a) more **nutrient enters** (as in the case of insulin) that (b) the enzyme **adenyl cyclase** is **activated,** leading to an increase in the intracellular concentration of cyclic AMP (as in the case of several other hormones), or that (c) **Ca^{2+} enters** (as in the case of these and other hormones). The outcome in each case is an increase or decrease in the rate of certain intracellular reactions, due to the activation or inactivation of specific enzymes.

Membrane movement

1. Substances are transported into and out of cells not only by movement through the plasma membrane, but by movement with the plasma membrane. Such movement is known as **endocytosis** and **exocytosis** respectively.

2. Endocytosis, which results in the uptake of high molecular weight compounds and particles such as viruses (a process also known as **phagocytosis**), is particularly marked in the case of certain white blood cells. The result can be beneficial, as when pathogenic bacteria or viruses are subsequently destroyed by the lysosomes within the cells, or it can be detrimental, as when such viruses are not destroyed, but replicate and spread to other cells.

3. Exocytosis is the mechanism by which many hormones, plasma proteins and other substances are secreted from cells.

4. During endocytosis and exocytosis, **fusion** of surface **membrane** occurs at certain points. Fusion of membranes takes place in other situations also; the mechanism is not clear.

FURTHER READING

General

For a general account of the biochemistry of the cell surface: see the books on membranes mentioned in the further reading section of Chapter 1.

A more detailed account of the cell coat:

G.M.W.Cook and R.W.Stoddart (1974). *Surface carbohydrates of the eukaryotic cell.* Academic Press, New York.

J.H.Luft (1976). The structure and properties of the cell surface coat. *Int. Rev. Cytol.* **45,** 291.

J.R.Clamp (ed.) (1978). Mucus. *Brit. med. Bull.* **34,** No.1.

Cell junctions:

C.Peracchia (1977). Gap junction structure and function. *Trends biochem. Sci.* **2,** 26.

L.A.Staehelin and B.E.Hull (1978). Junctions between living cells. *Scient. Am.* **238,** No. 5, 140.

Transport proteins:

G.Guidotti (1976). The structure of membrane transport systems. *Trends biochem. Sci.,* **1,** 11.

The metabolic consequences of calcium entry:

B.Gomperts (1976). Calcium and cell activation. In: *Receptors and recognition* (eds P.Cuatrecasas and M.F.Greaves) (Series A, Vol.2). Chapman and Hall, London.

H.Rasmusson and D.B.P.Goodman (1977). Relationship between calcium and cyclic nucleotides in cell activation. *Physiol. Rev.* **57,** 421.

For surface receptors:

P.Cuatrecasas and M.F.Greaves (eds) (1977). *Receptors and recognition* (Series A, Vol.4). Chapman and Hall, London.

M.F.Greaves (1975). *Cellular recognition* (Outline studies in Biology). Chapman and Hall, London.

S.Jacobs and P.Cuatrecasas (1977). The mobile receptor hypothesis for cell membrane receptor action. *Trends biochem. Sci.* **2,** 280.

T. Yamakawa and Y. Nagai (1978). Glycolipids at the cell surface and their biological functions. *Trends biochem. Sci.* **3,** 128.

Part Two

Cellular specialization

7

Vascular system

7.1 Introduction

The vascular system, which for the purpose of this discussion will be confined largely to **blood,** has several functions. The first is to keep cells supplied with **nutrients,** and the second is to remove **end-products** of **metabolism.** In the body, cells are grouped together in tissues and organs, each of which has a specialized function. The distribution of some of the major organs, and their relation to the vascular system, is illustrated in Figs 7.1 and 7.2. The metabolic activity of organs and tissues is co-ordinated by the action of hormones. The **transport** of **hormones** between tissues and organs is the third main function of the vascular system.

Nutrients and waste products are transported in one of two ways. Sugars, amino acids, fats, and vitamins — and hormones — are transported by the fluid part of blood, the **plasma.** Some end-products of metabolism, such as H_2O, CO_2, and urea are also transported by plasma. Oxygen and some carbon dioxide are transported by the **red cells.**

In addition to plasma and red cells, blood contains white cells and platelets (Fig. 7.3). The function of **white cells** is predominantly that of **protecting** the body against **infection.** In short, the neutralization and elimination of viruses, bacteria, and other microbes. The function of **platelets** is to prevent accidental **loss** of **blood.** A continual supply of blood is crucial to the functional maintenance of all organs, especially brain and heart. A disruption of the blood-supply, by mechanical stoppage or by excessive loss, rapidly leads to organ failure and death. By initiating the clotting of blood at sites of damage to the vascular system, and thus 'sealing' the injury, platelets prevent blood loss.

7.2 Plasma

7.2.1 Constituents

Plasma is defined as the fluid part of blood which is obtained by removing red cells, white cells, and platelets. This is most readily achieved by centrifuging whole blood in the presence of anticoagulants such as citrate or heparin, whereby plasma is found to account for half the volume of blood, with the various cells making up the other half. The total volume of blood in a 70 kg adult is some 6 litres; in addition there is some 12 litres of interstitial fluid (section 6.3.2), with another 30 litres of fluid within cells. In other words, approximately 70 per cent of the body is water.

The main constituents of plasma are proteins, fats, low molecular weight compounds, and ions. A typical distribution is shown in Fig. 7.4. If blood is allowed to

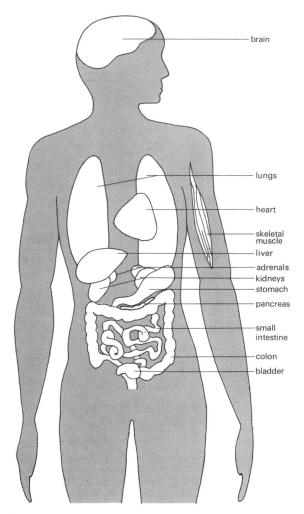

Fig. 7.1 Tissues and organs of the body

Labels: brain, lungs, heart, skeletal muscle, liver, adrenals, kidneys, stomach, pancreas, small intestine, colon, bladder

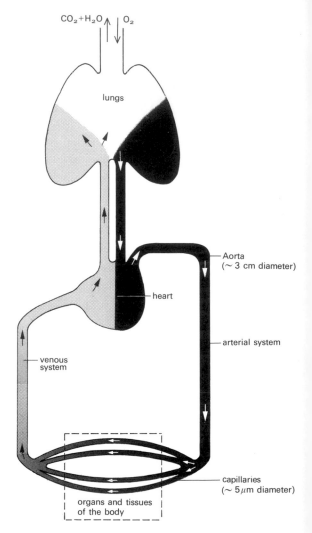

Fig. 7.2 Vascular system

Solid arrows indicate the flow of blood. The intensity of shading indicates the relative oxygenation of the blood. Note that this is a highly diagrammatic representation: in fact the heart lies between the lungs (Fig. 7.1).

Labels: $CO_2 + H_2O$, O_2, lungs, Aorta (~ 3 cm diameter), heart, arterial system, venous system, capillaries ($\sim 5\,\mu$m diameter), organs and tissues of the body

coagulate before it is centrifuged, one of the components of plasma, namely fibrinogen, is removed. In this case the supernatant fluid, that is, fibrinogen-free plasma, is termed **serum.**

7.2.1.1 Proteins and lipids

Plasma, like intracellular fluid, has a high concentration of protein. Interstitial fluid, that is the fluid in the spaces between cells and blood capillaries, contains little protein. Hence plasma (like intracellular fluid) has a high osmotic pressure in relation to interstitial fluid. It is this pressure that controls the movement of water into, and out of, cells (Fig. 7.5). The predominant protein in plasma is **albumin** (molecular weight 65 000). If its concentration falls, as in protein malnutrition and certain types of renal failure, water is retained in interstitial fluid and oedema develops.

Albumin has another role. Although the amino acid

Fig. 7.3 Constituents of blood

Constituent	Amount present
Plasma	55–60% by volume
Cells	40–45% by volume;
Red	4–7×10^{12} cells/l
White	4–11×10^{9} cells/l
Platelets (thrombocytes)	1–4×10^{11} cells/l

The values, which can vary considerably, refer to a normal adult. A value of $<4 \times 10^{12}$ red cells/l is referred to as anaemia (see Fig. 7.28); a value of $<4 \times 10^{9}$ white cells/l is referred to as leucopenia; and a value of $<1 \times 10^{11}$ platelets/l is referred to as thrombopenia.

Fig. 7.4   Constituents of plasma

Constituent	Amount present
Protein	69–85 g/l of which
	61% is albumin
	9% is α-globulins
	11% is β-globulins
	14% is γ-globulins
	4% is fibrinogen
Fat	3.5–8.5 g/l of which approx.
	30% is cholesterol
	30% is phospholipid
	30% is triglyceride
Inorganic ions	~6 g/l of which approx.
	140 meq/l is Na^+
	4 meq/l is K^+
	2.5 meq/l is Ca^{2+}
	1 meq/l is Mg^{2+}
	100 meq/l is Cl^-
	27 meq/l is HCO_3^-
	2 meq/l is phosphate
Carbohydrate	~0.8 g/l of which most is glucose
Urea	~0.3 g/l
Amino acids	~0.2 g/l of which
	alanine, glutamine, glycine, lysine, proline, and valine are the predominant species
Ketone bodies	~0.02–0.2 g/l

These values, which can vary considerably, especially after meals, refer to a normal adult. Only the major constituents are shown. There are in addition other low-molecular-weight compounds, ions, vitamins, proteins, hormones (see Fig. 9.24), enzymes (see Fig. 7.8), etc.

residues exposed at its surface are largely polar, that is forming many hydrogen bonds with the surrounding water and making the protein a soluble one, there is an area of apolar, hydrophobic amino acid residues. This allows the binding of free fatty acids and other lipids. Albumin is the main carrier of free fatty acids in plasma. The free fatty acids pass, rather like water, across blood-vessels and the cell plasma membrane, while albumin remains in the blood-stream. An albumin-like molecule, present at much lower concentration, has a specific site for the fat-soluble vitamin A.

Some of the **globulins** have binding sites for steroid hormones and other lipids. The majority of triglyceride and cholesterol is transported in the form of **lipoprotein complexes.** These are actually vesicles made up of a membrane containing α- and β-globulins surrounding a mixture of triglyceride and cholesterol. Depending on the relative amounts of triglyceride and cholesterol compared to protein, the vesicles are termed 'very low density lipoproteins' (VLDL), 'low density lipoproteins' (LDL), and 'high density lipoproteins' (HDL), (Fig. 7.6).

After a fatty meal, the plasma concentration of triglyceride rises (Fig. 7.7). Triglyceride is initially in the form of chylomicrons, absorbed into the lymphatic system and released into the blood-stream by way of the thoracic duct. Triglyceride is degraded to free fatty acids and glycerol by the enzyme 'clearing factor lipase' situated on the outside of certain cells. The enzyme is so called because it 'clears' the blood of its milky appearance, caused by the presence of triglyceride. Production of the enzyme is stimulated by heparin. The enzyme facilitates the uptake of some of the free fatty acids into adipose cells and muscle. The rest of the free fatty acids become bound to albumin. In the liver such free fatty acids are reconstituted into triglyceride and, in concert with phospholipid and protein, released in the form of VLDL. Triglyceride is again degraded to free fatty acids by clearing factor lipase of adipose cells, muscle, and other tissues such as heart and kidney cortex; most of the free fatty acids are taken up into the cells. The triglyceride-depleted VLDL are now LDL, carrying mainly cholesterol. Cholesterol is not utilized so effectively and much of it remains in the blood-stream. Exactly how the circulation of cholesterol, in LDL or HDL, is controlled is not clear.

The major outcome of the processes described above is to store triglyceride in adipose tissue; at the same time some of the free fatty acids are oxidized by muscle, heart and other tissues. During starvation, adipose tissue triglyceride is broken down to free fatty acids, which are utilized by muscle and other tissues either directly, or indirectly after conversion to ketone bodies by the liver (Fig. 8.18).

Globulins have other roles. The γ-globulins, or immunoglobulins, act as antibodies to neutralize foreign antigens. Their structure and functions are described in section 11.2.1.

Other globulins, such as prothrombin and various clotting factors, are, together with fibrinogen, involved in the blood coagulation. The coagulation of blood is closely associated with the function of platelets, and is discussed in section 7.5. Several other globulins, some with as yet undefined functions, exist. A very minor globulin, for example, specifically binds thyroxine (thyroxine-binding protein) and is responsible for its transport in the blood.

In addition to the proteins described above, which, with the exception of the immunoglobulins, are mostly synthesized in the liver, plasma contains circulating protein and peptide hormones, all in very low concentration, that are synthesized in the respective endocrine organ (section 9.2). Also present are several enzymes, synthesized in various tissues. Their presence in abnormally large amounts is indicative of damage to the cells of origin. Some examples are given in Fig. 7.8.

7.2.1.2 Low molecular weight compounds and ions

The low molecular weight compounds present in plasma comprise water-soluble and fat-soluble molecules. The fat-soluble molecules, cholesterol, steroid hormones, fat-soluble vitamins, and other lipids, are protein-bound and are transported, partly in

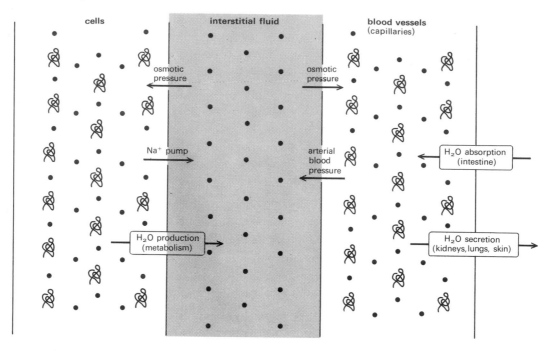

Fig. 7.5   Control of water movement

Interstitial fluid occupies the space between blood-vessels (capillaries) and cells. It contains virtually no proteins (or other high-molecular weight compounds). Cell plasma membranes and the walls of capillaries are 'semi-permeable'. That is, movement of water (and ions) is relatively unrestricted; proteins, however, do not cross. Hence water tends to move out of interstitial fluid into cells and blood vessels (by osmotic pressure). Interstitial fluid would thus 'dry up', were it not for the fact that the Na^+ pump tends to drive water out of cells (see section 12.3.1), and the arterial blood-pressure tends to drive water out of blood-vessels. Superimposed on these forces is a net movement of water (generated by cellular metabolism) out of cells, a movement of water (absorbed from the intestine) into the blood-stream and a net movement of water (secreted through kidneys, lungs, and skin) out of the blood-stream. Interstitial fluid itself drains into the lymphatic system, and hence back to the blood-stream. Under normal conditions all these movements (indicated by the arrows) tend to cancel each other out, and the volume of cells and interstitial fluid remains constant.

   Dots = low molecular weight compounds (mainly inorganic ions)
   Larger symbols = high molecular weight compounds (mainly proteins)

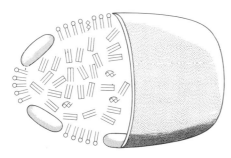

|| triglyceride   ⦂ cholesterol   ⋔ phospholipid

⬭ protein   ⦂ cholesteryl ester

Composition of lipoprotein vesicles (~ % of dry wt)

Type	Density	Protein	Phospholipid	Cholesterol*	Triglyceride
HDL	~1.1–1.2	50	30	20	<5
LDL	~1.0–1.1	25	20	45	10
VLDL	~0.9–1.0	10	15	15	60
Chylomicrons	<~0.9	<5	5	5	90

* The figures refer to total cholesterol (free plus esterified).

The composition of chylomicrons (containing entirely dietary-derived lipids) is shown for comparison. Note that all the values shown are approximate, and can vary between certain limits.

Fig. 7.6   Plasma lipoproteins

   Plasma lipoproteins circulate in the form of vesicles. A single layer of phospholipid, protein and cholesterol surrounds a hydrophobic interior containing cholesteryl esters and triglyceride. Unlike the double-layered membranes of cellular organelles, there are no hydrophilic 'channels' (Fig. 1.35): the proteins are folded so as to present an entirely hydrophilic exterior and an entirely hydrophobic interior. A high-density lipoprotein (HDL) vesicle, which is flat and disc-shaped, is shown. Low-density lipoprotein (LDL) and very low-density lipoprotein (VLDL) vesicles are progressively larger and more spherical, as they contain increasingly more triglyceride. Chylomicrons, which contain the most triglyceride, are the largest type of vesicle.

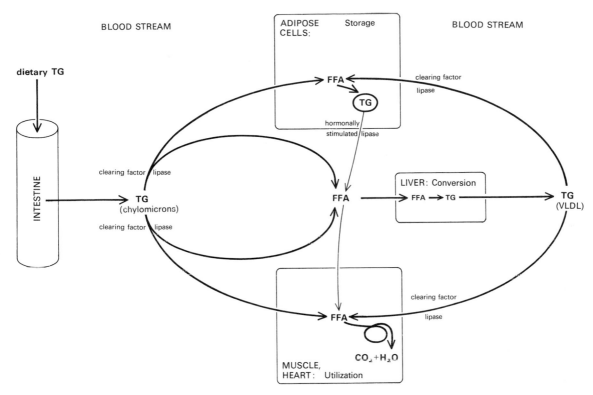

Fig. 7.7   Circulation of lipids

The major 'flow' of lipid following a fatty meal is shown in bold. TG = triglyceride FFA = free fatty acids; the release and metabolism of glycerol has been omitted for clarity.
The utilization of free fatty acids by liver (to form CO_2 and ketone bodies) is illustrated in Fig. 8.18. Note that triglyceride is synthesized and stored to a certain extent in muscle also; in this case the products of hydrolysis remain within the cell.

the form of vesicles, as described above. The water-soluble molecules include glucose and other sugars, amino acids, glycerol, urea, ketone bodies, and vitamins; some water-soluble molecules, such as certain vitamins

Fig. 7.8   Elevation of serum enzymes in disease

Enzyme	Cellular origin	Disease
Most; some transaminases up 10–200-fold	Liver	Acute hepatitis
Transaminases up 2–8-fold	Liver	Chronic hepatitis
Alkaline phosphatase	Liver	Obstructive jaundice
Creatine kinase, aldolase, aspartate transaminase, and LDH (type 1+2) up 2–10-fold	Heart	Myocardial infarct
Creatine kinase, aldolase, aspartate transaminase, and LDH (type 5) up 3–6-fold	Muscle	Muscular disease (e.g. muscle injury, muscular dystrophy)
Amylase, lipase	Pancreas	Acute pancreatitis

Under normal conditions, the above enzymes are present in serum at only 0.01–0.1 per cent of the amount present in tissues such as liver or muscle.

and hormones, are transported in a protein-bound form also. Following a meal, sugar and amino acids are absorbed by the intestine and their concentration in the blood-stream rises. They are utilized for storage, energy production and biosynthesis by liver, muscle, and other tissues. The deamination of amino acids by liver results in the formation of urea which is excreted by the kidneys. During starvation, the level of nutrients returns to normal, or a little below. Glucose is replenished largely by the liver: by glycogen breakdown and by gluconeogenesis. Amino acids are replenished, largely from muscle, by an excess of protein breakdown over protein synthesis. At the same time the output by liver of ketone bodies, formed from free fatty acids derived from adipose tissue triglyceride, increases (section 8.4.3.2). Ketone bodies are a major energy source for muscle, heart, and other tissues.

Carbon dioxide, derived from cells as the end-product of sugar, amino acid, and fat breakdown, is transported largely as bicarbonate ion. The equilibrium position is rapidly reached by participation of the enzyme carbonic anhydrase (Fig. 7.9), present in red blood cells. Red blood cells are freely permeable to

149

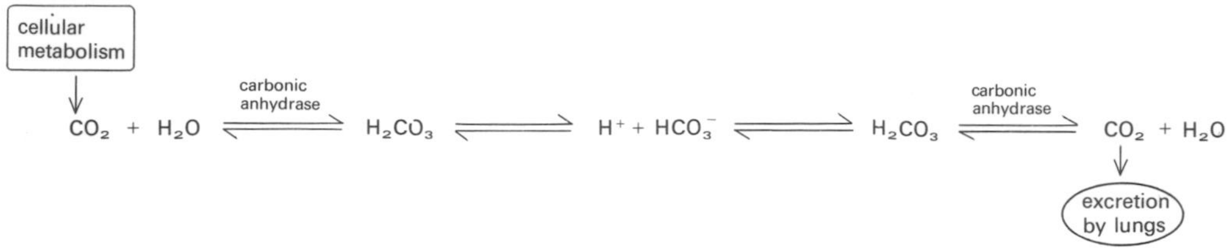

Fig. 7.9   Function of carbonic anhydrase

carbon dioxide, which enters, and to bicarbonate, which leaves (section 7.3.2). Transient lowering of pH (acidosis) due to excessive production of carbon dioxide, or other cause (Fig. 7.12) is avoided by the buffering capacity of plasma proteins. In addition, the kidney controls plasma pH (section 7.2.2.2).

The concentration of the major inorganic ions in the blood-stream is like that of bicarbonate, linked to cellular metabolism. Failure to absorb Na^+ and K^+ across the epithelial cells of intestine or renal tubules, or failure of the Na^+ pump to maintain ionic asymmetry in other cells, leads to an imbalance of plasma Na^+ and K^+. Ca^{2+} levels are controlled by the action of two hormones: calcitonin, which stimulates Ca^{2+} removal by deposition as calcium phosphate in bone, and parathyroid hormone, which has the reverse effect (section 9.2.4). Vitamin D stimulates Ca^{2+} absorption by the small intestine (Fig. 8.5).

7.2.2 Circulation: kidney

The circulation of the blood-stream is depicted in outline in Fig. 7.2. That diagram does not take account of the fact that certain constituents of plasma, chiefly water, ions, carbon dioxide, and urea, are continuously being lost from the body. There are three main sites of elimination: lungs, skin, and kidney. The **lung** is a specialized tissue, allowing **gaseous exchange** to take place (Fig. 7.10). Oxygen is absorbed; CO_2 and H_2O (both in the gaseous form) are eliminated. The **sweat glands** in the skin cells secrete **water** and **ions** (chiefly Na^+ and Cl^-). The **kidney** secretes **water, ions, urea,** and other **waste products.** Under certain conditions some loss of water and ions through the intestine also occurs, as in diarrhoea. This is the result of failure to reabsorb the water and ionic content of food and gastrointestinal secretions by lower portions of intestine. Normally the faeces contain predominantly solid waste matter (largely the products of haem degradation).

7.2.2.1 Excretion of waste products

The main function of the kidneys, two organs each approximately 150 g in weight, is the formation of **urine.** Another function is the secretion of hormones (Fig. 9.12). The part of the kidney that is con-

cerned with urine formation is known as a **nephron.** Approximately 1 000 000–1 500 000 nephrons make up each kidney. The nephron consists of two separate components: glomerulus and tubule (Fig. 7.11). The function of **glomerulus** is to **filter** the blood and thus to retain cells, proteins and other high molecular weight substances. The function of the **tubule** is to **reabsorb** most of the water, ions, sugars, and amino acids that are present in the glomerular filtrate, back into the blood-stream. What is then left is a concentrated solution of urea, uric acid, creatinine, and other excretory products, that is passed via a collecting duct to the bladder.

The absorptive capacity of the tubule is enormous. Out of some 160–180 litres of fluid that may pass through the glomerulus in one day, only 1.5 litres is excreted. The rest, about 99 per cent of the glomerular filtrate, is reabsorbed. This is achieved largely as a result of the increased osmotic pressure of the protein-rich fluid leaving the glomerulus, which flows through blood-vessels that are entwined around the tubule. Additional water is reabsorbed by active transport, linked to the Na^+ pump and controlled by the antidiuretic hormone (ADH or vasopressin); moreover the anatomical arrangement of the tubulues is such that a counter-current system of absorption is set up, which serves to absorb more water. Failure to synthesize and to secrete antidiuretic hormone (from the posterior pituitary gland) results in diuresis, that is, the production of too watery a urine. This is a characteristic of the type of diabetes known as diabetes insipidus.

Like water, much of the sugar, amino acids, and ions are reabsorbed by active transport. The **Na^+ pump** is the main driving force. It is situated within the epithelial tubule cells, so as to pump Na^+ from cell into the blood-stream. The cells resemble epithelial cells of the small intestine in having the outer (absorptive) part of their plasma membrane elaborated into a mass of microvilli ('brush border') in order to maximize the surface area for absorption (Fig. 6.4). In other words, Na^+, and hence Cl^- and water, as well as other ions and low molecular weight compounds, diffuse from the glomerular filtrate into the epithelial tubule cell, and are actively extruded at the other end of the cell by operation of the Na^+ pump. In order to

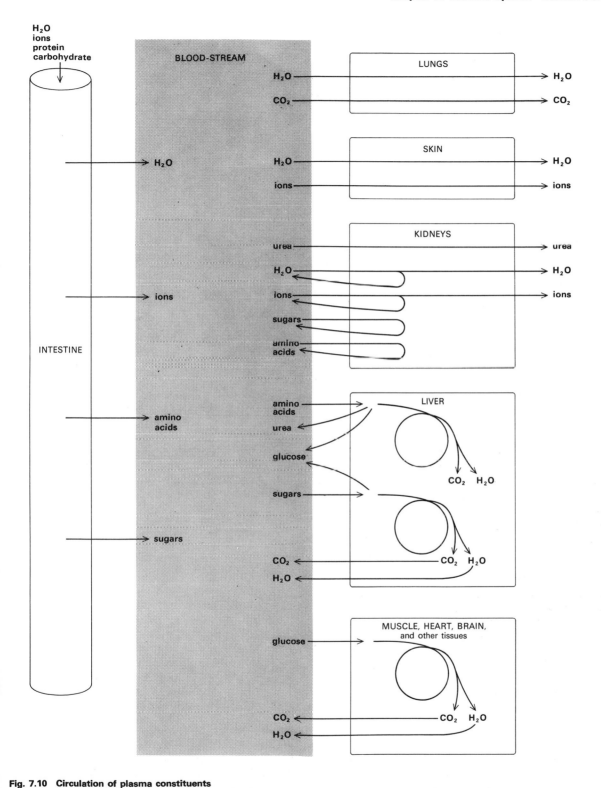

Fig. 7.10 Circulation of plasma constituents

Circulation of lipids is shown in Fig. 7.7.

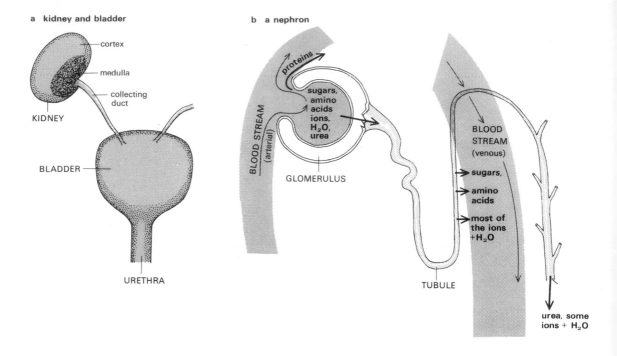

a   kidney and bladder

cortex

medulla

collecting
duct

KIDNEY

BLADDER

URETHRA

b   a nephron

proteins

sugars,
amino
acids
ions,
H₂O,
urea

BLOOD STREAM
(arterial)

GLOMERULUS

BLOOD
STREAM
(venous)

sugars,

amino
acids

most of
the ions
+H₂O

TUBULE

urea, some
ions + H₂O

Fig. 7.11   The kidney

Each kidney consists of ~1 500 000 nephrons; these are so arranged that the glomeruli are in the cortex region, and the tubules in the medulla region. A simplified diagram of a nephron is shown. Arterial blood flows through the glomerulus: proteins are retained; low molecular weight compounds, ions and water are removed by filtration. These constituents pass into the tubule, from which most of the constituents are absorbed back into the blood-stream, through venous capillaries wound round the tubule. A concentrated solution of urea and some ions is left and is passed into the bladder. Actual blood-vessels have been omitted from the diagram for clarity.

reabsorb something like 1.4 kg of NaCl per day, the pump utilizes most of the ATP generated by oxidative phosphorylation within the kidney. The hormone aldosterone potentiates the reabsorption of Na^+ from the glomerular filtrate, but not of K^+. Moreover, as Na^+ is secreted from the tubule cell into the blood-stream, it exchanges with K^+. As a result almost as much Na^+ as water is reabsorbed (98–99 per cent) whereas only some 90 per cent of K^+ is reabsorbed. In other words urine is relatively richer in K^+ than is plasma (Fig. 7.13).

The plasma membrane at the brush border of the epithelial cells is rather impermeable to urea, uric acid, creatinine, and other waste products, so that these are not reabsorbed.

The urine provides a useful diagnostic system for many diseases affecting metabolism. Diseases resulting from lack of a particular degradative enzyme, such as phenylketonuria or galactosaemia are characterized by the presence of phenylpyruvate or galactose in the urine.

7.2.2.2 Maintenance of plasma pH

The kidney performs a second role, which is related to its function as an excretory organ. This is the mainte-

nance of plasma pH near neutrality. Basal metabolism releases H^+ (Fig. 7.12), due to:

i.   the ionization of carbon dioxide produced by respiration,

ii.   the ionization of acids such as lactic and acetoacetic, produced by glycolysis and ketogenesis respectively, and

iii. the ionization of sulphuric acid, produced by the oxidation of cysteine and methionine (although the amount of cysteine and methionine present in dietary and other proteins is relatively small, the strong dissociation of sulphuric acid rather compensates for this).

It is true that reaction (a) of Fig. 7.12 is largely reversed when carbon dioxide is released from the lungs, but the net effect of the metabolic breakdown of foods is nevertheless the production of an excess amount of hydrogen ions. While the buffering ions in the blood, especially phosphate, haemoglobin, and some plasma proteins, are able to counteract small changes in pH, a continued release of hydrogen ions has to be balanced

H⁺-producing processes

a oxidation of nutrients $\longrightarrow$ carbon dioxide (all tissues)

$$CO_2 + H_2O \rightleftharpoons H_2CO_3$$
$$H_2CO_3 \rightleftharpoons HCO_3^- + H^+$$

b glycolysis $\longrightarrow$ lactic acid (skeletal muscle; red cells; other tissues)

$$CH_3CH(OH)COOH \rightleftharpoons CH_3 \cdot CH(OH)COO^- + H^+$$

ketogenesis $\longrightarrow$ ketone bodies (liver)

$$CH_3COCH_2COOH \rightleftharpoons CH_3COCH_2COO^- + H^+$$
acetoacetic acid

$$CH_3CH(OH)CH_2COOH \rightleftharpoons CH_3CH(OH)CH_2COO^- + H^+$$
β–hydroxybutyric acid

c oxidation of sulphur-containing amino acids (liver and other tissues)

cysteine
$$\searrow$$
$$\rightarrow H_2S \rightarrow \rightarrow H_2SO_4$$
methionine $\nearrow$

$$H_2SO_4 \longrightarrow SO_4^{2-} + 2H^+$$

H⁺-removing process

d amino acids $\longrightarrow$ keto acids + NH_3
(liver and other tissues)

$$NH_3 + H^+ \rightleftharpoons NH_4^+$$

Fig. 7.12   Control of plasma pH

by an equivalent removal if neutrality in body fluids is to be maintained.

The kidney achieves the **maintenance** of **plasma pH** near 7·4 by excreting a slightly **acid urine** (normal pH range between 5·3 and 7·0, with an average value of about pH 6·0). This is brought about by several factors (Fig. 7.13). First the renal tubule reabsorbs bicarbonate very effectively, whereas chloride is relatively less well absorbed, and phosphate and sulphate hardly at

Fig. 7.13   Ionic content of plasma and urine

Ion	Plasma meq/l	Urine meq/l
Cations		
Na⁺	140	150
K⁺	4	30
NH₄⁺	<1	**30**
Anions		
Cl⁻	100	140
HCO₃⁻	**30**	<1
phosphate	2	30
sulphate	1	40
protein	**14**	<1

Only the major ionic species are shown. The ions that are concentrated in urine, compared to plasma, are K⁺, NH₄⁺, phosphate, and sulphate. The ions that contribute to the greater acidity of urine compared to plasma, i.e. cations of weak base in urine and anions of weak acid in plasma, are shown in bold.

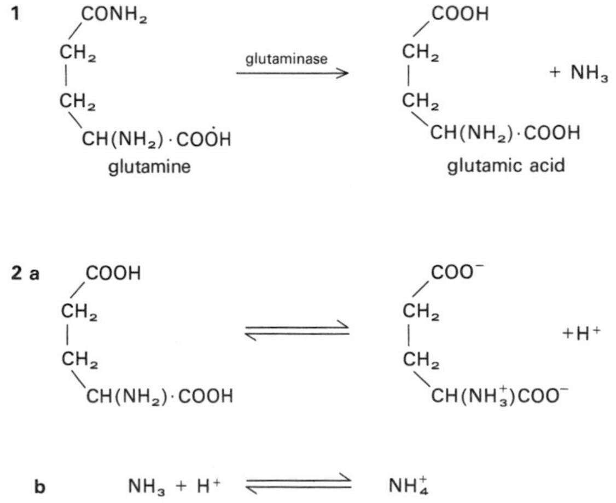

all. As a result phosphate, sulphate, and chloride (the last two of which are the dissociation products of a strong acid), are selectively excreted over bicarbonate (the dissociation product of a weak acid). Secondly, the plasma proteins, which carry a negative charge (that is, which are the dissociation products of weak acids) are retained in plasma by virtue of their lack of entry into the glomerular filtrate. Thirdly, the renal tubule produces ammonia by the action of glutaminase on glutamine (Fig. 7.14). This does not in itself contribute to acidity since reactions 2(a) and (b) of Fig. 7.14 counterbalance each other. But the NH₄⁺ which appears in the extracellular fluid is relatively less well reabsorbed than is Na⁺. As a result NH₄⁺ (which is the dissociation product of a weak base) appears in urine instead of Na⁺ (which is the dissociation product of a strong base). Fourthly, the renal tubule cells reabsorb a certain amount of Na⁺ in exchange for H⁺. Although the H⁺ is produced intracellularly, by hydrolysis of ATP linked to the Na⁺ pump, the overall effect is to remove H⁺ from plasma. In the same way the secretion of HCl in the stomach raises the pH of plasma, though in this case the effect is largely offset by the secretion of bicarbonate from the pancreas.

The kidney is able to control the rate at which Na⁺ bicarbonate, phosphate, and other anions are reabsorbed. It is therefore able to control the pH of plasma: decreased reabsorption of bicarbonate, for example, leads to a fall in plasma pH; increased reabsorption has the opposite effect. The kidney is also able to control the action of glutaminase, that is, the rate at which ammonia is produced. This too controls the pH of plasma: a fall in plasma pH is corrected for by increased activity in glutaminase, whereas a rise in plasma pH is corrected for by a decrease in activity. Damage to the renal tubule and failure to secrete

Fig. 7.14   Production of ammonia in kidney

ammonia therefore has a direct effect on the maintenance of plasma pH, and leads to an acidosis (and excessive loss of Na^+).

The glutamine that enters the kidney is synthesized largely in liver, from glutamate and ammonia produced by deamination of amino acids (Fig. 2.18). In other words the amino groups of dietary amino acids are excreted in the urine both as urea (the major form) and as ammonium ions. It is the relative amount that is excreted as ammonium ions, by the combined action of glutamine synthesis in liver and glutamine degradation in kidney, that exerts a control over the pH of body fluids. The exact mechanism by which the kidney reacts to an altered plasma pH is not clear. The important point is that there exists a type of 'negative

feedback' (sections 2.6 and 9.1) such that an increase in plasma pH leads to conditions that operate to lower plasma pH, and *vice versa*.

7.2.3 Disease

An abnormal concentration of plasma constituents generally reflects either a nutritional deficiency, or an impairment in the function of organs such as liver or the endocrine glands (Fig. 7.15). Liver disease (section 8.4.5) affects the levels of plasma proteins such as albumin, α- and β-globulins; endocrine disease (section 9.5) affects the levels of the circulating hormones. In each case the concentrations may be reduced or

Fig. 7.15 Abnormal levels of plasma constituents

Constituent	Alteration	Cause
Albumin; proteins in general	Decrease	Nutritional deficiency (Kwashiokor); liver disease
β lipoproteins	Decrease	Defective synthesis of β lipoproteins by liver (abetalipoproteinaemia)
	Increase	Increased synthesis of β lipoproteins in liver
γ globulins	Decrease	Defective synthesis of immunoglobulins by immune system (agammaglobulinaemia)
	Increase	Increased synthesis of immunoglobulins by immune system (infections and other causes)
Fibrinogen	Decrease	Defective synthesis of fibrinogen by liver (afibrinogenaemia)
Prothrombin and other clotting factors	Decrease	Defective synthesis of specific proteins by liver (certain types of haemophilia and other diseases)
	Incomplete structures	Lack of vitamin K (Fig. 8.6)
Tissue enzymes	Increase	Damaged tissue of origin (Fig. 7.8)
Cholesterol	Increase	Nutritional and other causes, including excessive lipoprotein synthesis by liver (hyperlipidaemia; cholesterolaemia)
Triglyceride	Increase	Nutritional and other causes (hyperlipidaemia)
Glucose	Increase	Defective utilization by muscle and other tissues (diabetes mellitus)
Glucose, amino acids	Increase	Defective tubular reabsorption (renal disease, e.g. chronic nephritis)
Urea, creatinine	Increase	Defective glomerular filtration (renal disease, e.g. acute nephritis)
Na^+	Decrease	Lack of aldosterone (Addison's disease); excessive loss from gastro-intestinal tract (e.g. vomiting) or from kidney
	Increase	Excess of aldosterone (certain adrenal tumours); excessive water loss
K^+	Decrease	Excessive loss from gastro-intestinal tract (e.g. vomiting) or from kidney
	Increase	Lack of aldosterone (Addison's disease)
Ca^{2+}	Decrease	Lack of parathyroid hormone (hypoparathyroidism)
	Increase	Excessive vitamin D intake; excessive bone destruction; hyperparathyroidism
Hormones	Increase and decrease	Various causes (see Fig. 9.33)
Vitamins	Decrease	Nutritional deficiencies (see Fig. 8.2)

Note that defective synthesis of specific proteins is often due to a genetic defect.

raised. The latter situation is often the result of a malignant tumour.

Changes in plasma proteins affect those substances that are transported by them. In abetalipoproteinaemia, for example, a lowered concentration of the proteins responsible for transporting triglycerides and cholesterol results in an altered deposition of fat. Conversely an increase of β-lipoproteins can result in an increased level of cholesterol (cholesterolaemia) and triglyceride in plasma, which may be a contributing factor to the development of atherosclerosis. For atherosclerosis involves the deposition of lipid 'plaques' containing cholesterol, triglyceride, and protein on the walls of blood-vessels. Such plaques tend to form a clot or thrombus with platelets and red blood cells. Thrombi present a danger as they may detach and block a vessel in the heart (leading to angina pectoris and myocardial infarction, that is, coronary thrombosis) or the brain (leading to a cerebral infarct, that is, stroke).

Diseases of the kidney have two consequences:

i. failure to retain substances such as proteins, amino acids, sugars, water, and ions and

ii. failure to excrete substances such as urea, creatinine, and other waste products.

The first consequence is generally not so serious as the second.

In chronic nephritis in which the reabsorptive capacity of the kidney is impaired, for example, there is a loss of nutrients in the urine, and mild symptoms of nutritional deficiency may develop. The same occurs in rare inherited disorders in which there is a failure to reabsorb specific amino acids or sugars from the urine; the resulting high concentrations in urine are known as aminoaciduria and glycosuria respectively. A quite different reason for glycosuria occurs in diabetes mellitus. Here the glucose content in blood is so high that the reabsorbing capacity of the tubule is exceeded. Assay of urine for the presence of glucose is the primary test in the case of suspected diabetes mellitus.

In acute nephritis, on the other hand, the glomerulus is often affected (perhaps by an autoimmune reaction) and ceases to function. No fluid is filtered, with the result that all the toxic components of blood are retained and accumulate; there is also a retention of water, leading to oedema. Such defects of glomerular filtration are the most common forms of renal disease.

Other diseases of the kidney are those that reflect an endocrine imbalance, such as an insufficiency or excess of vasopressin (ADH; Fig. 9.4) or aldosterone (Fig. 9.12). A malfunctioning kidney is also a common site for the start of bacterial infections; blockage due to precipitation of calcium oxalate or other material ('renal calculi') is often the cause.

7.3 Red blood cells

The main function of red blood cells (erythrocytes) is to **carry oxygen** from the lungs to all the other organs of the body. At the same time red cells accelerate the removal of carbon dioxide from the tissues back to the lungs. The constituents and metabolism of red blood cells reflect these two functions. In order to bring red cells, and the other constituents of blood, as close as possible to the cells that make up lung and other organs, all tissues contain a network of blood-vessels so that as many cells as possible are in contact with the blood. Blood-flow through tissues is designed so that incoming blood (containing oxygen) is kept separate from outgoing blood (containing carbon dioxide). This is achieved by having separate networks of incoming (arterial) and outgoing (venous) blood within tissues (Fig. 7.2). Within tissue the blood-vessels become progressively smaller, until they are some 5–10 μm in diameter. Such capillaries then form a network of ingoing (arterial) and outgoing (venous) vessels for the supply and removal of nutrients from cells.

7.3.1 Constituents

The chief constituent of red blood cells is **haemoglobin.** It is the compound that transports oxygen, and about 15 per cent of the carbon dioxide in blood. Haemoglobin accounts for some 90 per cent of the dry weight of red blood cells. The remainder, which is largely plasma membrane (red cells have no nucleus, mitochondria, endoplasmic reticulum or other internal membrane system) consists of other proteins, glycoproteins, phospholipids, cholesterol, inorganic ions (K^+, Na^+, Mg^{2+}, Cl^-), and low molecular weight compounds such as sugar phosphates and other metabolic intermediates.

7.3.1.1 Haemoglobin

Haemoglobin is a protein made up of four subunits. In adult haemoglobin the subunits are two α-chains and two β-chains ($\alpha_2\beta_2$) (Fig. 1.12). Each subunit, called globin (Fig. 1.4), contains a haem group (Fig. 3.10). The essential component of haemoglobin is the iron atom in each of the four haem groups. Of the six coordinate valencies of the iron, only one is linked to globin; four are linked to the pyrrole group of haem; and the last is free to bind oxygen, carbon monoxide, or other ligand. Carbon monoxide binds 500 times more tightly than oxygen (to yield an inactive haemoglobin–CO complex), which explains its great toxicity.

The structure of iron in haemoglobin may be contrasted with the structure of iron in cytochromes, in which two of the six valencies are linked to protein (Fig. 7.16). Also in contrast to cytochromes is the fact

haemoglobin

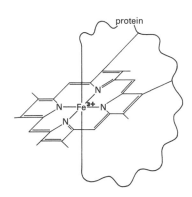

cytochromes

that in haemoglobin the iron remains in the Fe^{2+} state, whether it is carrying oxygen or not. Were haemoglobin to become oxidized by the oxygen it carries, its function would be nullified, since oxygen would be used up and not be available for use by the tissues. Oxidation of haemoglobin by oxidizing agents *other* than molecular oxygen, however, does occur; the resulting Fe^{3+} compound, known as methaemoglobin, is unable to bind oxygen. The red cell possesses a powerful reductase, linked to glutathione, that maintains haemoglobin in the Fe^{2+} state (Fig. 7.17). The state of the haem groups in haemoglobin is, as in cytochromes, readily ascertained from the absorption spectrum (Fig. 7.18).

The **binding** of **oxygen** by haemoglobin is characterized by **cooperativity** between the four haem groups. That is to say, the affinity of any one haem group for oxygen is affected by the presence of an oxygen molecule on one of the other haem groups. As a result, a plot of the amount of oxygen bound against the concentration of oxygen is sigmoidal (Fig. 7.19). In

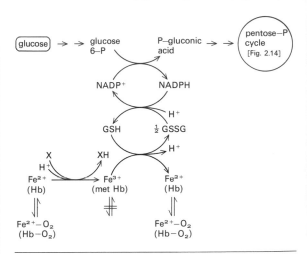

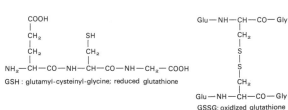

Fig. 7.17 Role of glutathione reductase in red cells

Several compounds that are oxidizing agents, such as the sulphonamides and other drugs, oxidize haemoglobin (Fe^{2+}) to methaemoglobin (Fe^{3+}). Methaemoglobin, which binds H_2O in preference to O_2, is reduced back to haemoglobin by a number of intracellular reducing agents. One such compound is glutathione (GSH); GSH is regenerated by the action of glutathione reductase, which is linked to the oxidation of glucose via the pentose phosphate pathway. Glutathione is important also for preventing the oxidation of fatty acids and proteins in the red cell plasma membrane.

Fig. 7.16 Comparison between haemoglobin and cytochromes

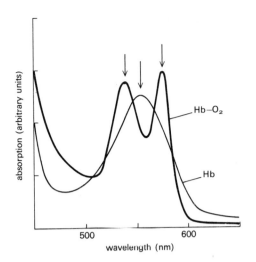

Fig. 7.18   Absorption spectra of haemoglobin

The absorption spectra for haemoglobin (Hb) and oxygenated haemoglobin (Hb–O₂) are shown. Partially oxygenated haemoglobin has a spectrum intermediate between the two; by measuring the absorption at the wavelengths shown by the arrows, the degree of oxygenation can be assessed

contrast, the binding of oxygen to a haem group that is present as a single chain only, as is the case for the muscle protein myoglobin, is characterized by the more usual rectangular hyperbola.

The physiological advantage of cooperativity in oxygen binding by haemoglobin is as follows. In the lungs, where oxygen tension is high, haemoglobin is fully saturated with oxygen. As it passes through the arterial system to the tissues, the oxygen tension, and hence

the amount of oxygen bound, drops; at this point myoglobin (which is also present in tissues other than muscle) has a relatively greater affinity for oxygen than haemoglobin (Fig. 7.19), thus facilitating transfer. Once this has occurred, haemoglobin tends to bind carbon dioxide (at free amino groups of globin, not at the iron atom of haem), and the haemoglobin–CO_2 is carried by the venous system to the lungs. Many other factors, such as the concentration of chloride, affect the transport of carbon dioxide.

The mechanism whereby in haemoglobin the affinity of one haem group for oxygen is modulated by the presence or absence of oxygen on another haem group is due to the quaternary structure of haemoglobin. Although the four haem groups are not in direct contact, the binding of oxygen by one haem group distorts the globin chain to which it is attached in such a manner that a neighbouring globin chain, and hence its haem grouping, is distorted, with the result that oxygen is able to bind more easily. The distance moved is some 0·7 nm (7Å) (Fig. 7.20). This type of cooperativity is seen also in enzymes that possess quaternary structure; in this instance binding of

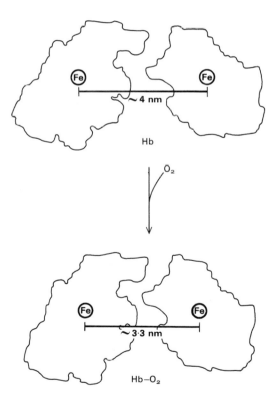

Fig. 7.20   Allosteric change in haemoglobin

Diagrammatic representation of the two β-chains in haemoglobin; the α-chains have been omitted for clarity (see Figure 1.12). The distance between the respective Fe^{2+} atoms decreases by some 0.7 nm (7 Å) when haemoglobin is oxygenated.

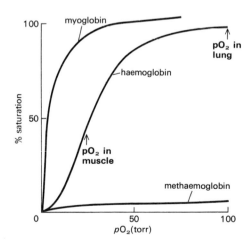

Fig. 7.19   Binding of oxygen to haemoglobin

The binding of oxygen to haemoglobin is characterized by a S-shaped curve (cf. Fig. 1.26), indicative of co-operativity between the 4 sub-units.

substrate to one subunit, and hence the catalytic activity, is modulated by the presence of substrate on another subunit (Fig. 1.27). Such cooperative phenomena are termed **allosteric** ('other shape'). The affinity of haemoglobin for oxygen is modulated by a variety of factors. Increase in pH favours binding (known as the Bohr effect); increase in the concentration of 2,3-diphosphoglycerate, present in relatively high concentrations in red blood cells, has the opposite effect.

From what has been said it is likely that changes in the composition of the globin chain can affect the binding of oxygen by the haem group. This is what happens in inherited disorders of haemoglobin synthesis, the **haemoglobinopathies.** In sickle-cell anaemia, for example, the glutamate at position 6 of the β-chain of haemoglobin is replaced by valine. In this instance the resulting molecule is altered to such an extent that its solubility, especially in the non-

oxygenated state, is reduced to the point of precipitation. As a result the red cells are distorted in a characteristic 'sickling' manner (Fig. 7.21). The anaemia results from the fact that sickled cells are unable to bind oxygen normally; moreover the cells are recognized as 'foreign' or 'aged' by the spleen, and hence degraded faster than normal.

The haemoglobin genes are on autosomal chromosomes (the α and β genes are actually on different chromosomes), and hence an individual expresses the genes of both his parents (see Fig. 4.12). There are thus two forms of a haemoglobinopathy such as sickle cell disease: the heterozygous state in which the disease is not serious, and the homozygous state, in which it is (Fig. 4.12).

Many other types of haemoglobinopathy have now been recognized. Anaemia and muscular fatigue, due to the failure of proper oxygenation, are common symptoms. In each case, haemoglobin has an abnormal

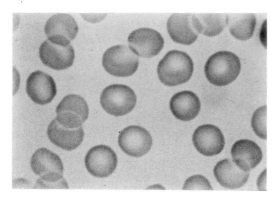

Normal red cells, containing HbA.

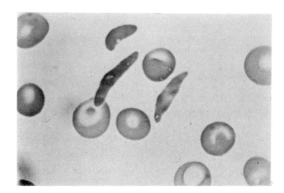

'Sickled' red cells, containing HbS. Some unsickled cells are also present.

Courtesy of Professor P. T. Flute.

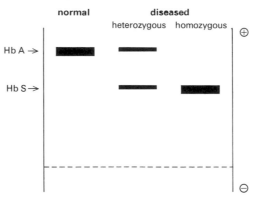

In sickle cell haemoglobin (HbS), glutamate at position 6 of the β-chain (Fig. 1.4) is replaced by valine. As a result, the electrophoretic mobility of Hb is reduced. In the heterozygous condition ('sickle cell trait'), approximately equal amounts of normal haemoglobin (HbA) and HbS are present; in the homozygous condition ('sickle cell disease'), only HbS is present (see Fig. 4.12).

Fig. 7.21 Sickle cell disease

electrophoretic mobility, due generally to the replacement of a particular amino acid in one of the two globin chains (Fig. 7.22). In some instances, an entire globin chain is missing. This could be due to a 'nonsense' mutation, so that an mRNA that is untranslatable into protein is produced, or it could be due to a mutation affecting the translational ability of an otherwise normal mRNA. It could also, of course, be due to lack of the globin gene itself. All three types of mutation have been observed.

The haemoglobinopathies provide an excellent example of the molecular basis of a particular disease. The reason why they are so well documented is twofold. First, diseases of the blood are easy to investigate because specimens can readily be obtained. Secondly, haemoglobin is the major constituent of red blood cells, and can readily be purified and analysed. Once methods are developed for the analysis of enzymes and other proteins present in small amounts, and in less accessible tissues such as liver, heart, or brain, it is likely that a host of other molecular-disease relationships will become apparent. Just as the change in haemoglobinopathies involves a part of the molecule *other* than the haem group (which can be described as the 'active site' of haemoglobin), so the change in enzymes is likely to affect amino acids other than those directly participating in the catalytic site;

the increased sensitivity to proteolytic degradation of the enzyme glucose 6-P dehydrogenase in G6PDH deficiency (section 2.2.1.4) is an example that has been documented. It is because the efficiency of the altered molecules is *reduced* not *lost*, that recognition is difficult to achieve.

7.3.1.2 Membrane components

The membrane of a red cell is similar to the plasma membrane of any other cell: it contains protein, phospholipids, and cholesterol, as well as glycoproteins and glycolipids. The asymmetric distribution of phospholipids across the bilayer (Fig. 6.2) is particularly marked in red cells, since there is no 'flow' of membrane between plasma membrane and other cellular organelles (Fig. 2.64) to disturb it. Some of the glycolipids have been characterized and found to fall into various groups, such as the ABO or the Rh system, known as the **blood groups.**

In the **ABO system,** glycolipids having either the A configuration, the B configuration, both configurations, or neither configuration, are present. Persons having the A type of glycolipid on their blood cells possess an antibody in plasma that is specifically directed against the B type of glycolipid (anti-B serum); the converse is also true (Fig. 7.23). If an anti-B serum is mixed with the red cells having B specificity, red cells agglutinate (haemagglutination). In the presence of complement factors

Fig. 7.22 Some haemoglobinopathies

Disorder	Type of Hb	Subunits	Abnormality	Result	Clinical manifestation
Change in gene structure					
Sickle cell disease	HbS	$\alpha\alpha\beta\beta$	Glu → Val at position 6 of β chain	Decreased sol. of Hb; sickling	Haemolytic anaemia
No particular name	HbC	$\alpha\alpha\beta\beta$	Glu → Lys at position 6 of β chain	Decreased sol. of Hb	Mild anaemia
No particular name	Hb Freiburg	$\alpha\alpha\beta\beta$	Val deleted at position 23 of β chain	Unstable Hb	Haemolytic anaemia
No particular name	Hb Riverdale–Bronx	$\alpha\alpha\beta\beta$	Gly → Arg at position 24 of β chain	Unstable Hb	Haemolytic anaemia
No particular name	Hb Savannah	$\alpha\alpha\beta\beta$	Gly → Val at position 24 of β chain	Very unstable Hb	Severe haemolytic anaemia
Change in gene expression					
α thalassaemia	HbH	$\beta\beta\beta\beta$	Defective synthesis of α chains	Insol. Hb; decreased red cell survival	Mild anaemia
β thalassaemia	HbF*	$\alpha\alpha\gamma\gamma$	Failure to switch from synthesis of γ chains to β chains at birth	Abnormal O₂ affinity	Anaemia

* Normal Hb prior to birth.

Changes in gene structure give rise to haemoglobin of altered *structure*. Over 200 types of amino acid substitution or deletion, involving either the α or the β chain, are known; the results vary, depending on the effect that the change has on the tertiary and quaternary structure of haemoglobin.

Changes in gene expression give rise to haemoglobins containing subunits in altered *amount*. Two extreme types are illustrated; many others, involving partial deficiency of α or β chain, are known.

Fig. 7.23   The ABO blood group system

Blood group	Glycolipid (antigen) on red cell	Antibody in serum	Enzyme present in cells[†]
A	A	Anti-B	UDP-GalNAc transferase
B	B	Anti-A	UDP-Gal transferase
AB	A and B	—	UDP-GalNAc and UDP-Gal transferases
O	—	Anti-A and anti-B	—

Structure of A:
GalNAc-Gal-GlcNAc-Gal-Glc-ceramide
|
    Fuc

Structure of B:
Gal-Gal-GlcNAc-Gal-Glc-ceramide
|
Fuc

[†] Specific for transferring to precursor glycolipid

Fig. 7.24   Relation between ABO system and incidence of disease

Blood group	Predisposition to disease
A	Carcinoma of stomach; pernicious anaemia; coronary thrombosis (? related to increased serum cholesterol)
O	Duodenal ulcer

In no case is the relation a direct one; that is, persons of a particular blood group merely have a statistically increased predisposition to certain disease. (For example, if the risk of an O or B person developing cancer of the stomach is taken as 1, that of an A person is 1.2).

(section 11.2.1.3), lysis (haemolysis) then occurs. When cells agglutinate, there is a danger of blood clot formation (section 7.5); also, agglutinated cells are destroyed by the spleen. Haemolysed cells lose haemoglobin, and are therefore non-functional. Haemagglutination and haemolysis form the basis of ABO blood group assays. It is important to perform such assays prior to blood transfusion, as an unmatched transfusion (that is, between persons of differing blood groups) can result in haemagglutination and haemolysis.

A and B glycolipids have been purified and analysed. The differences in composition are shown in Fig. 7.23. The AB specific carbohydrate groupings occur not only in red blood cells, but also on other cells, and in mucous secretion such as those of the alimentary tract. In these cases the carbohydrates are present as terminal groupings on glycoproteins rather than on glycolipids. The genetic basis underlying the ABO blood groups is the presence or absence of a specific glycosyl transferase enzyme (Fig. 7.23). The enzymes are not cell-specific, which accounts for the fact that a person secreting A type glycoprotein in the alimentary tract is also A positive on his red blood cells. Whether a person secretes a blood group-specific glycoprotein at all is under separate genetic control.

The physiological function of the ABO glycolipids and glycoproteins is not clear. What has been established is that there is a significant correlation between the possession of a particular blood grouping and an increased sensitivity to certain diseases (Fig. 7.24).

7.3.2 Metabolism

The human red cell has no nucleus, mitochondria, endoplasmic reticulum, or other organelle. It therefore cannot synthesize DNA, RNA, protein, phospholipid, or carbohydrate, nor can it obtain ATP by oxidative phosphorylation. Its major metabolic transformation is the breakdown of glucose to lactate by the pathway of **glycolysis.** Glucose is therefore the major metabolic fuel required by red cells. The ATP that is generated (Fig. 2.28) is used primarily to drive the Na^+ pump. Its main function in red cells is to maintain a high internal K^+ concentration (conducive to glycolysis) and to prevent excessive water entry (section 6.3.3.1).

The glycolytic sequence is used also to generate 1,3-diphosphoglycerate, from which 2,3-diphosphoglycerate is formed by a mutase enzyme; normally 1,3-diphosphoglycerate reacts with ADP to form 3-phosphoglycerate and ATP (Fig. 2.8). 2,3-diphosphoglycerate is important in modulating the oxygen affinity of haemoglobin.

In addition to being broken down to lactate by glycolysis, glucose is oxidized to CO_2 and H_2O by the pentose phosphate shunt. As in other cells, the enzymes of this pathway (Fig. 2.14) are all cytoplasmic. NADPH, which is generated by the pathway, is oxidized back to $NADP^+$ by coupling with glutathione reductase, the reduced glutathione (GSH) so formed being used to maintain haemoglobin in the Fe^{2+} form (Fig. 7.17). When there is a defect of the pentose phosphate pathway, as in glucose 6-phosphate dehydrogenase deficiency, haemoglobin tends to become oxidized to methaemoglobin (Fe^{3+}); the result is a failure to transport oxygen as well as the premature destruction of red blood cells (section 2.2.1.4).

Red cells contain a high concentration of carbonic anhydrase (Fig. 7.9). By accelerating the formation of bicarbonate ion, which is more water-soluble than carbon dioxide, the enzyme promotes the passage of CO_2 from tissues to lungs as follows. CO_2 is absorbed by red cells and the resulting H_2CO_3 dissociated into H^+ and HCO_3^-; HCO_3^- is exchanged for plasma Cl^- (a process referred to as the chloride shift), while H^+ is largely buffered by the histidine residues of haemoglobin; since reduced haemoglobin has a greater buffering capacity than haemoglobin–oxygen, uptake of H^+ is particularly favoured in venous blood where carbon dioxide is high and oxygen is low. In addition, carbon dioxide is bound

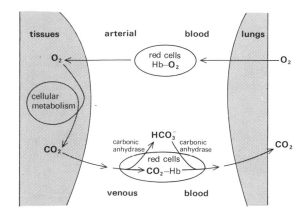

Fig. 7.25   Movements of oxygen and carbon dioxide in blood

Note that in Hb–O_2, the O_2 is bound to Fe^{2+}; in CO_2–Hb, the CO_2 is bound to globin.

directly by haemoglobin; again, reduced haemoglobin has a greater affinity than haemoglobin–oxygen. The result is efficient transport of CO_2 and HCO_3^- to the lungs, where the processes are reversed, and CO_2 is released (Fig. 7.25).

7.3.3 Formation and destruction

Red cells have a life span of some 120–130 days. This means that approximately 2×10^{11} red cells, or 1 per cent of the total red cell population, is formed (and destroyed) every day. In consequence, haemoglobin synthesis and degradation are processes of considerable metabolic importance. Since a red cell contains some 3×10^8 molecules of haemoglobin, the rate of synthesis is 6×10^{19} molecules, or about 8 grams of

haemoglobin per day; this corresponds to about 5 per cent of the total weight of proteins turned over per day. During pregnancy the demand for haemoglobin synthesis increases, as a result of which an extra intake of iron is required.

The formation of red cells takes place in the bone marrow. The precursor cells are known as erythroblasts, and contain nucleus, mitochondria, and endoplasmic reticulum. Haemoglobin begins to be synthesized at this stage. As the erythroblast matures, it loses its nucleus to become a reticulocyte (Fig. 7.26). This contains sufficient of the haemoglobin-mRNA synthesized in the erythroblast (and not subject to rapid turnover), to continue haemoglobin synthesis. The pathway for haemoglobin synthesis is outlined in Fig. 7.27. Failure to absorb sufficient dietary iron prevents haemoglobin synthesis and leads to anaemia. Anaemia also results from failure to absorb sufficient vitamin B_{12} or folic acid, both of which are required for the rapid DNA synthesis that accompanies the formation and proliferation of erythroblasts.

In foetal development, during which a blood-supply is initiated long before the establishment of bones, haemoglobin synthesis takes place in the liver. The type of haemoglobin that is synthesized differs from adult haemoglobin in having two γ-chains instead of two β-chains. The resultant haemoglobin is known as foetal haemoglobin (HbF). In β thalassaemia (Fig. 7.22), foetal haemoglobin synthesis continues into adult life, even though production shifts from liver to the bone marrow.

In adults, red cells are destroyed at the same rate as they are produced, by the combined functioning of two organs: spleen and liver. In spleen, macrophages ingest red cells by phagocytosis. Haemoglobin is degraded to haem and free amino acids. Since newly

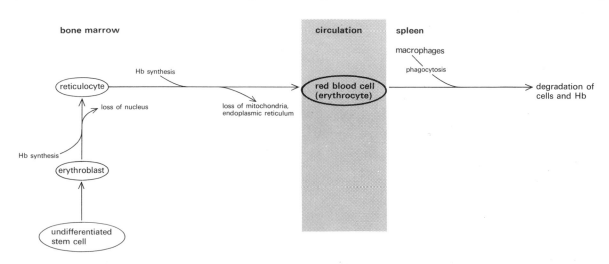

Fig. 7.26   Formation and destruction of red cells

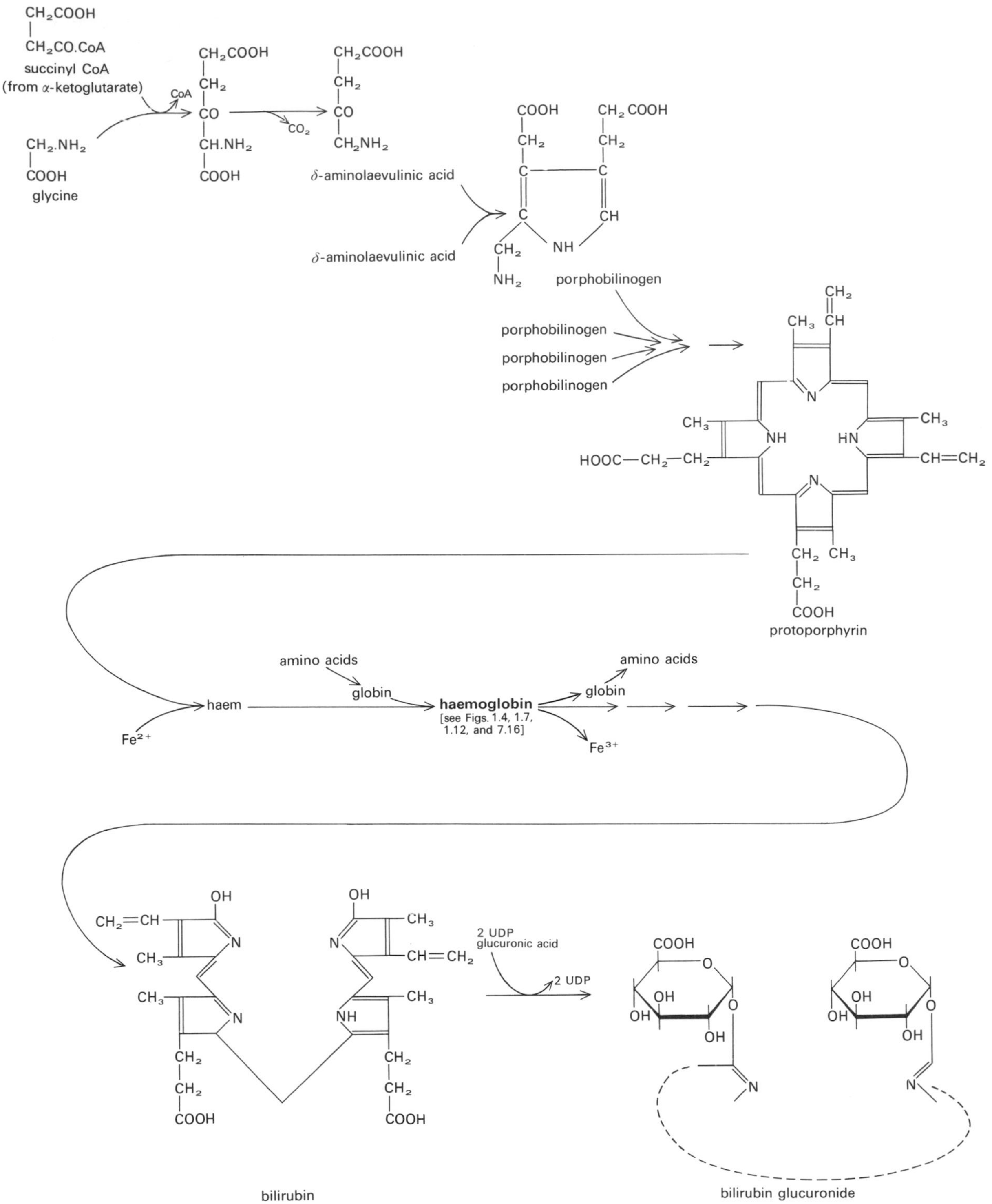

Fig. 7.27 Biosynthesis and degradation of haemoglobin

Haemoglobin (Hb) synthesis takes place in erythroblasts and reticulocytes; (haem synthesis takes place in most cells, since it is the precursor of cytochromes (Fig. 3.10) and of myoglobin, as well as of Hb). Hb degradation begins in the spleen and is continued in the liver. The product (bilirubin glucuronide) is secreted in the bile.

synthesized, as well as 120 day-old, red cells continuously pass through the spleen, it is clear that macrophages somehow distinguish between aged and non-aged red cells and attack only the former. The exact nature of the difference between an aged and a non-aged red cell is not yet clear.

The haem that is released from the spleen passes to the liver, where it is taken up by the Kupffer cells. Any haem originating from cellular cytochrome and myoglobin also passes to the liver. Haem is degraded to free Fe^{2+} and bilirubin (Fig. 7.27). Bilirubin, which is rather insoluble in water, is conjugated with glucuronide, by reaction with UDP glucuronic acid, on the endoplasmic reticulum of the liver cell (Fig. 5.4). Bilirubin glucuronide and free Fe^{2+} then pass through the biliary system into the small intestine. Fe^{2+} is reabsorbed, but bilirubin is excreted, after further metabolism, in the faeces. In jaundice in which there is an impairment of liver function, the degradation products of haem accumulate in the blood-stream, giving the affected individual a yellowish appearance

7.3.4 Disease

Diseases of red blood cells can be considered in three categories (Fig. 7.28):

i.   production of insufficient red cells (**anaemia**)
ii. production of faulty red cells, and
iii. production of an excess of red cells

Category (i) is by far the most common: the cause may be nutritional, as in iron, folate, or vitamin B_{12} deficiency (section 8.2.3.1); it may be infectious, as in malaria or infectious mononucleosis; or it may be due to a diseased bone marrow (section 10.3.1).

Category (ii) includes the haemoglobinopathies (Fig. 7.22) and other hereditary disorders such as G6PDH deficiency (section 2.2.1.4). These disorders may themselves lead to anaemia.

Category (iii), known as polycythaemia, is much less serious than category (i). It occurs in situations in which an excessive demand for red blood cells has arisen: loss of blood or anoxia are the most common causes. Only where excessive production fails to return to normal, does polycythaemia become a hazard.

7.4 White blood cells

The function of white blood cells, or **leucocytes**, is to combat and prevent infection. There are two types of leucocytes: granulocytes and agranulocytes, so classified because of the presence or absence of granules (Fig. 7.29). The granules are best revealed by staining with specific dyes, which is also the basis of the classification. **Eosinophils** stain with eosin (an acid dye), **basophils** with toluidine and other basic dyes, and **neutrophils** stain only slightly with either basic or acidic dyes. All three classes of granulocyte have a bizarre-shaped nucleus which results in their alternative name, polymorphonuclear leucocytes.

Different types of white cells participate in different aspects of the defence mechanism. Neutrophils are phagocytic, that is to say they ingest bacteria and other foreign agents. They also produce lysozyme, an enzyme which attacks bacterial cells by degrading their outer wall.

Eosinophils play a role in allergy, by neutralizing the histamine that is released in allergic conditions

Fig. 7.28   Diseases of red blood cells

Cause	Result
1 Insufficient red cells: anaemia	
Abnormal structure of Hb (see Fig. 7.22)	Increased loss of red cells due to haemolysis
Abnormal structure of enzymes (e.g. glucose 6–P dehydrogenase)	Increased loss of red cells due to haemolysis
Infections, immune disorders, etc.	Increased loss of red cells due to haemolysis
Abnormal synthesis of haemoglobin (e.g. Fe deficiency)	Decreased production of red cells
Abnormal synthesis of nucleic acids in stem cells (e.g. B_{12} or folate deficiency)	Decreased production of red cells
Other defects of erythropoiesis (e.g. various diseases, protein deficiency)	Decreased production of red cells
Invasion of bone marrow (e.g. leukaemia)	Decreased production of red cells
2 Abnormal red cells	
Abnormal structure of Hb (see Fig. 7.22)	Abnormal O_2 transport by red cells
Abnormal structure of enzymes (e.g. glucose 6–P dehydrogenase)	Abnormal metabolism of red cells
3 Excessive red cells: polycythaemia	
Anoxia; loss of blood	Increased production of red cells; generally only temporary

Anaemia is by far the most common disorder of red blood cells; some 20 per cent of hospital admissions are due to anaemia.

granulocytes:
(polymorphonuclear
leucocytes)

neutrophils

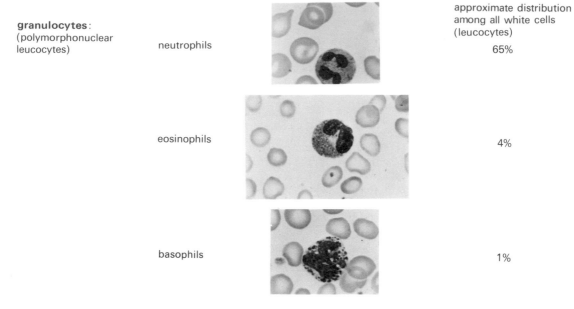

approximate distribution
among all white cells
(leucocytes)

65%

eosinophils

4%

basophils

1%

Note the densely staining granules, and the bizarre-shaped nucleus.

agranulocytes:

lymphocytes

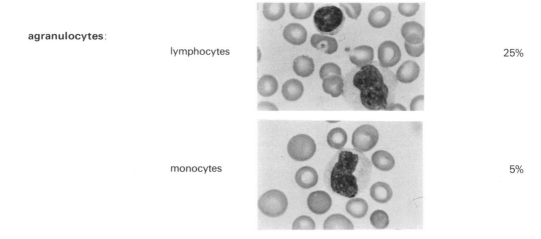

25%

monocytes

5%

The larger of the two lymphocytes is a blast cell (transformed B (or T) lymphocyte; see section 11.2.1.2).

platelets:
(thrombocytes)

The photographs are stained blood smears, taken from normal adults; the size of the various white cells, and of platelets, may be compared with the surrounding red cells (which are approximately 8 μm in diameter).

Courtesy of Professor P. T. Flute.

Fig. 7.29 The major types of white blood cells

(section 11.2.2). Eosinophils also contribute to the lysis of blood clots (section 7.5), by releasing an enzyme that is otherwise contained within the granules.

Basophils have the opposite function, namely the release of histamine. Histamine, as well as heparin, is contained in the granules. On stimulation (Fig. 11.13), the granules, which are actually membrane-coated vesicles, fuse with the plasma membrane and release their contents by exocytosis.

Lymphocytes are even more directly associated with the immune system; their functions are described in Chapter 11. **Monocytes** are weakly phagocytic. Their main role appears to be that of precursors of macrophages. **Macrophages** are the most potent phagocytic cells in the body. They are present not in free circulation, but in organs such as spleen, liver (Kupffer cells may be regarded as 'fixed' macrophages of liver), lymph nodes, and other sites. The whole system of macrophages and other phagocytic cells is known as the reticuloendothelial system.

All white cells contain a nucleus, mitochondria, and endoplasmic reticulum and possess the corrresponding biosynthetic machinery. Mature, circulating white cells, however, do not divide and hence do not synthesize DNA; there is also relatively little RNA synthesis. Only when lymphocytes are activated by an immunological stimulus (section 11.2.1.2) does DNA synthesis and cell division, as well as RNA and protein synthesis, take place.

The main metabolic activity of neutrophils, monocytes, and macrophages is related to their phagocytic function. Thus these cells contain active **lysosomes,** which degrade ingested material to low molecular weight compounds. In particular such lysosomes contain peroxidases which serve to degrade foreign compounds. During phagocytosis, part of the plasma membrane becomes ingested. This is subsequently made good by extra membrane synthesis. Such cells therefore have relatively high rates of phospholipid and membrane protein turnover.

All white cells are synthesized in the bone marrow. Lymphocytes and monocytes are in addition synthesized, or at least further developed, in other organs such as thymus, tonsils, lymph nodes, and spleen. The lymph nodes form an important part of a vascular system distinct from blood, known as the **lymphatic** system or lymph (Fig. 7.30). Lymph contains lymphocytes, water, proteins, fats, low molecular weight substances, and ions. Its main function is as an additional drainage for tissue fluids; interstitial fluid, for example, drains into lymph. The lymphatic system is a major site for the absorption of dietary fat in the intestine. The contents of lymph are passed into the blood-stream by way of the thoracic duct, which empties into the venous system.

The average life span of white cells is rather short. Eosinophils for example, have a half-life as short as 12

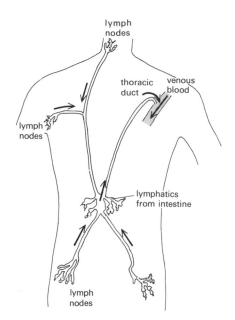

Fig. 7.30   The lymphatic system

The flow of lymph is indicated by the arrows.

hours. Hence some two-thirds of the bone marrow is engaged in synthesis of white cells, even though there are only approximately one-thousandth as many white cells as red cells. A very small number of lymphocytes, on the other hand, has an extremely long life span, approaching a year or longer. These cells play a part in the phenomenon of acquired immunity, and have therefore been termed 'memory' cells (see Chapter 11).

The number of white cells in blood is often higher (leucocytosis) than normal. During many infections, for example, the total white cell count increases markedly. More rarely, it is lower than normal (leukopenia). This is in contrast with the red cell count which is often depressed (anaemia), but rarely elevated (polycythaemia). In certain types of cancer affecting the bone marrow or other site of white cell production, the number of white cells is increased. This is known as leukaemia; according to which cells proliferate the most, leukaemias are defined as granulocytic, lymphocytic, and so forth. Unlike non-leukaemic cells, leukaemic cells divide indiscriminantly and invade other tissues to form malignant growths.

7.5 Platelets

The major function of platelets, or thrombocytes, is to prevent loss of blood through damaged blood-vessels. This is achieved by the coagulation or clotting of blood.

7.5.1 Structure

Platelets are the smallest cells in the body, their diameter being approximately 2–3 μm. They resemble red cells in lacking a nucleus, though they contain the other intracellular organelles. Like red cells and white cells, platelets are made in the bone marrow. The mechanism, however, is different. Whereas red cells *mature* from an intact nucleated precursor cell (the erythroblast), platelets arise by the *pinching off* of a piece of cytoplasm from the precursor cell (a megakaryocyte). This process is one of the few examples of a kind of cytokinesis that takes place in the absence of mitosis.

7.5.2 Function

The mechanism of **blood coagulation** is complex and not completely understood; moreover, several confusing nomenclatures exist. In the present discussion, only the bare essentials will be presented. The key, and final, step of blood coagulation involves the trapping of red blood cells in a network of strands composed of the protein **fibrin**. Fibrin is produced from a soluble precursor protein present in plasma called fibrinogen. The change, which involves the hydrolytic removal of part of the fibrinogen molecule, is catalysed by the enzyme thrombin (Fig. 7.31); like chymotrypsin and several other hydrolytic enzymes, the catalytic activity of thrombin involves a specific serine residue (section 1.2.3). Thrombin is itself produced by hydrolytic cleavage of a precursor plasma protein called prothrombin. The enzyme catalysing this step (autoprothrombin C or Factor Xa), is itself derived from a precursor plasma protein. Blood coagulation is thus a cascade process rather like glycogen degradation (Fig. 2.4) or the complement system (Fig. 11.11): relatively few molecules of Factor XII cause the conversion of a large number of molecules of fibrinogen to fibrin.

The coagulation process requires Ca^{2+} at several stages (Fig. 7.31). Removal of Ca^{2+} by binding to citrate or other chelating agent prevents coagulation. Two other types of anticoagulant are heparin and dicoumarols such as warfarin. Heparin, which is present in mast cells and basophils, is secreted into the blood-stream during anaphylactic shock. It prevents the formation of thrombin from prothrombin. The

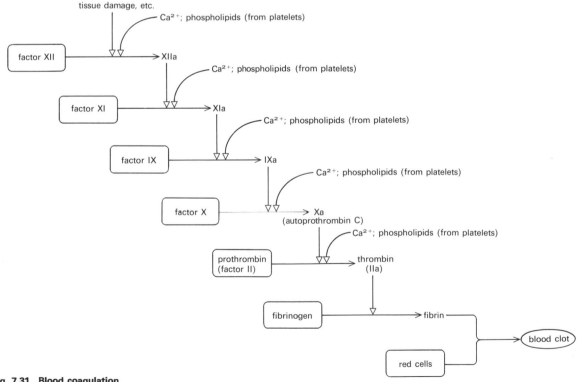

Fig. 7.31 Blood coagulation

Factors XIIa, XIa, IXa, Xa, and thrombin are proteolytic enzymes. Like chymotrypsin (Fig. 1.29), each is derived (by proteolysis) from an inactive precursor; also like chymotrypsin, their action involves acylation of a serine residue at the active site (Fig. 1.29). The factors are all plasma proteins; their concentration in the blood stream reflects the fact that each acts catalytically for the next step (i.e. the concentration of XII<XI<IX, etc).
Note that other factors, in addition to Ca^{2+} and phospholipids, are involved at the various stages.

other action of heparin, that of clearing the blood-stream of chylomicrons after a fatty meal, occurs at the sites of fat absorption, and the concentration of heparin in plasma is normally very low. Dicoumarols act by preventing the synthesis of prothrombin. This occurs in the liver and requires the participation of vitamin K. Dicoumarols, which resemble vitamin K in structure, thus act as inhibitors of vitamin K. Their action is long-term, rather than immediate, and for this reason they have proved useful in preventing further blood clotting following an initial clot in a pathological situation, such as a cardiac embolism (that is, heart attack).

What initiates blood coagulation, and how are platelets involved? The answer depends to a certain extent on the cause of coagulation; that is, whether coagulation is precipitated by damage to a blood-vessel, by the action of snake venom or, in the case of blood that has been removed from the body, by contact with the vessel in which it is stored. Certainly in the latter case, platelets initiate the process by themselves clumping together to form a clot. The surface of platelets is intrinsically 'sticky' and so an additional substance such as fibrin is not required to cause aggregation. All that is needed is the presence of a foreign substance such as glass or other solid. Should such a substance find itself in the blood-stream, aggregation of platelets occurs, and this becomes the stimulus for blood coagulation. In other instances the aggregation of platelets is stimulated by compounds released from the site of tissue injury, or by a component of the snake venom.

Once a **platelet clot** has formed, it acts in three ways. First, it becomes a focus for the coagulation of red blood cells. An interaction with the collagen fibres of blood-vessel walls helps to anchor the ensuing blood clot. Secondly, agglutinated platelets begin to liberate serotonin, which causes constriction of damaged blood-vessels and thus helps to prevent blood loss; serotonin also causes platelets to release ADP, which stimulates clumping of unagglutinated platelets. These then release more ADP, and in this manner a 'wave' of agglutination is set up. Thirdly, agglutinated platelets secrete other factors which act as cofactors at various stages of the coagulation process. Whatever the exact nature of the cell–cell interaction in an aggregate of platelets is, it is a good example of a surface-mediated effect resulting in a change in the internal environment of the cell (Fig. 6.15).

The platelet factors which potentiate blood clotting act at two distinct stages. First, they play a role in the formation of clotting factors leading to thrombin (Fig. 7.31). The most important platelet factor in the latter reaction is phospholipid. Whether phospholipid is actually secreted, or whether it acts as part of the platelet membrane, is not clear. The second stage in which platelets act is in stabilizing the fibrin clot. This is achieved partly as a result of cross-linking of the fibrin molecules enzymatically.

Once a blood clot has stabilized to prevent blood loss, and the lesion has repaired itself, the continued presence of the clot is no longer required. In fact it becomes a hazard. For if a clot detaches from its site of action, reaches the vessels of the heart or brain, and becomes established so that further blood-flow is halted, cardiac or cerebral failure ensues. In short, a clot is beneficial at one site, but detrimental at another. There is therefore a mechanism for removing clots once formed and in this process too, platelets probably play a part. The mechanism involves lysis of the fibrin meshwork by an enzyme called fibrinolysin or plasmin, which is produced from an inactive precursor by another activation process. The source of the inactive precursor is eosinophils.

Blood coagulation and lysis together help to keep blood inside the vascular system and to maintain it in the fluid state. The two processes are interconnected by feedback mechanisms, so that either coagulation or lysis can be initiated when the need arises. Defects can occur at several places. In a genetic disorder known as haemophilia, one of the clotting factors is defective, with the result that thrombin formation is retarded. Afflicted persons are therefore at risk of extensive internal or external bleeding (e.g. when an injury is sustained). Disorders of the liver, in which the proteins involved in blood coagulation are synthesized, likewise affect the process (Fig. 7.15). Disorders of the bone marrow, in which platelets are formed, have a similar effect.

7.6 Summary

Nature of vascular system

1. The functions of the vascular system (that is, **blood**) are to supply cells with **nutrients,** to remove **end-products** of **metabolism,** and to transport **hormones** between specific types of cell. All three functions are carried out by the fluid part of the blood, the **plasma;** oxygen (a nutrient) and some carbon dioxide (an end-product of metabolism) are transported by the **red cells.**

2. Blood also contains **white cells** and **platelets.** White cells function largely in the **protection** of the body against **infection;** platelets function to prevent accidental **loss** of **blood** from the circulation.

Plasma

1. The main constituents of plasma are proteins, lipids, low molecular weight compounds, and ions. The

major proteins are albumin, globulins, and fibrinogen. Hormones and enzymes are also present. The role of **albumin** and **α-** and **β-globulins** (which are all synthesized in the liver) is:

i. to maintain osmotic pressure of plasma (and thus prevent infiltration of fluid into tissues) and

ii. to transport lipids and other substances between cells.

γ-globulins (which are synthesized in lymphocytes) function as antibodies in the immune system.

2. The major lipids of plasma are triglyceride, cholesterol, phospholipids free fatty acids, and steroid hormones. Triglyceride and cholesterol (derived largely from the diet) are transported in the form of fat droplets (chylomicrons) or membrane vesicles (lipoprotein complexes); the latter also contain varying amounts of protein and phospholipid. Free fatty acids (derived largely from adipose cells), fat-soluble vitamins, and steroid hormones are transported bound to albumin, α- or β-globulins.

3. The major low molecular weight compounds, other than lipids, are sugars and amino acids (derived from the diet or from liver and other cells), ketone bodies and urea (derived from liver), water-soluble vitamins, and hormones. The major ions are Na^+, K^+, Mg^{2+}, Ca^{2+}, Cl^-, HCO_3^-, and PO_4^- ions.

4. Urea and other water-soluble waste products are excreted by the **kidney.** This is achieved as follows. Blood entering the kidney is **filtered** (through a system called the **glomerulus**) so as to retain cells and plasma proteins in the blood-stream. The major constituents of the filtrate (water, sugars, amino acids, Na^+, K^+, Mg^{2+}, Ca^{2+}, Cl^-, and HCO_3^-) are actively **reabsorbed** into the blood-stream (through a system of **tubules**), leaving a concentrated solution of urea and other waste products, that is excreted by the bladder in the form of **urine**. The energy for the reabsorptive processes is supplied largely through the operation of the **Na^+ pump.**

5. The pH of urine is slightly acid compared with that of the plasma. As a result, the blood leaving the kidney is slightly alkaline; this counteracts the acidity produced by the carbon dioxide (which forms H^+ in water by the dissociation of carbonic acid) that is formed by the oxidation of foodstuffs. By variation of the acidity of urine, the kidney is able to **control** the **pH** of **plasma.**

6. An abnormal composition of plasma or urine is generally an index of some disease process. An increase or decrease in the synthesis of plasma proteins may lead to an alteration in lipid transport, which may result in atherosclerosis. An accumulation of toxic compounds may result from a failure of kidney function (for example, nephritis).

Red blood cells

1. The main constituent of red blood cells (or erythrocytes) is **haemoglobin.** Haemoglobin consists of four subunits; each sub-unit is made up of an Fe^{2+}-containing **haem group** linked to a protein called **globin.** The function of haemoglobin is to transport **oxygen** from the lungs to the other tissues of the body; the Fe^{2+} atoms of haemoglobin bind oxygen in such a manner that in the lungs (where oxygen concentration is high) binding is favoured, whereas in tissues (where oxygen concentration is low) release is favoured.

2. Red blood cells are unlike other cells in that they contain no nucleus, no mitochondria, no endoplasmic reticulum or other organelle. They cannot divide or synthesize macromolecules; once formed (from precursor cells in the bone marrow), red blood cells remain in the circulation for approximately 120–130 days before they are destroyed (by liver and spleen). As in other cells, the ionic composition of the cytoplasm is controlled largely by the Na^+ pump. The ATP required to drive the pump is derived predominantly from glycolysis (anaerobic breakdown of glucose to lactate).

3. The plasma membrane of red blood cells contains different glycolipids in different persons. The differences fall into categories, such as the ABO **blood group** system. Blood from a person of one type (for example group A) must not be used for transfusion into a person of another type (for example group B).

4. Lack of red blood cells (**anaemia**) results in failure to maintain an adequate oxygen supply to tissues. The cause may be **nutritional** (such as a deficiency of iron or certain vitamins) or it may be **hereditary** (such as the possession of an abnormal haemoglobin, or an abnormal or missing cytoplasmic enzyme).

White blood cells

1. There are several types of white blood cell (leucocyte) in the vascular system. Their functions are

i. **phagocytosis** (that is, the ingestion and destruction of foreign and other materials) and

ii. the establishment of **immunity.**

Phagocytosis is carried out largely by **neutrophils** (polymorphonuclear leucocytes), **monocytes,** and **macrophages** (which are tissue cells derived from monocytes). The establishment of immunity is carried out by **lymphocytes.**

2. In certain situations excessive production of white cells that divide indiscriminately occurs; the result is a type of cancer known as **leukaemia.**

Platelets

1. The main function of platelets, which are the smallest cells in the body, is to promote the **coagulation** of blood. This is achieved by the release of certain factors from platelets that activate a series of enzymes present in plasma, which results in the formation of a **fibrin clot.**

2. Failure (due to hereditary or nutritional causes) to coagulate blood when necessary, leads to diseases such as haemophilia. Excessive coagulation may lead to coronary or cerebral thrombosis.

FURTHER READING

Good introduction to all the systems described in Part 2:

J.H.Green (1976). *Introduction to human physiology* (4th S.I. edition). Oxford University Press.

General account of the vascular system:

E.Neil (1975). *The mammalian circulation* (Oxford Biology Reader no.82). Oxford University Press.

General account of the kidney:

D.B.Moffat (1971). *The control of water balance by the kidney* (Oxford Biology Reader no.14). Oxford University Press.

Plasma Components:

R.L.Jackson, J.D.Morrisett, and A.M.Gotto (1976), Lipoprotein structure and metabolism. *Physiol. Rev.* **56,** 259.

F.W.Putnam, ed. *The plasma proteins* (2nd ed.) (Vols 1 and 2, 1975; vol 3, 1977). Academic Press.

The function of red cells:

H.Lehmann and R.G.Huntsman (1974). *Man's haemoglobins.* North-Holland, Amsterdam.

N.MacLean (1978). *Haemoglobin* (Studies in Biology no.93). Edward Arnold, London.

The function of phagocytic white cells:

D.Roos (1977). Oxidative killing of micro-organisms by phagocytic cells. *Trends biochem. Sci.,* **2,** 61.

The function of platelets:

R.G.Mason and K.M.Brinkhouse (1976). The platelet. *Ann. Rev. Physiol.* **38.**

8

Alimentary system

8.1 Introduction

The supply of nutrients that is available to cells depends on three factors:

i. the amount of food that is eaten,

ii. the efficiency with which it is digested and absorbed, and

iii. the ability to convert one type of nutrient to another.

The factors, which are the function of the alimentary system, are conveniently discussed under the headings of nutrition (section 8.2), digestion and absorption (section 8.3), and liver function (section 8.4).

8.2 Nutrition

The **nutritional requirements** of a person are three-fold: **water, calories,** and **specific molecules.** Water is the most important substance. Lack of water for more than 24 hours leads to severe dehydration and eventually to death, whereas lack of food can be tolerated for several weeks.

8.2.1 Water

Under normal conditions, approximately 3 litres of water are lost daily. Half of this is excreted in the urine; the other half is excreted through the skin, lungs, and faeces. Hence approximately 3 litres of water has to be replaced daily. This is subject to wide variation. The urinary output alone can vary from less than 500 ml to over 20 litres per day under abnormal conditions. Excessive sweating and diarrhoea result in an increased demand for water.

The water that is taken into the body comes partly from liquids and partly from food. All food contains a certain amount of water; in the case of vegetables and fruit, it can be as high as 90 per cent. In addition water is generated by cellular oxidation of food (e.g. Fig. 3.2).

8.2.2 Caloric requirements

The amount of food that is required to be eaten depends on the rate at which it is metabolized in the body (Fig. 8.1). The basal metabolic rate (BMR) defines the amount of energy that is expended by a human being under resting, non-eating conditions. It is determined from the amount of oxygen that is absorbed over a given period of time.

Conversion of the amount of oxygen into units of energy requires knowledge of the type of molecules that are being broken down (that is, oxidized; anaerobic degradation of glucose to lactate may be

Fig. 8.1 Caloric requirements

Activity	Metabolic rate[1]
Resting	6900 kJ/day (Basal level)
Additional energy required for:	
Walking slowly for 1 h	800 kJ
Swimming for 1 h	2000 kJ
Running for 1 h or walking upstairs for $\frac{1}{2}$ h	2400 kJ
Typing, ironing or dish washing for 5 h	3000 kJ
Carpentry or metal working for 5 h	5000 kJ
Sawing wood for 3 h	5800 kJ
Mental activity for 5 h	<100 kJ
An average working person might expend some	12 000 kJ/day

[1] Calculated from O_2 absorbed and from values of RQ (see text)

Caloric content of food

Fat	~40 kJ/g
Carbohydrate	17 kJ/g
Protein	~17 kJ/g

A daily diet of:

100 g fat	(4000 kJ)
370 g carbohydrate	(6300 kJ)
100 g protein	(1700 kJ)
thus provides	12 000 kJ/day

ignored, since the lactate is not excreted but is converted back to glucose before being broken down by oxidation). For example, if carbohydrate is being oxidized, the reaction is

$$C_6H_{12}O_6 + 6O_2 \rightarrow 6CO_2 + 6H_2O;$$

$$\Delta H \simeq -3200 \text{ kJ/mol of solid carbohydrate.}$$

That is, 3200 kJ are released for every 6 gmol of oxygen utilized; this is equivalent to 21 kJ/l O_2, or 17 kJ/g of carbohydrate. If triglyceride (for example 1,2-dipalmityl, 3-stearyl glycerol) is being oxidized, the reaction is

$$C_{53}H_{108}O_6 + 77O_2 \rightarrow 53CO_2 + 54H_2O;$$

$$\Delta H \simeq -33\,600 \text{ kJ/mol of solid fat.}$$

That is, 33 600 kJ are released for every 77 gmol of oxygen utilized; this is equivalent to 20 kJ/l O_2, or about 40 kJ/g of fat.

The type of molecule that is being degraded is most easily ascertained by measurement of the carbon dioxide that is expired, relative to the oxygen that is utilized. This ratio is known as the respiratory quotient, or RQ. In the case of carbohydrate oxidation, the RQ is 1·0; in the case of triglyceride oxidation it is 0·7. In the case of protein degradation, it is intermediate between these two values. Since the molecules that are being broken down are generally a mixture of carbohydrate, triglyceride, and protein, the RQ is generally somewhere between 0·7 and 1·0. From a knowledge of the RQ and the caloric yield per

litre of oxygen inhaled (varying between 19 and 21 kJ for different types of food), the caloric yield of the food that is being eaten can be calculated (an alternative method is to burn the food in a bomb calorimeter). It varies somewhere between 17 and 40 kJ/g.

From Fig. 8.1 it is seen that muscular activity increases the caloric requirement considerably, whereas mental activity has virtually no effect. A man whose BMR is 6900 kJ/day might expend half as much again in a normal working day. Growth requires relatively little extra caloric intake. An adolescent child, for example, uses only some 170 additional kJ/day.

The caloric value of foods depends on their content of carbohydrate, fat, and protein. As has been mentioned, a given weight of triglyceride releases more energy than a given weight of carbohydrate. Approximate values are given in Fig. 8.1. The aim of a proper diet should be to supply as much energy as is required. Lack leads to weight loss; excess to weight gain. A glance at Fig. 8.1 shows how difficult it is to prevent weight gain by exercise. If a person whose daily requirements are some 12 000 kJ starts to eat 50 per cent more than this, he would have to spend some 3 hours per day carrying out vigorous exercise such as sawing wood or swimming in order to remain the same weight. Avoiding an excessive caloric intake is by far the most effective way to prevent gaining weight.

An optimal diet should contain a proper balance between carbohydrate, fat, and protein. If no protein at all is eaten, the body will go into negative nitrogen balance. This is because a certain amount of body protein is continuously being broken down by turnover, and the nitrogen lost from the body as urea. Sufficient protein should be eaten at least to compensate for this; that is, to maintain an even nitrogen balance. A positive nitrogen balance (excess retention over excretion) is not serious, unless it is due to renal damage or other pathological cause.

If no carbohydrate at all is eaten, the body has difficulty in breaking down the fat, since carbohydrate is the main source of the oxaloacetate (via pyruvate, Fig. 2.27) that is necessary for maintaining optimal activity of the tricarboxylic acid cycle. Some Eskimos, who have adapted to a diet consisting solely of fat and protein have an abnormally high BMR, which is related to this nutritional imbalance.

8.2.3 Specific molecules

Certain molecules that are not synthesized within the body are required for the proper functioning of cells. These are the vitamins and the essential minerals. In addition certain fatty acids, amino acids, and other molecules that are the constituents of fat and protein become essential if the type of fat and protein that is eaten is unusual.

8.2.3.1 Vitamins

The vitamins have been classified into **fat-soluble** and **water-soluble** molecules. All are organic compounds required in relatively small amounts (Fig. 8.2). This is because the degradation of vitamins in the body is relatively slow so that the requirement of a grown adult is minimal; it is during pregnancy (vitamins are effectively transferred to the foetus) and childhood that vitamins are most required. Moreover some vitamins, such as A, B_{12}, C, and D are stored in considerable amounts, which helps to delay the onset of deficiency symptoms. A child born of a well-fed mother, for example, is unlikely to develop a requirement for vitamin A until many years after birth.

The **B vitamins** (thiamin, riboflavin, niacin, pyridoxine, pantothenic acid, folic acid, and vitamin B_{12}) and lipoic acid each play a basic role in intermediary metabolism; this is described in section 2.4. Deficiency leads to the diseases listed in Fig. 8.2, which reflect the tissues that show the initial signs of deficiency. Complete lack of any B vitamin over a sufficiently long period leads to death.

The other vitamins play more diverse roles in specific cell types, and for this reason may be considered to be functionally akin to hormones. Note, however, that the turnover of vitamins is quite different to that of hormones. All vitamins are degraded and excreted very slowly, so that the requirement for replenishment is low. Hormones are degraded and excreted rapidly, so that the requirement for replenishment is high.

Vitamin A, retinene or retinol, has two main roles: in vision and in the maintenance of epithelial cell function. In the retina the aldehyde form, retinal, in its *cis* configuration, binds to a protein called opsin to form a compound called rhodopsin or visual purple. This is one of the main receptors for light reaching the eye (Fig. 8.3). Light causes retinal to change to the *trans* configuration; *trans*-retinal binds to opsin less tightly than does *cis*-retinal, and is accordingly detached. The process generates a nerve impulse. *Cis*-retinal is regenerated from *trans*-retinal by an isomerase that acts on the reduced form, retinol. The cycle is driven clockwise by the energy contained in the light, not all of which is dissipated in the nerve impulse. Lack of vitamin A causes a gradual failure of vision, the first symptoms of which are manifest as night blindness.

The second role of vitamin A is concerned with the maintenance of epithelial cell function. The epithelial (outermost) cells of tissues such as gastrointestinal tract secrete a proteoglycan-rich mixture of compounds called mucus (section 6.2.1), which aids in digestion and prevents damage to the gastrointestinal tract. Epithelial cells of glands such as the salivary, lacrimal, and sweat glands likewise secrete proteoglycans, as does the epithelial layer of the cornea. Lack of vitamin A changes the function of all these cells so that they no longer synthesize proteoglycan. The cells instead begin to synthesize collagen; they become more fibroblast-like and begin to degenerate. In the cornea this leads to blindness (xerophthalmia); in other words vitamin A has two quite distinct effects on

Fig. 8.2   Vitamin requirements

Vitamin	Alternative name	Property	Approximate daily requirement	Deficiency symptoms
A	Retinol; retinene	Fat-soluble	1.5 mg; 5 μmol	Xerophthalmia; night blindness; keratinization of soft tissues
B_1	Thiamin	Water-soluble	1.5 mg; 4.5 μmol	Beri-beri
B_2	Riboflavin	Water-soluble	1.8 mg; 5 μmol	Various skin lesions
B_6	Pyridoxin	Water-soluble	2 mg; 10 μmol	Various skin lesions
B_{12}	Cobalamin	Water-soluble	3 μg; 2 nmol	Pernicious anaemia
	Niacin	Water-soluble	20 mg; 160 μmol	Pellagra
	Folic acid	Water-soluble	0.4 mg; 0.9 μmol	Anaemia (macrocytic and megaloblastic)
	Biotin	Water-soluble	(?) 0.2 mg; 0.8 μmol	Various skin lesions
	Pantothenic acid	Water-soluble	?	None described for humans
C	Ascorbic acid	Water-soluble	45 mg; 250 μmol	Scurvy
D	Calciferol	Fat-soluble	10 μg; 30 nmol	Rickets (children); osteomalacia (adults)
E	Tocopherol	Fat-soluble	(?) 15 mg; 35 μmol	None described for humans
K	Phylloquinone	Fat-soluble	—	Tendency to haemorrhage

The recommended intake for young adults is shown. During growth or pregnancy the requirements are increased.

Note that folic acid and vitamin K are generally synthesized by the intestinal bacteria in amounts sufficient to prevent deficiency syndromes. In any case, most of the vitamins are present in common foods, and additives are seldom necessary.

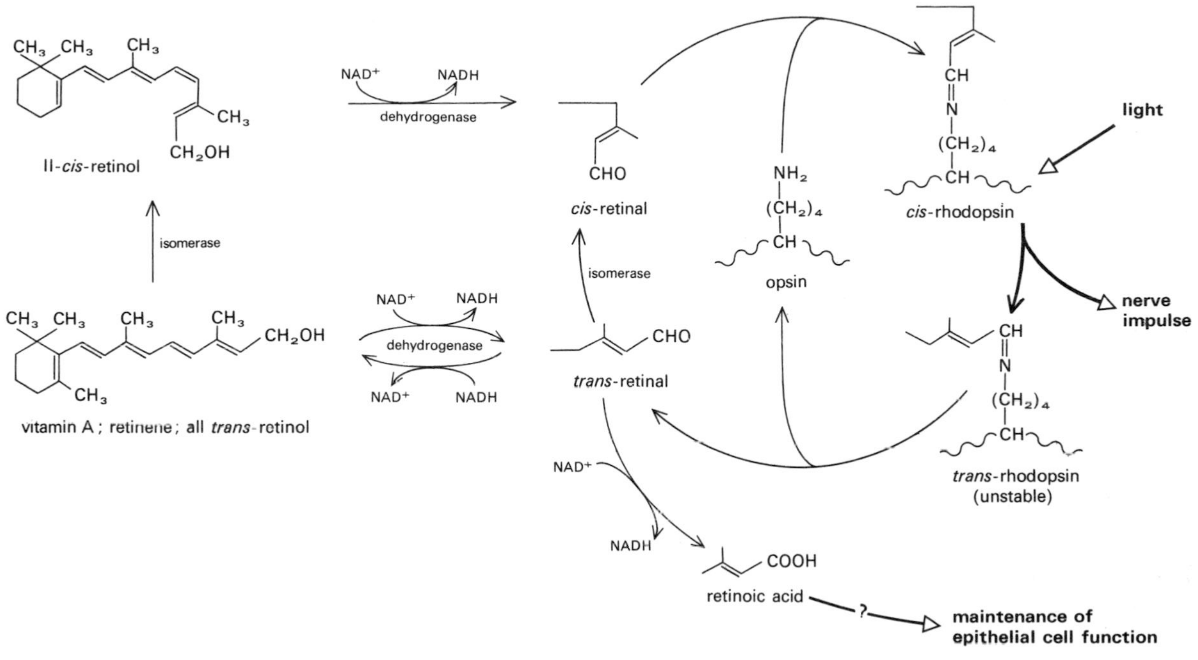

Fig. 8.3 Role of vitamin A

It is not clear whether isomerization is predominantly between *cis*- and *trans*-retinol, or between *cis*- and *trans*-retinal. Note that both isomerizations are reversible; the main pathway is indicated in the diagram.

the eye: on the retina and on the cornea. Vitamin A, probably in the form of retinoic acid, prevents the epithelial changes by a mechanism which is not yet clear.

Vitamin C, ascorbic acid, acts as a cofactor for the hydroxylation of certain proteins. Collagen (section 10.3.2.1) and one of the components of complement (section 11.2.1.3) are particularly rich in proline residues. Many of these become hydroxylated immediately after translation (section 2.3.3), to form hydroxylated proline residues. Some lysine residues also become hydroxylated to form hydroxylysine. Vitamin C, which readily undergoes oxidation and reduction, somehow functions in the hydroxylation mechanism (Fig. 8.4).

Vitamin D, cholecalciferol, promotes the absorption of calcium in the upper part of the small intestine. The vitamin is first activated by hydroxylation in the kidney cortex to form the 1,25-dihydroxy derivative. This passes via the blood-stream to the intestine, where it activates the synthesis of a specific calcium-transporting protein (Fig. 8.5). Lack of vitamin D prevents the absorption of calcium, which leads to a fall in blood calcium levels and a resultant defect in bone formation. The action of vitamin D may be compared with that of aldosterone (section 9.2.2). Each compound (a steroid) is taken up and concentrated in the nucleus of a specific target cell: aldo-

sterone in the kidney; 1,25-dihydroxy chloecalciferol in the intestine. Synthesis of a specific protein is stimulated, presumably at the level of transcription. That protein then functions in the transport of a cation: aldosterone-mediated sodium absorption in the kidney; vitamin D-mediated calcium absorption in the intestine (and bone). The result is an increase in the plasma concentration of sodium or calcium.

A certain amount of vitamin D can be synthesized from cholesterol in the presence of light. Hence there is some conversion of cholesterol to vitamin D in the skin of individuals exposed to bright sunshine.

Vitamin E, tocopherol, plays a role in the reproduction of experimental animals; its lack leads to sterility. A requirement for vitamin E by humans, however, is doubtful.

Vitamin K is required for the coagulation of blood (section 7.5). It has been termed the anti-haemorrhagic factor. Its action is in the liver, where it stimulates the synthesis of prothrombin (Fig. 8.6). The mechanism is likely to be at the translational stage, and to involve the carboxylation of certain glutamic acid residues in the protein. A derivative is formed that has a high affinity for Ca^{2+}, and thus localizes Ca^{2+} at the site of blood coagulation. Vitamin K is rather effectively formed by the intestinal flora and its requirements in the food are therefore negligible.

173

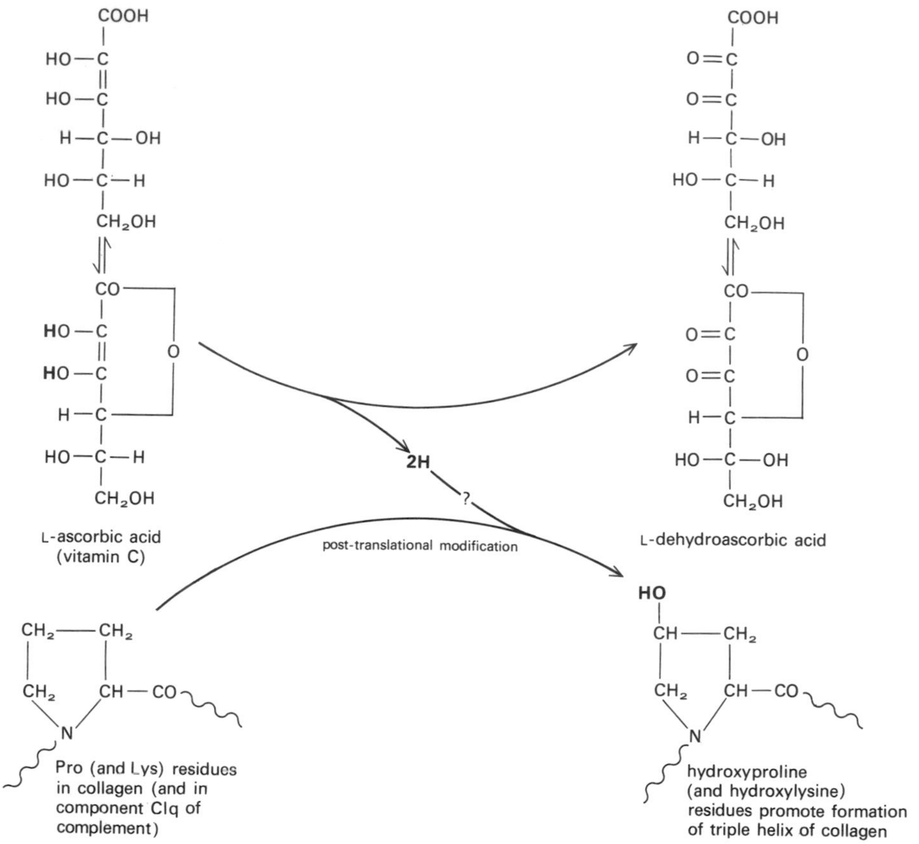

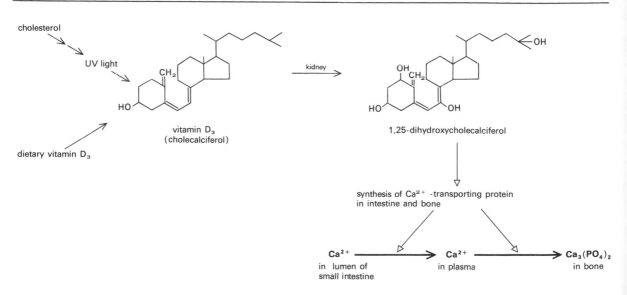

Fig. 8.4   Role of vitamin C

The exact mechanism by which ascorbic acid promotes hydroxylation is not clear.

Fig. 8.5   Role of vitamin D

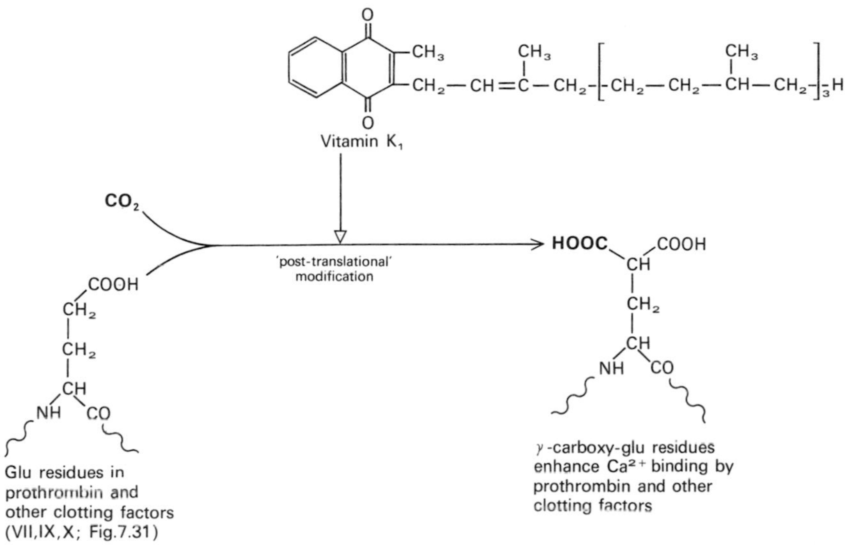

Fig. 8.6   Role of vitamin K

8.2.3.2 Minerals

The main **minerals** that are required are listed in Fig. 8.7. Since none of the ions is metabolized in any way, except by oxidation or reduction, the amount that is required is equivalent only to the amount that is excreted. For example the requirement for Na^+ and Cl^- reflects the fact that these are the main ions excreted in urine and sweat; requirement for iron is determined largely by the rate of its excretion in the faeces. The ions are ingested in a variety of ways. The bulk of Na^+ and Cl^- is eaten in the form of salt. The other ions are present either as free salts in food and water, for example K^+, Mg^{2+}, Ca^{2+} and F^-, or bound to other molecules, for example Fe^{2+}, Zn^{2+}, PO_4^-, and I^-.

Each ion has a number of functions in the body. Na^+ and Cl^- are the main ions of plasma and other extracellular fluids and maintain osmolarity at approximately 0·3 mol/l (made up approximately of 0·15 mol/l of Na^+ and 0·15 mol/l of Cl^-; see Fig. 7.13). Exactly *why* approximately 0·3 mol/l has evolved to be the molarity of most biological fluids is not clear. Presumably the requirement for specific ions within cells is such that the total adds up to approximately 0·3 mol/l. Hence this has to be the molarity of the fluid surrounding cells, otherwise swelling or shrinkage of cells would occur.

K^+ is required by several enzymes of glycolysis. It may be required for nuclear activity and for protein synthesis also. Chemically Na^+ and K^+ are very

Fig. 8.7   Mineral requirements

Ion	Approx. daily requirement	Deficiency symptoms
Na^+ (sodium)	5 g; 200 mmol	Dehydration; acidosis
K^+ (potassium)	3 g; 75 mmol	Acidosis; renal damage; cardiac arrest
Mg^{2+} (magnesium)	0.3 g; 15 mmol	Muscular tremor; mental depression
Ca^{2+} (calcium)	0.8 g; 20 mmol	Rickets (children); osteomalacia (adults)
$Fe^{2+(3+)}$ (iron)	10 mg; 200 μmol	Anaemia
Zn^{2+} (zinc)	15 mg; 200 μmol	Anaemia; stunted growth
Cu^{2+} (copper)	3 mg; 50 μmol	Anaemia (hypochromic); various nervous lesions
Cl^- (chloride)	7 g; 200 mmol	Alkalosis
Phosphate	0.8 g; 30 mmol	'Renal rickets'
iodine	0.1 mg; 1 μmol	Endemic goitre (hypothyroidism)
F^- (fluoride)	2 mg; 100 μmol	Dental caries

The recommended intake for young adults is shown. During growth or pregnancy the requirements, especially for Fe, are increased. Most of the ions are present in common foods, and additives are seldom necessary.

similar. Their ionization in aqueous media is so strong that few insoluble salts are known. In other words Na^+ and K^+ bind water in preference to other ligands. Nevertheless proteins that bind K^+ (that is, some glycolytic enzymes) do exist and one protein, at least, binds Na^+ and K^+ at different sites. This is the membrane ATPase, or the Na^+ pump, that maintains the ionic asymmetry across the plasma membrane (Fig. 6.12).

In the nerve cells, a transient diminution of the ionic asymmetry results in the generation of an impulse (section 12.3.1). Nervous activity, without which death ensues, is dependent on such impulses and hence on the presence of Na^+ and K^+ in body fluids. K^+ is toxic in high amounts by virtue of the fact that it stops the heart beat and thus leads to death.

Mg^{2+} is required by many enzymes. It binds strongly to phosphate groups. The folding of ATP and other compounds into a particular structure by Mg^{2+} is the basis of its action in stimulating enzymes that metabolize ATP. The most important of such enzymes are the kinases.

Ca^{2+} has three main functions:

i. it is required for muscle concentration (Fig. 10.5);

ii. together with phosphate it is the major constituent of bone (Fig. 10.10);

iii. it is required for blood coagulation (Fig. 7.31).

In high amounts Ca^{2+} is toxic and, like K^+, slows the heart. Because the balance between excess and deficiency of Ca^{2+} in the blood-stream is a delicate one, its concentration is controlled by the antagonistic action of two hormones, parathormone and calcitonin (section 9.2.4).

The function of iron in the body is two fold. As the active centre of the haem group, it is required:

i. for the function of haemoglobin and

ii. for the function of cytochromes.

The magnitude of haemoglobin turnover, with resultant excretion of iron, makes an adequate iron intake important. This is especially so when blood is being excessively lost, as during menstruation, or excessively required, as during pregnancy. Iron undergoes a series of oxidations and reductions during its absorption and transport in the body (Fig. 8.8). Lack of iron results in anaemia.

Zinc is required by several enzymes, of which the most important are carbonic anhydrase (Fig. 8.11) and carboxypeptidase (section 8.3.1.1). It is also bound by insulin, though this is not essential for biological activity. Other cations are required in such small amounts that their nutritional need is insignificant. Cobalt is absorbed already bound to the molecule with which it reacts (vitamin B_{12}).

The reader should not forget that several inessential cations are toxic if present in the diet. Mercury is one of the most common of these. Its toxicity, like that of arsenic, is due to its affinity for binding to essential SH groups.

The role of Cl^- is largely that of counter-ion in the maintenance of electrical neutrality. This is seen in the fact that most of its movements are passive; in other words it largely follows Na^+ movements.

PO_4^{3-} has three main functions:

i. it participates in intermediary metabolism, in the form of organic phosphates; the most important of these is ATP;

ii. it is a constituent of nucleic acids, of phospholipids, and of some proteins; and

iii. as calcium phosphate or hydroxyapatite, it is the major constituent of bone.

Since most foods of natural origin contain phosphorylated compounds of one sort or another, phosphate deficiency is rare.

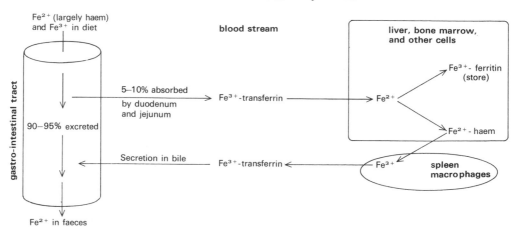

Fig. 8.8   Metabolism of iron

I^- is required for the maintenance of thyroid function (Fig. 9.20). It is absorbed as I^-, I_2, or bound to organic molecules.

F^- is a desirable, rather than an essential, mineral. It prevents the onset of dental decay.

8.2.3.3 Other compounds

There are a number of other compounds that the body cannot synthesize. These are the essential amino acids, the essential fatty acids, and inositol.

Most of the proteins that are eaten, whether of animal, plant, or microbial origin, contain the 20 amino acids in roughly the same proportion as the proteins of the human body. Hence the fact that the body cannot synthesize some 8 of the 20 amino acids, termed the essential amino acids, is usually irrelevant. An exception is in the case of those cereal proteins that lack tryptophan and lysine, which are essential amino acids. In populations in which little protein other than that in cereals is eaten, a high incidence of protein malnutrition is common.

On the other hand the content of fatty acids is not the same in all sources of fat. Hence those fatty acids that the body cannot synthesize become essential if the wrong type of fat is eaten. These fatty acids are the **polyunsaturated acids** such as linoleic, linolenic, and arachidonic acid (Fig. 9.21). Surprisingly the essential fatty acids are more common in vegetable fat than in animal fat. The function of essential fatty acids is two fold: as precursors of phospholipids and as precursors of prostaglandins. The exact role of phospholipid-containing polyunsaturated fatty acids is not clear. A deficiency of fatty acids has been linked to a tendency for atheroma formation, that is, atherosclerosis, to occur. Prostaglandins are polyunsaturated, hydroxylated fatty acids containing 20 carbon atoms. They have several pharmacological actions, all at extremely low concentration, the major one being to relax smooth muscle. The turnover of prostaglandins, which are synthesized in most cell types, is rapid. This fact, coupled with the nature of their action, has caused prostaglandins to be considered as a class of hormones (section 9.3.3).

Inositol is a constituent of certain phospholipids. If for some reason insufficient inositol-containing phospholipids are ingested, a deficiency develops. The same is true of choline and choline-containing phospholipids, except that choline can be synthesized from glycine, provided a sufficient amount of methionine is available.

8.2.4 Disease

The major nutritional diseases may be summarized as follows (Fig. 8.9). Lack of water leads to dehydration

Fig. 8.9   Diseases of nutrition

Nutrient	Alteration	Outcome
Water	Lack	Dehydration; renal failure
Calories	Lack	Weight loss (e.g. anorexia nervosa)
	Excess	Obesity and associated complications, e.g. atherosclerosis
Protein	Lack	Weight loss; oedema (kwashiokor)
Vitamins	Lack	Various symptoms (Fig. 8.2)
	Excess	Hyperkeratinization (vitamin A); hypercalcaemia (vitamin D)
Minerals	Lack	Various symptoms (Fig. 8.7)
	Excess	Liver damage; haemochromatosis (Fe); Wilson's disease (Cu).

and death. Lack of caloric intake leads to breakdown of the body reserves of carbohydrate, fat, and protein, to the point where the supply of oxidizable substances is too low for the function of brain or heart to continue. Lack of protein leads to an excess of breakdown of body proteins over synthesis, with resulting failure to maintain the osmotic pressure of plasma and in severe cases oedema develops. Excess of caloric intake can be as damaging as deficiency, and can lead to obesity, hypertension, and diabetes, as well as to atherosclerosis and heart failure.

Lack of vitamins is initially a much less serious problem. Only after considerable time are the deficiency syndromes listed in Fig. 8.2 manifest. Lack of minerals is a more immediate problem. Na^+, K^+, and Ca^{2+} are required for nervous and muscular activity, and are not stored or bound in the way that vitamins are. A sufficient intake of salt is especially important in situations in which excessive loss through sweating or through renal failure occurs. Lack of iron results in anaemia, lack of I^- in endemic goitre. Excess of minerals, vitamins, or water is in general not as serious as excess of calories, though exceptions exist. Thus in certain cases excess copper or iron absorption leads to liver damage; and excess vitamin D can be fatal.

8.3 Digestion and absorption

The major sites of digestion are the stomach, the duodenum, and the upper part of the small intestine. The major site of absorption is the small intestine, though considerable absorption of water and ions also occurs in the colon (Fig. 8.10).

8.3.1 Digestion

Before food can be digested it has to be broken into small pieces. This is achieved largely by chewing, and

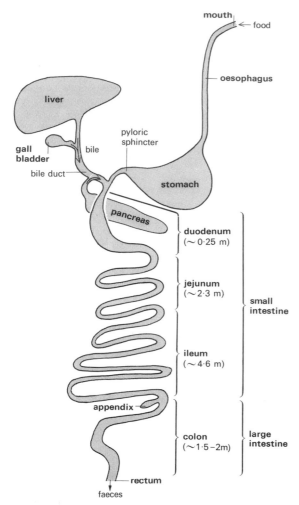

Fig. 8.10   The alimentary tract

Diagrammatic representation; the organs actually all lie quite close together (see Fig. 7.1).

by the physical movement of food through the alimentary tract. **Digestion** involves the **hydrolysis** of the components of food into their low molecular weight building blocks (section 2.2): proteins to amino acids, carbohydrates to sugars, and fats (in part) to free fatty acids and glycerol; nucleic acids, phospholipids, and other molecules that are present in smaller amounts in food are likewise degraded by hydrolysis at susceptible bonds.

The movement of food and its digested products through the alimentary tract is achieved through a wave-like contraction of muscles lining the walls of the tract. Such movement, which is known as peristalsis, occurs in an involuntary manner (section 10.2.1); it is maintained, as is the secretion of mucus, by nerve impulses that form an important part of the autonomic nervous system (section 12.1).

The presence of potent degradative enzymes within the gastrointestinal tract could be a hazard to the epithelial cells that line it. They are prevented from being digested by the enzymes they secrete by the fact that they are covered with a layer of mucus, which is a mixture of compounds rich in proteoglycans. Hence the epithelial cells that line the entire alimentary tract, from mouth to anus, are known as mucosal cells. Mucus, which helps to prevent access of infectious viruses and microbes as well as of degradative enzymes, also acts as a lubricant for the passage of food. Despite the protection by mucus, epithelial cells are nevertheless liable to physical and chemical damage. To safeguard against this, they are regenerated at a relatively rapid rate, by being sloughed off the surface and replaced by newly synthesized cells at their base. The rate of cell division is as high as once every 24 hours; only the haemopoietic cells of the bone marrow (Fig. 10.8) have division times (that is, cell cycle times) as high as this.

The main enzymes that are responsible for the degradation of food are proteases, glycosidases, and lipases, hydrolysing peptide, glycoside, and ester bonds respectively (Fig. 8.11).

8.3.1.1 Proteins

Protein degradation begins in the stomach. First, proteins are denatured by the presence of **hydrochloric acid;** denatured proteins are more susceptible to hydrolysis than are highly folded, globular proteins. Hydrochloric acid is secreted by the epithelial cells that line the upper part of the stomach (the fundus); secretion is triggered by the hormone gastrin and by stimulation of the vagus nerve (section 9.2.6.3). Secretion is an active process, in that hydrogen ions are transported against a concentration gradient of some 10^6 fold (internal concentration about 10^{-7} M; external concentration about 10^{-1} M; pH about 1). ATP is required (Fig. 6.13), but the coupling mechanism, which probably involves carbonic anhydrase, is not clear.

Secondly, proteins are hydrolysed by various **proteases.** Since the cells in which the proteases are synthesized would themselves be liable to proteolytic attack, the enzymes are secreted as inactive precursors (called zymogens; see Fig. 1.29). The mucosal cells in the lower part of the stomach secrete pepsinogen. This is broken down by the acid conditions of the stomach into the active enzyme pepsin, with release of a small (42 amino acid long) polypeptide. Pepsin, which is stable at pH 1, degrades proteins into small fragments, chiefly attacking bonds at an aromatic, methionine, or leucine (i.e. hydrophobic) residue. The partially degraded proteins pass into the duodenum, where most of the degradation, by proteins secreted from the pancreas, takes place.

Fig. 8.11   Digestive enzymes

Enzyme	Site of synthesis	Precursor	Substrate	Products	Predominant site of action
Pepsin	Stomach	Pepsinogen	Protein (at hydrophobic residues)	Peptides	Stomach
Trypsin	Pancreas	Trypsinogen	Protein (at basic residues)	Peptides	Small intestine
Chymotrypsin	Pancreas	Chymotrypsinogen	Protein (at hydrophobic residues)	Peptides	Small intestine
Elastase	Pancreas	Proelastase	Protein (rich in Gly and Pro)	Peptides	Small intestine
Carboxypeptidase	Pancreas	Procarboxypeptidase	Protein and peptides (at C-terminal)	Amino acids; peptides	Small intestine
Peptidase	Small intestine	—	Peptides	Amino acids	Small intestine
Amylase	Pancreas	—	Glycogen; starch	Maltose	Small intestine
Maltase	Small intestine	—	Maltose	Glucose	Small intestine
Lactase	Small intestine	—	Lactose	Glucose + galactose	Small intestine
Sucrase	Small intestine	—	Sucrose	Glucose + fructose	Small intestine
Lipases	Pancreas	—	Triglyceride, phospholipids	Diglyceride, monoglyceride, free fatty acids, etc.	Small intestine

Pancreatic juice is alkaline (pH 8–9), due to a high content of bicarbonate. This raises the pH of the gastric contents as they enter the duodenum to approximately 5–6. The major pancreatic proteases, which are active at this pH, are trypsin, chymotrypsin, elastase, and carboxypepsidase; each is stored as an inactive precursor (trypsinogen, chymotrypsinogen, etc.), from which the active enzyme is released by hydrolytic removal of a small peptide. Hydrolysis of trypsinogen is achieved by an enzyme called enterokinase, which is secreted by the mucosal cells of the duodenum; trypsin then hydrolyses chymotrypsinogen (Fig. 1.29), and so forth. Chymotrypsin and trypsin attack proteins at hydrophobic and basic bonds respectively. Elastase attacks particularly the proline–glycine bonds of collagen and elastin. The secretion of pancreatic juice, like that of gastric juice, is controlled by a series of hormones (Fig. 9.13).

The partially degraded peptides are hydrolysed further by the enzyme carboxypeptidase, which splits amino acids one by one from the COOH-terminal end of the molecule. The end-product of protein digestion is a mixture consisting largely of free amino acids and some dipeptides, tripeptides, etc. The digestion of proteins is entirely non-specific: any protein that enters the alimentary tract is degraded. In new-born infants digestion of proteins is not complete. As a result, some maternal proteins present in milk, such as immunoglobulin A (section 11.2.1.1) reach the small intestine intact. They become absorbed by a mechanism that is not yet clear. In this manner, some immunity is passed from mother to offspring (even more immunoglobulins are passed from mother to foetus by way of the placenta).

8.3.1.2 Carbohydrate

The main degradative enzyme is **amylase.** This breaks the α 1:4 bonds of starch or glycogen, the major polysaccharides present in food, as far as the disaccharide maltose; oligosaccharides containing the 1:6 linkage that remain are broken down to glucose by other glycosidases. Note that amylase cannot hydrolyse β 1:4 links. That is why cellulose is undigestable. Although some amylase is secreted by the salivary glands, relatively little hydrolysis takes place until the food reaches the duodenum. Here more amylase is secreted by the pancreas, and the majority of starch and glycogen is degraded. The mixture of maltose and glucose that reaches the small intestine is degraded further by a maltase localized in the brush border of the epithelial cells (Fig. 6.10); some maltase is secreted into the intestine, and this helps in the degradation of glycogen and starch. The brush border also contains sucrase and lactase, which hydrolyse the sucrose and lactose present in foods. Failure to produce lactase is not uncommon; the main symptom is diarrhoea following the ingestion of milk (due to the fact that the lactose, which is not absorbed, retains water). As with the more serious, but much less common, disease of galactosaemia (section 2.2.1.3), the remedy is to avoid drinking milk.

8.3.1.3 Fats

Before fats are degraded they are **emulsified** by the action of **bile.** The process is in some ways analogous to the denaturation of proteins by hydrochloric acid, in that the purpose is to make the fat more accessible to degradative enzymes. The active components of bile, discharged from the gall bladder as a result of hormonal stimulation (Fig. 9.13) are the conjugated bile acids, taurocholate and glycocholate (Fig. 2.17). They act as detergents and disperse the fat so as to facilitate contact with the water-soluble enzymes present in the duodenum. These are lipases secreted by the pancreas. Lipase action breaks triglyceride down to a mixture of diglyceride, monoglyceride, glycerol, and free fatty acids. Degradation is incomplete, and some triglyceride, containing medium-chain fatty acids, is absorbed intact.

Minor components of foods, such as phospholipid, nucleic acid, proteoglycans, glycoproteins, and glycolipids are degraded by concerted action of the various enzymes described above, as well as by specific enzymes such as phospholipases, DNAases, RNAases, etc.

8.3.2 Absorption

Absorption takes place chiefly in the upper part of the **small intestine,** though some substances, mainly bile acids and vitamin B_{12}, are absorbed in the lower part of the small intestine, and others, chiefly water and ions, are absorbed in the colon as well. It must be remembered that not only is the bulk of dietary material absorbed, but the bulk of the digestive secretions also; the total amount of material secreted by the salivary glands, stomach, pancreas, and gall bladder alone is some 5 litres a day, approximately 90 per cent of which is reabsorbed in the gastrointestinal tract. In addition mucus is secreted all the way down to the caecum and most of this is reabsorbed also.

The surface in the small intestine is folded into a series of protuberances known as **villi.** This increases the area across which absorption takes place. The villi beat in a waving motion, thereby stirring the digested contents and aiding in absorption. The epithelial cells lining each villus are themselves elaborated into microvilli, the so-called **brush border** (Figs 6.4 and 8.12)

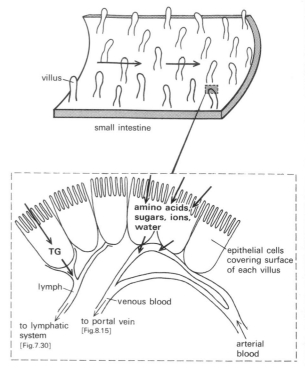

Fig. 8.12 Intestinal absorption

Note: the absorption of nutrients into the lymphatic system and into the blood stream is shown occurring in separate epithelial cells, for clarity. In fact every cell supplies both systems.

8.3.2.1 Amino acids

As indicated above, proteins are broken down as far as amino acids and low molecular weight peptides. Amino acids are absorbed by active transport (section 6.3.1). This is linked, at least for some amino acids, to the passive influx of Na^+; Na^+ is itself actively extruded from the epithelial cell into the blood-stream by the ATP-driven Na^+ pump (Figs 6.9 and 8.13).

Peptides are absorbed at the same time as they are broken down by hydrolysis to free amino acids. The **peptidases** that catalyse the hydrolysis are positioned at the brush border of the epithelial cells in such a manner that substrate binding occurs on the outside of the plasma membrane, while release of products occurs on the inside. The energy that is required for the accumulation of amino acids is provided not by ATP, but by hydrolysis of the peptide bond ($\Delta G^{\ominus} \simeq -10 \text{ kJ/mol}$) (Fig. 6.10).

Amino acids leave the epithelial cell by passive diffusion and enter the venous capillaries of the blood-stream.

8.3.2.2 Sugars

The products of starch digestion that are absorbed by the epithelial cells of the small intestine are maltose and glucose. Dietary sucrose and lactose are also absorbed. As for amino acid absorption, two processes operate (Fig. 8.13). First, glusoce is absorbed by an active transport process that is linked to the Na^+ pump. Some other monosaccharides, such as galactose, which are present as degradative products of glycoproteins and glycolipids, are absorbed by the same mechanism. Secondly the disaccharides maltose, sucrose, and lactose are absorbed by specific **disaccharidases** (maltase, sucrase, and lactase) situated at the brush border of the epithelial cell. Hydrolysis is simultaneous with uptake, as described for the absorption of peptides (Fig. 6.10). The products, glucose, fructose, and galactose, are released into the blood-stream by passive diffusion.

8.3.2.3 Fat

The products of fat digestion, largely monoglycerides, glycerol, and free fatty acids, with minor amounts of triglycerides and diglycerides (largely those composed of medium-chain fatty acids) are absorbed by passive diffusion across the brush border. The driving force for uptake in this case is biosynthesis. That is, triglyceride is resynthesized from its components within the epithelial cell (Fig. 8.13). Note that the components of bile are not absorbed at the same time as the fat. Absorption occurs lower down the intestinal tract.

Triglyceride is released not into the venous capillaries, as are amino acids and sugars, but instead is released into the capillaries of the **lymphatic system** (Fig. 8.12). Prior to release, triglyceride is complexed with newly synthesized protein and phospholipid to form chylomicrons; these eventually appear in the blood-stream by virtue of the fact that the contents of the lymphatic system are discharged into the venous blood-supply at the thoracic duct.

8.3.2.4 Other compounds

The major material absorbed by the intestine is water. The process is largely passive; that is, water accompanies the active transport of Na^+, amino acids, and

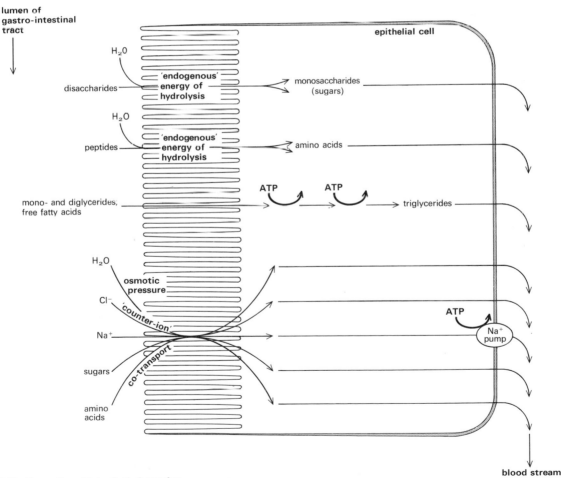

Fig. 8.13 Energetics of intestinal absorption

sugars across the epithelial cells (Fig. 8.13). The osmolarity remains approximately the same in epithelial cells as in the blood-stream. This is in contrast to the absorption of water in the kidney, in which a hyperosmolar secretion, namely the urine, is formed.

The active absorption of Na^+ is one of the main energy-consuming reactions of epithelial cells of the renal tubule and small intestine. Na^+ transport is coupled to the utilization of ATP by means of a specific Na^+K^+-ATPase, the so-called **Na^+ pump** (Fig. 6.12). In cells such as liver or blood cells, in which there is no net movement of Na^+, the Na^+ pump expends energy merely to *maintain* the asymmetric distribution of Na^+ between interior and exterior of the cell; in cells such as nerve or muscle cells, in which there is transient movement of Na^+ into the cell, the Na^+ pump expends energy to *restore* the asymmetric distribution following a series of impulses. In the epithelial cells of the renal tubule and small intestine, in which there is a net movement of Na^+ across the cell layer, the Na^+ pump expends energy to *drive* Na^+ continuously out of the cell: 24 mol of Na^+ (1.4 kg of NaCl) per day in the case of the renal tubule, and some 50–250 mmol of Na^+ (3–15 g of NaCl) per day in the small intestine.

The effect of an active transport of Na^+ is to cause a passive movement of Cl^- (as counter-ion) and water (as the result of osmotic pressure created by the NaCl) to occur (Fig. 8.13). The active movement of Na^+ is linked to the uptake of glucose and amino acids, by a mechanism that is not yet clear.

The actual entry of Na^+ into the epithelial cells is passive. A sufficiently rapid rate of entry is maintained by making the surface area as large as possible in the form of microvilli. The Na^+ pump is situated on the lower part of the epithelial cell, away from the brush border (Figs 8.12 and 8.13).

The manner in which K^+ is transported across the epithelial cells is not clear. Following absorption by the microvilli, K^+ somehow 'leaks out' into the blood stream; but note that for every Na^+ ion pumped out of the epithelial cell, some K^+ *enters*.

The absorption of Ca^{2+} and iron is catalysed by specific transport proteins. A somewhat different mechanism operates for the absorption of the cobalt-containing vitamin B_{12}. A glycoprotein called 'intrinsic factor', secreted by the mucosal cells of the stomach, aids in the absorption of vitamin B_{12}. In other words the factor acts from the outside of the intestinal epithelial cell. Individuals who are defective in the production of intrinsic factor suffer from pernicious anaemia, a disease due to lack of vitamin B_{12} despite the presence of the vitamin in the diet. The remedy is to administer vitamin B_{12} intravenously; since it is degraded very slowly once it is inside the body, injections every month or so suffice. This may be contrasted with deficiency of a hormone such as insulin, which is degraded rapidly: in this case injections one or more times per day are necessary.

The absorption of the bile acids taurocholate and glycocholate takes place in the lower parts of the small intestine. Reabsorption is very efficient and relatively little is excreted in the faeces (95 per cent reabsorbed; 5 per cent in faeces, Fig. 8.20). On the other hand bilirubin and other breakdown products of haem that are present in bile, are not reabsorbed but are further metabolized and excreted in the faeces. Faeces also contain some 50 per cent of the cholesterol (as coprosterol, a reduced metabolite of cholesterol) that is present in bile. The other 50 per cent of the cholesterol is reabsorbed. The secretion and reabsorption of bile acids and cholesterol is referred to as the enterohepatic circulation.

8.3.3 Disease

Failure to secrete any of the digestive enzymes results in nutritional deficiency and an abnormal amount of faecal matter, with diarrhoea resulting from the retention of water (Fig. 8.14). In pancreatitis, for example (caused most commonly by gall stones or by excessive alcohol), the digestion and absorption of protein, fat, and carbohydrate is impaired. Gastric malfunction, on

Fig. 8.14 Some diseases of digestion and absorption

Component	Alteration	Cause	Outcome
Pepsin and HCl	Increased secretion	Unknown	Gastric ulcer
Pancreatic lipases	Decreased secretion	Pancreatitis	Steatorrhea; nutritional deficiency
Pancreatic proteases	Decreased secretion	Pancreatitis	Not serious
H_2O	Decreased absorption (continued secretion of mucus, etc.)	Inflammation of epithelial cells due to bacterial or viral infection (e.g. cholera, typhoid)	Diarrhoea; dysentery
Ca	Decreased absorption	Lack of vitamin D	Rickets (children); osteomalacia (adults)
Vitamin B_{12}	Decreased absorption	Lack of intrinsic factor	Pernicious anaemia
Fe	Decreased absorption	Lack of transferrin	Anaemia

the other hand, is less serious, and achlorhydria (failure to secrete hydrochloric acid) can be tolerated remarkably well. If bile is not secreted, due to blockage of the bile duct or other cause, fat digestion is impaired and a nutritional deficiency results. This includes the lack of essential fatty acids and fat-soluble vitamins; the faeces become pale coloured due to the content of fat (steatorrhea).

Excessive secretion of hydrochloric acid and pepsin can overcome the protective ability of mucus, and the epithelial cells become damaged: gastric ulcer may result. Failure to secrete sufficient bicarbonate in the duodenum, and hence to neutralize the hydrochloric acid produced in the stomach, may be a contributing factor to the development of duodenal ulcer. A correlation exists between secretion of mucus containing particular blood group substances and a predisposition towards gastric and duodenal ulcer (collectively known as peptic ulcers), as well as towards gastro-intestinal cancer (Fig. 7.24). The molecular basis of these correlations is not known, despite the fact that cancers of the alimentary tract are one of the most common forms of the disease.

Pathogenic microbes and viruses that escape degradation can give rise to gastrointestinal infections. Water is generally retained, resulting in diarrhoea. The diseases range in severity from food poisoning (various strains of *Salmonella*) to typhoid and cholera. Failure to degrade certain viruses, followed by absorption of sufficient molecules to initiate an immune response, has been turned to advantage in the case of oral vaccines such as that for poliomyelitis.

Diseases due to faulty absorption include rickets, due to lack of vitamin D-dependent Ca^{2+} uptake, and anaemia, due to lack of vitamin B_{12} uptake protein (intrinsic factor) or to lack of iron transport proteins.

8.4 Liver function

The type of food with which caloric requirements are met can vary widely. A diet that is rich in either carbohydrate, fat, or protein satisfies the needs of the body equally. Yet the major energy-consuming cells, namely muscle, cannot metabolize amino acids completely; and when utilizing fat, are more effective with ketone bodies than with fatty acids. The cells that transform amino acids into glucose, and fatty acids into ketone bodies, are those of liver. **Liver** alone possesses the enzymes necessary for these two transformations, that is **gluconeogenesis** and **ketogenesis;** kidney has a slight capacity for gluconeogenesis also.

The cells of the liver perform four other important biochemical functions: the formation of **urea,** the synthesis of **plasma proteins,** the formation of **bile acids,** and the **detoxification** of endogenous and foreign

compounds. All these and other functions of the liver are conveniently described under the headings of amino acid metabolism (section 8.4.1), carbohydrate metabolism (section 8.4.2), fat metabolism (section 8.4.3), and detoxification (section 8.4.4). Failure of liver function is treated in section 8.4.5.

The anatomical structure of the liver confirms its role as an intermediary between the digestion and absorption of food on the one hand, and the metabolism of the resulting nutrients on the other. For unlike other organs and tissues that are perfused by an inflow of blood from the heart (arterial blood) alone, liver directly receives blood from the gastrointestinal tract (stomach, duodenum, small intestine, and colon), spleen, and pancreas (all venous blood), in addition to arterial blood from the heart (Fig. 8.15). The liver is also the first organ through which the contents of the

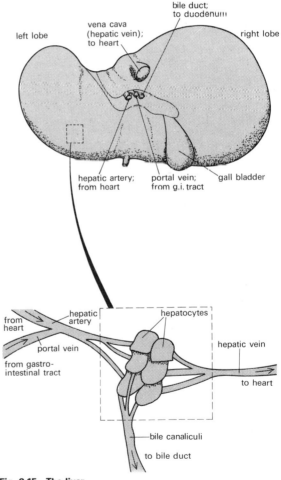

Fig. 8.15   The liver

The liver consists of two large lobes, shown from the underneath in this diagram, and two much smaller lobes (not shown). The important blood-vessels are indicated.

The lower figure shows the special capillary system for liver.

lymph pass. It is in liver that chylomicrons are broken down and the constituent molecules processed and resecreted in the form of lipoprotein vesicles (Fig. 7.7). The close connection between spleen and liver may reflect the fact that haemoglobin degradation, which is begun in the spleen, is completed in the liver.

The liver is made up of four lobes, which together weigh approximately 1·7 kg. It is thus one of the largest organs in the body. Although the cells of the liver normally divide rather slowly, they have a high capacity for rapid growth following an injury. If the greater part of one of the lobes is surgically removed, for example, the residual cells regenerate the original mass of tissue within a matter of weeks.

Two types of cell constitute the functional mass of liver tissue: hepatocytes or parenchyma cells and Kupffer cells. In addition, cells lining the blood-vessels (endothelial cells) are present. Hepatocytes, which are responsible for all the functions to be discussed except that of haem degradation, comprise the major portion. In general when reference to liver is made, the hepatocyte population is implied.

8.4.1 Amino acid metabolism

8.4.1.1 Urea formation

The liver is the major organ in which amino acid breakdown occurs. It is the only organ in which the ammonia that is released is condensed with carbon dioxide to form urea. In the sense that free ammonia is toxic, the formation of urea is a detoxification reaction, and might equally be discussed in section 8.4.4. In the sense that the formation of urea is a key reaction in amino acid metabolism it is just as appropriately discussed here.

The breakdown of amino acids leads to their oxidation to CO_2 and H_2O, when carbohydrate is in plentiful supply, or to their conversion into glucose and ketone bodies, when carbohydrate is limiting. The first reaction of the two conversions is the same: the removal of the amino group by transamination (Fig. 2.18). The amino groups that are removed react with the CO_2 to form urea.

The formation of urea is complex, and involves a cyclic mechanism (Fig. 3.17), rather like the oxidation of acetyl CoA. Like the oxidation of acetyl CoA, the formation of urea occurs in the mitochondria. The mechanism is described in section 3.6.

8.4.1.2 Protein synthesis

Apart from synthesizing the proteins required for intracellular function, liver cells synthesize proteins for export to other parts of the body. The major class of such proteins are the plasma proteins. Albumin and α-, and β-globulins, including lipoproteins, clotting

factors and complement factors, and fibrinogen, are all synthesized in the liver (Fig. 8.16). Of the major plasma proteins only the γ-globulins (section 11.2.1) are synthesized outside the liver.

The mechanism of protein release is by exocytosis. The proteins are synthesized by the ribosomal machinery of the hepatocyte, utilizing the mRNA molecules corresponding to each of the proteins. In other words the genes for plasma proteins are specifically activated in the nuclei of liver cells. The finished protein migrates via the inside of the endoplasmic reticulum to the Golgi vesicles; these then fuse with the plasma membrane to release their contents.

8.4.2 Carbohydrate metabolism

8.4.2.1 Gluconeogenesis

The liver is the major site of glucose formation from non-sugar precursors. The process is known as gluconeogenesis. Three types of non-sugar precursor participate:

i.   glycogenic amino acids,

ii. lactate; and

iii. glycerophosphate.

The pathways are as follows.

Glycogenic amino acids are first transaminated, that is, the α-amino groups removed. The corresponding α-keto acids then enter the gluconeogenic pathway (Fig. 2.27) directly (for example pyruvate and oxaloacetate from alanine and aspartate), or after further metabolic conversion (for example α-ketoglutarate and β-hydroxypyruvate from glutamate and serine). Glycogenic amino acids undergo gluconeogenesis when dietary carbohydrate is insufficient and dietary protein is high. The utilization of glycogenic amino acids is stimulated by glucocorticoid hormones (Fig. 9.11).

Lactate is oxidized to pyruvate. Gluconeogenesis from lactate occurs when violent muscular activity has given rise to an accumulation of lactate in the blood-stream. The accumulation is due to the fact that oxygen consumption is insufficient to keep pace with glucose utilization; in other words anaerobic glycolysis rather than oxidation to CO_2 and H_2O occurs.

Fig. 8.16   Proteins secreted by liver

Albumin

α and β globulins
including: lipoproteins (Fig. 7.6); prothrombin and other clotting factors (Fig. 7.31); complement factors (Fig. 11.11); transport proteins (e.g. for Fe, for thyroxine and other hormones, etc)

Fibrinogen

Glycerophosphate is derived from triglyceride or phospholipid. Hydrolysis of triglyceride leads to free glycerol. A specific kinase, present in few tissues other than liver, converts glycerol to glycerophosphate at the expense of ATP. In the case of phospholipids, hydrolysis directly yields glycerophosphate. Glycerophosphate is oxidized by NAD^+ to dihydroxy acetone phosphate (an isomer of glyceraldehyde 3-phosphate; Fig. 2.7). Gluconeogenesis from glycerophosphate takes place when dietary carbohydrate is insufficient, but fat intake is high. The rest of the triglyceride molecule, that is to say fatty acids, are broken down to ketone bodies in this situation.

8.4.2.2 Interconversion of sugars

The conversion of dietary monosaccharides, such as galactose and fructose, to glucose takes place in the liver. In the case of galactose, for example, the pathway is as shown in Fig. 2.13.

8.4.3 Fat metabolism

8.4.3.1 Utilization of chylomicrons

Dietary fat enters the blood-stream in the form of **chylomicrons.** These are utilized as follows. Triglyceride is degraded by 'clearing factor lipase', and some of the resulting free fatty acids are absorbed by adipose cells; the remainder are passed to the liver, where they are resynthesized into triglyceride and secreted back into the blood stream in the form of very low density lipoproteins (VLDL). Triglyceride in VLDL is subsequently degraded by clearing factor lipase just as it is in chylomicrons (Fig. 7.7).

8.4.3.2 Ketogenesis

When fatty acids are oxidized by the liver, CO_2 and H_2O are not the only end-products. Instead, some of the acetyl CoA derived by β-oxidation is converted into acetoacetate; part of the acetoacetate becomes reduced to β-hydroxybutyrate (Fig. 8.17). These two substances, which are known as ketone bodies, diffuse out of liver into the blood-stream; their production is referred to as **ketogenesis.**

The formation of ketone bodies, like β-oxidation of fatty acids and the tricarboxylic acid cycle, occurs within mitochondria. Unlike β-oxidation of fatty acids or the operation of the tricarboxylic acid cycle, ketone body formation is restricted to the mitochondria of liver (hepatocyte) cells. The pathway is somewhat complicated (Fig. 8.17). Acetoacetyl CoA, the penultimate stage in the β-oxidation of fatty acids, is not simply hydrolysed to acetoacetate, as might be expected. Instead, acetoacetyl CoA condenses with another molecule of acetyl CoA to form β-hydroxy β-methyl glutaryl CoA. This compound then breaks down to yield acetoacetate, at the same time regenerating a molecule of acetyl CoA. β-hydroxybutyrate is formed by the reduction of acetoacetate. In the blood-stream, some of the acetoacetate is non-enzymically decarboxylated to yield acetone, which is excreted by the lungs and in the urine. Note that β-hydroxy β-methyl glutaryl CoA is also the starting-point for the synthesis of cholesterol (Fig. 2.34). Cholesterol synthesis is not, however, increased or decreased during ketogenesis.

Under normal conditions, the concentration of ketone bodies in plasma is relatively low (0.2–2 mM).

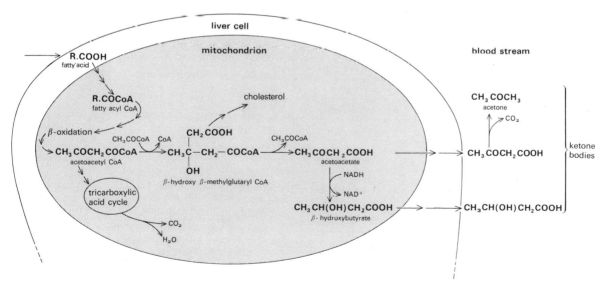

Fig. 8.17 Pathway of ketogenesis

See Fig. 3.14 for the oxidation of fatty acids to acetoacetyl CoA.

Consumption by tissues such as skeletal and heart muscle, fat, and to a certain extent nervous tissue balances output. It must be appreciated that ketone bodies form a very ready source of acetyl CoA for ATP production by cells. Unlike glucose or fatty acids, which form acetyl CoA only after a lengthy metabolic sequence, and which can be metabolized to products other than acetyl CoA such as glycogen or triglyceride, ketone bodies form acetyl CoA by only a few steps; moreover acetyl CoA formation is the only metabolic route available to ketone bodies in cells. Ketone bodies are thus the most useful form in which 'metabolic energy' circulates in the blood-stream.

Under conditions in which the blood glucose concentration falls, as in starvation, the output of ketone bodies by the liver in increased. That is, instead of converting free fatty acids predominantly into triglyceride, secreted in the form of VLDL, the liver switches to converting free fatty acids to ketone bodies. The plasma concentration of ketone bodies rises only slightly, since uptake by tissues is correspondingly increased; that is, tissues compensate for the lack of glucose by metabolizing relatively more ketone bodies. The stimulus for the increased ketogenesis is two-fold (Fig. 8.18). First, a fall in plasma glucose concentration leads to a decrease in the secretion of insulin, and an increase in the secretion of glucagon, from the pancreas (Fig. 9.10). The

result of these changes is to stimulate the degradation of triglyceride in fat cells, and thus to release fatty acids. Fatty acids enter the liver; here they are either resynthesized into triglyceride and secreted (in the form of lipoprotein vesicles), or they are broken down to acetyl CoA and hence to ketone bodies, CO_2, and H_2O. The second stimulus, which is in the liver, somehow acts to promote fatty acid breakdown, rather than triglyceride synthesis. The carnitine-mediated entry of fatty acids into mitochondria appears to be the site at which the stimulus acts, but how it does so is not clear.

Under pathological conditions such as diabetes, ketogenesis is increased, again by a fall in the concentration of plasma insulin; in this case the fall is a permanent one due to an inability to produce insulin. The release of free fatty acids from adipose tissue, and their breakdown in liver, is more extensive than during starvation, with the result that ketone bodies accumulate in plasma. Their concentration can rise as high as 20 mM. Under these conditions, an acidosis (due to the acidic nature of the ketone bodies) results. The same events appear to occur in alcoholics, for reasons that are not yet clear.

8.4.3.3 Bile formation

Bile is a mixture of compounds. The most important are the bile acids, taurocholate and glycocholate, and the degradation products of haem, chiefly bilirubin conjugated to glucuronic acid. Phospholipids and cholesterol are also present. The bile acids are synthesized from cholic acid, which is itself derived from cholesterol.

The liver is one of the major sites of cholesterol synthesis; other cell types are intestinal mucosa, fat cells, and the endocrine organs. Cholesterol is metabolized in one of three ways (Fig. 2.17):

i. to pregnenolone, the starting-point for the synthesis of the steroid hormones,

ii. to cholecalciferol (vitamin D), and

iii. to cholic acid.

The pathway to cholic acid is quantitatively the most important one and represents one way by which excess cholesterol is excreted; the other way is as unchanged cholesterol. In fact excretion of bile acids is extremely inefficient, in that 95 per cent of the bile acids are reabsorbed in the intestine. The importance of bile acid formation is clearly not to produce a more readily excretable form of cholesterol; rather it is to facilitate the excretion of intact cholesterol, as well as to aid in the digestion of dietary fat (section 8.3.1.3).

The pathway of cholic acid synthesis is shown in Fig. 2.17. Cholic acid is conjugated with taurine and glycine by enzymes located on the endoplasmic reticulum. From here the bile acids are secreted directly into the bile duct by way of the bile canaliculi; these

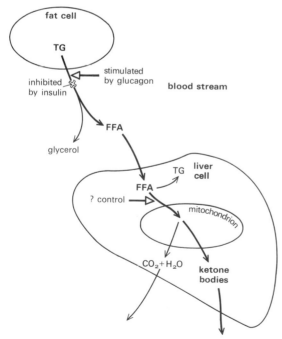

Fig. 8.18 Control of ketogenesis

TG = triglyceride; FFA = free fatty acid. (see Fig. 8.17 for details of pathway).

are capillary-sized vessels that connect directly with the hepatocyte. The plasma membrane of hepatocytes is thus differentiated into three distinct areas (Fig. 8.19). One area through which plasma protein, urea, glucose, lipoproteins, and ketone bodies are secreted, and through which nutrients are absorbed, is in contact with blood capillaries. A second area, through which bile acids are secreted, is in contact with the bile canaliculi. A third part of the plasma membrane is elaborated into intercellular junctions, which allow direct passage of regulatory molecules from cell to cell and which maintain the structural integrity of liver tissue. Exactly how the products of liver metabolism are channelled into one or other of these areas is not clear.

The bile acids secreted by hepatocytes mix with the bilirubin glucuronide (Fig. 7.27) secreted by the Kupffer cells, to form the mixture known as bile. Bile is secreted directly into the duodenum, or stored during fasting in the gall bladder. The overall circulation of bile is depicted in Fig. 8.20.

8.4.4 Detoxification

The liver is the major organ in which compounds are detoxified. Compounds that are produced within the body, such as ammonia, as well as compounds that enter the body from outside, such as alcohol, drugs, and other substances, are detoxified. The process is generally one of making substances more **water soluble,** and hence more readily excreted by the kidney. The cells in which detoxification takes place are the hepatocytes; many of the detoxifying enzymes are situated on the endoplasmic reticulum. The reactions are oxidation (largely hydroxylation), reduction (via NADPH), hydrolysis, and conjugation (including methylation). Some of the more important conjugation reactions are listed in Fig. 5.4.

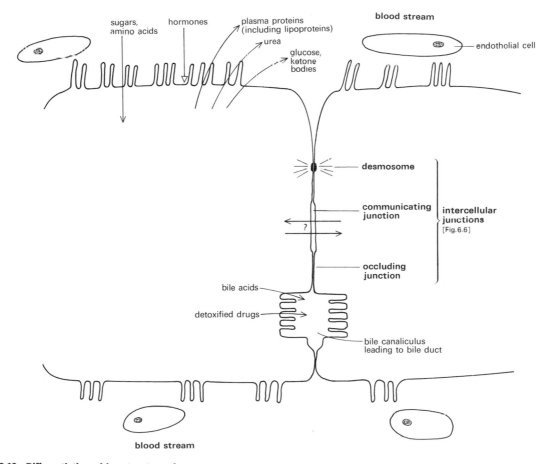

Fig. 8.19 Differentiation of hepatocyte surface

Three distinct types of plasma membrane surround the hepatocyte: plasma membrane in contact with the blood-stream, plasma membrane between adjacent cells (containing the three types of intercellular junction) and plasma membrane in contact with bile canaliculi.
Adapted from W. H. Evans (1977) *Trends biochem. Sci.* **2**, 169–71, with permission.

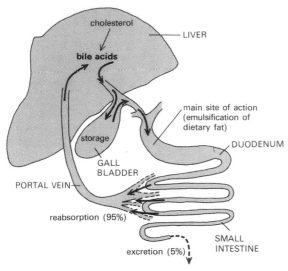

Fig. 8.20   Circulation of bile

Note that the reabsorption of bile acids is a more efficient process than the reabsorption of cholesterol (~50 per cent reabsorbed, ~50 per cent excreted).

The inactivation of hormones and other pharmacologically active substances can be considered as a type of detoxification. For the reactions by which biological activities are terminated are in several instances the same as those which cause the detoxification of foreign compounds; that is, hydroxylation, methylation and other forms of conjugation (Fig. 9.19). Indeed, the enzymes have presumably evolved primarily for the latter purpose, and not for that of eliminating ethyl alcohol, phenobarbitone, or other drug.

The conjugation of bilirubin with glucuronic acid is another type of detoxification, in the sense that the molecule is metabolized to make it more readily excreted; in this case the route of excretion is through the faeces, by way of the bile and the gastrointestinal tract.

8.4.5 Disease

It is clear from the nature of the functions that have been described that diseases of the liver lead to a disturbance of several metabolic pathways (Fig. 8.21). Failure to synthesize specific enzymes is often the result of a hereditary defect; failure in the overall metabolism of liver is usually the result of an infection or of the accumulation of toxic compounds.

Note that faulty secretion of specific proteins (Fig. 8.16) is as much a defect of the vascular system (Fig. 7.15) as of liver. The reason for inclusion in Fig. 8.21 is that the primary defect is a failure to synthesize a liver-specific protein; that is, a defect in a gene that is expressed only in liver.

Infections such as viral, bacterial, or protozoal hepatitis, or an excessive intake of alcohol or other toxic substance affect most liver functions simultaneously. Cells break down and release some of their contents into the blood-stream. The appearance of certain enzymes is used for the diagnosis of liver disease (Fig. 7.8). Another diagnostic change that may occur is the appearance of degradation products of haem in the blood-stream, instead of in the bile. The resultant condition, jaundice, is characterized by a yellow tint to the skin. Jaundice can also result from an anatomical defect, such as blockage of the bile duct (obstructive jaundice).

Alcohol taken in excess over a number of years gives rise to cirrhosis of the liver. In the early stages of this disease, the liver becomes 'fatty' as a result of an accumulation of triglyceride, which may be due to a failure of lipoprotein secretion. Eventually the liver becomes fibrous and unable to function normally.

Fig. 8.21   Some diseases of the liver

Function	Alteration	Cause	Outcome
Protein synthesis	Decreased synthesis of albumin	Chronic liver disease e.g. cirhosis	Oedema
	Decreased synthesis of fibrinogen, prothrombin or other clotting factor	Chronic liver disease or genetic defect in synthesis of specific proteins	Haemorrhage
	Decreased synthesis of complement factors	Chronic liver disease or genetic defect in synthesis of specific proteins	Decreased immune response (Fig. 11.16)
Carbohydrate metabolism	Lack of glycogen-metabolizing enzymes	Genetic defects	Glycogen storage diseases (Fig. 2.57) (rare)
	Failure of gluconeogenesis	Ingestion of alcohol following fasting	Hypoglycaemic coma
Urea synthesis	Decreased synthesis of specific enzymes	Chronic liver disease	Ammonia toxicity
Detoxification	Decreased excretion of bilirubin	Biliary obstruction; damaged liver due to infection or drugs.	Jaundice

Fatty liver also results from ingestion of toxic chemicals such as carbon tetrachloride, or from a deficiency of choline in the diet.

8.5 Summary

Nutrition

1. A person requires **water, calories,** and **specific molecules,** water being the most important.

2. The number of calories required varies according to the size and activity of the individual; an average person might require some **12 000 Kj** during a normal working day. The number of calories derived from food depends on its composition. Carbohydrate and protein yield approximately 17 kJ/g, fat approximately 40 kJ/g. In other words approximately half a kilogram of food might be required by an average person per day. This should be in the form of a balanced mixture of carbohydrate, fat, and protein.

3. The specific molecules that are required are those that cannot be synthesized in the body from carbohydrate, fat or protein. These are the **vitamins** and the essential **minerals.** Depending on the nature of the fat and protein that is eaten, certain **polyunsaturated fatty acids,** amino acids, and other molecules might in addition be required.

4. The major vitamins that are required are thiamin (B_1), riboflavin (B_2), niacin (B_5), pyridoxine (B_6), pantothenic acid, folic acid, vitamin B_{12}, vitamin A, vitamin C (ascorbic acid), vitamin D, and vitamin K. Each is required in trace amounts only.

5. The major minerals that are required are common salt (NaCl) and trace amounts of K^+, Mg^{2+}, Ca^{2+}, Fe^{2+}, Zn^{2+}, PO_4^- and I^-.

6. An insufficient intake of water, calories, or specific molecules leads to a variety of nutritional diseases.

Digestion and absorption

1. The bulk constituents of food (carbohydrate, fat, and protein) are **hydrolysed** (that is, digested) to their constituent 'building blocks' (sugars, fatty acids, amino acids, and so forth) in the small intestine; some digestion of protein occurs in the stomach. Hydrolysis is achieved by enzymes that are secreted into the gastrointestinal tract: **pepsin** (which hydrolyses protein), as well as **hydrochloric acid,** is secreted into the stomach; **amylase** (which hydrolyses glycogen and starch), **chymotrypsin** and **trypsin** (which hydrolyse protein), **lipases** (which hydrolyse triglyceride and phospholipids) and **bicarbonate** (which neutralizes the hydrochloric acid) are secreted from the pancreas into the duodenum; also secreted into the duodenum (from the liver) are **bile acids,** which help to emulsify dietary fat.

Hydrolysis within the lumen of the intestine is incomplete. Some triglyceride is absorbed intact, and some disaccharides and peptides are hydrolysed during the process of absorption. Throughout the alimentary tract, **mucus** is secreted, to protect the outermost (epithelial) cells and to aid in the passage of food.

2. Absorption of amino acids and sugars across the epithelial cells of the small intestine and into the blood-stream is an **active process.** Absorption is linked either to hydrolysis (in the case of disaccharides and peptides) or to the Na^+ pump (in the case of free amino acids and monosaccharides). The **Na^+ pump,** which actively transports Na^+ across the epithelial cells, also transports (in a passive manner), Cl^-, and water. Fatty acids are passively absorbed; they are resynthesized in the epithelial cells into triglyceride, which (together with triglyceride absorbed intact) is secreted into the **lymphatic system** in the form of **chylomicrons;** the lymphatic system empties its contents into the blood-stream by way of the thoracic duct.

3. Incomplete digestion occurs if the secretion of hydrolytic enzymes (as in pancreatitis) or of bile acids (as in biliary or liver failure) is impaired; nutritional deficiency, accompanied by diarrhoea and steatorrhea, ensues. Faulty secretion of bicarbonate or mucus may lead to the formation of intestinal ulcer. The presence of pathogenic microbes and viruses in the gastrointestinal tract leads to a number of diseases.

Failure of specific uptake systems in the small intestine, such as that for Ca^{2+} or for vitamin B_{12}, results in nutritional disease, such as that of rickets or of pernicious anaemia.

Liver function

1. Several pathways of intermediary metabolism occur only in liver.

i. Ammonia, which is produced during the breakdown of amino acids, is converted (by the mitochondria of liver cells) into **urea,** which is less toxic than ammonia, and which is eliminated by the kidney.

ii. **Gluconeogenesis,** that is the synthesis of glucose from non-sugar precursors (such as certain amino acids, lactate or glycerol) takes place in liver.

iii. Free fatty acids (released from triglyceride stores in fat cells) are broken down to **ketone bodies** in the liver, to be used as an energy source by muscle and other cells.

iv. The liver converts cholesterol to **bile acids,** which are required for digestion.

v. Many foreign compounds and other substances, including breakdown products of cytochromes and haemoglobin, are **detoxified** in the liver.

2. Liver synthesizes and secretes important macromolecules and other substances. Apart from glycogen, which is retained within liver cells, the liver is the site of synthesis of most of the **plasma proteins** and cholesterol.

3. Failure of liver function, due to infections (as in various types of hepatitis) or to the accumulation of toxic substances (as in alcoholism), leads to defects of intermediary metabolism and of plasma protein synthesis; the effects of such abnormalities cover a wide range of diseases.

FURTHER READING

General account of nutrition:

G.T.Taylor (1978). *The principles of human nutrition* (Studies in Biology No. 94). Edward Arnold, London.

More specific aspects of nutrition:

Symposium on fat soluble vitamins (1974). *Vitam. Horm.* **32,** 131–545.

S.Lewin (1976). *Vitamin C: its molecular biology and medical potential.* Academic Press, London.

J.Stenflow (1976). Vitamin K-dependent carboxylation of blood coagulation proteins. *Trends biochem. Sci.* **1,** 256.

General account of the gastrointestinal tract:

R.M.H.McMinn (1974). *The human gut* (Oxford Biology Reader No.56). Oxford University Press.

Specific aspects of the liver cell surface:

W.H.Evans (1977). Resolving biochemically the hepatocyte's multifunctional pericellular membrane. *Trends biochem. Sci.* **2,** 169.

Diseases associated with the alimentary system:

R.H.S.Thompson and I.D.P.Wootton (eds) (1970). Chapters on Nutritional diseases, Disorders of the gastrointestinal tract, Diseases of the liver and biliary tract. In *Biochemical disorders in human disease* (3rd edition). J. and A.Churchill, London.

9

Endocrine system

9.1 Nature of endocrine function

The endocrine system is concerned with control. It is a means of coordinating the metabolism of the organs of the body. For example, when food is plentiful, liver, muscle, and adipose tissue each stores the excess as glycogen or fat. When energy is required rapidly, for example for running away, glycogen and fat are simultaneously broken down in liver, muscle, and adipose tissue. How is this coordination of function achieved, and what is the link between metabolism and the intake of food, or between metabolism and the desire to run away?

Coordination is achieved by the action of specific compounds called **hormones,** which pass to the various organs through the blood-stream. The link is achieved by the fact that hormones are produced in specialized cells, called endocrine cells, which are sensitive to changes in the internal and external environment. The stimulus may be chemical, such as the presence of food, or electrical, such as a nervous impulse from the brain.

Endocrine cells are either grouped together in secretory glands such as pituitary, thyroid, adrenal cortex, or adrenal medulla, or they are part of organs having other functions also, such as kidney, pancreas, or the intestinal tract. The major hormone-producing sites are shown in Fig. 9.1. It will be noted that the sexual organs, testis and ovary, have an endocrine function. This is almost as important as their production of sperm or ova, for the whole metabolism of males and females — metabolic rate, the number of red cells in blood, or the concentration of cholesterol in plasma, as well as the amount of facial hair or body fat — differs. All these functions are controlled to a certain extent by hormones produced in the gonads (testes or ovaries).

Just as the production of hormones is a function of specific cells, so is the responsiveness to hormone action. The degree of specificity, however, varies. For example, thyroid cells are the only cells sensitive to thyroid stimulating hormone (TSH). Muscle, liver, and adipose cells, on the other hand, are all sensitive to adrenalin. But muscle, liver, and adipose cells are also sensitive to insulin and to other hormones. It is not so much the cell *type* that is important, as whether it possesses the relevant *receptor* for a particular hormone (see sections 6.3.4.1 and 9.4.1).

Hormone-secreting cells respond not only to a stimulus from the environment, but to a stimulus from within the body also. Often the internal stimulus has an effect opposite to that of the external one. In other words, one stimulus turns hormone secretion *on*, another turns it *off*. The secretion of adrenocorticotrophic hormone (ACTH) by the anterior lobe of the pituitary gland may be cited as an example (Fig. 9.2).

An environmental stimulus, such as traumatic or

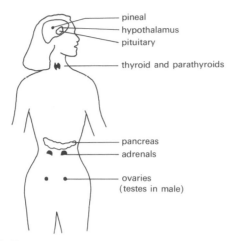

Fig. 9.1   Endocrine organs

The kidneys (Figs. 7.1 and 7.11) and alimentary tract (Figs. 7.1 and 8.10) are also important hormone-producing organs.

psychological stress, causes secretion of ACTH; the stimulus is not direct, but is transmitted by way of an ACTH-releasing hormone secreted by the hypothalamus. ACTH production is turned *off* by the presence of cortisol, secreted by the adrenal cortex gland. However the secretion of cortisol occurs only in response to ACTH. A **negative feedback** system is thus set up. Should too much ACTH be released by the pituitary, it will result in an increased release of cortisol from the adrenal, which will diminish the release of ACTH from the pituitary. Conversely, a decrease in release of ACTH will result in a decrease in the level of cortisol, which will tend to stimulate the pituitary to secrete more ACTH.

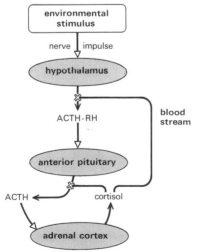

Fig. 9.2   Feed-back control of hormone release

The control of cortisol levels by ACTH (adrenocorticotrophic hormone) and ACTH-RH (ACTH releasing hormone) is illustrated. The concentrations of other hormones are controlled in a similar way.

Inhibition of ACTH release is not, of course, the main function of cortisol. One of its main functions is to stimulate the conversion of protein into carbohydrate. The feedback system serves to insure that if for some reason the rate of cortisol production begins to exceed the rate of its removal, or vice versa, the original level is restored. In short, the plasma concentration of a hormone such as cortisol is controlled in two ways: by negative feedback and by hypothalamic stimulation. The situation is somewhat analogous to that of a constant temperature bath. There is a feedback control such that the temperature is maintained at a particular level (plus or minus a degree or so). But the level itself — whether it is 37 °C or 25 °C — is determined by another control. It is the latter type of control that is exercised by the hypothalamus in the case of cortisol: a 'setting' of the mean, around which the plasma concentration of cortisol is allowed to fluctuate. Negative feedback is a feature typical of endocrine control and is shown by several other hormones (Fig. 9.4); it operates also in non-hormonal control (see for example Figs 1.27 and 2.66).

The major function of the adrenal cortex is to secrete cortisol and other steroids. The metabolism of its cells is directed to that end, and the amount of steroids that are secreted is proportional to the number of cells that make up the gland. In fact ACTH acts not only to secrete more steroids, but also to stimulate the division of adrenal cortex cells; it is a mitogen (section 6.3.4) as well as a hormone. Thus ACTH controls the size as well as the activity of the adrenal cortex. Other hormones react in the same way on the cells of their respective target organs, and are accordingly termed **trophic** (or 'nourishing') hormones (Fig. 9.3).

Insufficiency of a trophic hormone leads to atrophy, that is, wasting away, of the target organ. Excess leads to hyperplasia, that is, excessive cell division.

9.2 Endocrine organs

9.2.1 Pituitary

The pituitary or hypophysis, situated in a bony cavity below the hypothalamic part of the brain, is one of the most important endocrine glands. More than a dozen different hormones are secreted by it. The gland (approximately 1 cm long) is divided into an anterior and a posterior lobe, with an intermediate area, the pars intermedia, in between (Fig. 9.4). The **anterior** lobe secretes **growth hormone** (GH), **thyroid stimulating hormone** (TSH), **adrenocorticotrophic hormone** (ACTH), **luteinizing hormone** (LH), **follicle stimulating hormone** (FSH), and (in females) **prolactin.** The

Fig. 9.3 Trophic hormones

Hormone	Target organ	Response
Adrenocorticotrophic hormone (ACTH)	Adrenal cortex	Maturation of adrenals and secretion of glucocorticoids (e.g. cortisol)
Thyroid stimulating hormone (TSH)	Thyroid	Secretion of thyroid hormones (e.g. thyroxine)
Luteinizing hormone (LH) Follicle stimulating hormone (FSH)	Gonads (ovaries or testes)	Maturation of gonads and secretion of oestrogens (e.g. oestradiol) or androgens (e.g. testosterone)
Growth hormone (GH)	Connective tissue (cartilage and bone)	Maturation of connective tissue; growth
Glucagon	β cells of endocrine pancreas	Maturation of endocrine pancreas and secretion of insulin

The function of some of the major trophic hormones is indicated.

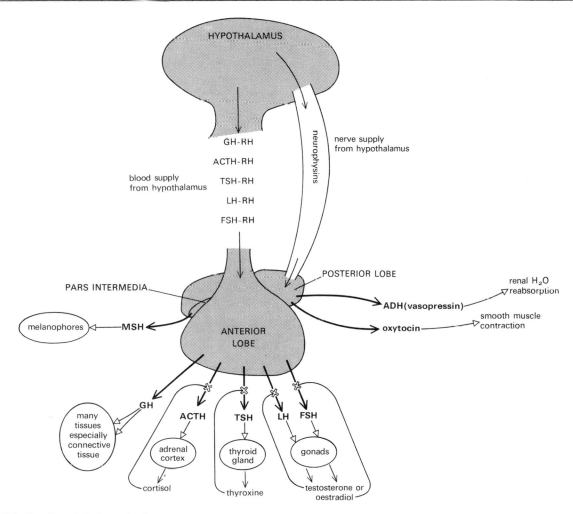

Fig. 9.4 Function of pituitary gland

The major hormones secreted by the pituitary gland (hypophysis) are shown: GH = growth hormone; ACTH = adrenocorticotrophic hormone; TSH = thyroid stimulating hormone; LH = luteinizing hormone; FSH = follicle stimulating hormone; GH-RH, ACTH-RH, etc. = GH-releasing hormone, ACTH-releasing hormone, etc; ADH = antidiuretic hormone; MSH = melanophore-stimulating hormone.
Note that several hormones inhibit the production of releasing hormones by the hypothalamus also (see Fig. 9.2).

posterior lobe secretes **antidiuretic hormone** (ADH or vasopressin) and **oxytocin.** The **pars intermedia** secretes **melanophore stimulating hormone** (MSH).

The secretions of the anterior lobe are stimulated by hormones secreted by the **hypothalamus.** Each hormone is stimulated by its respective **releasing hormone** (RH): growth hormone by GH-RH, ACTH by ACTH-RH, TSH by TSH-RH, and so forth. Release-inhibiting hormones, which counteract the effect of releasing hormones, are also secreted by the hypothalamus. Release of the pituitary hormones is inhibited by negative feedback; that is, by the hormones that are secreted in response to stimulation by the pituitary hormones. There are thus two mechanisms by which secretion of anterior pituitary hormones is controlled by feedback: by direct inhibition of hormonal release at the anterior pituitary, and by inhibition of the releasing hormone at the hypothalamus. Thyroxine, for example, inhibits TSH release predominantly by the first mechanism; cortisol inhibits ACTH release predominantly by the second.

It may be wondered how synthesis and release of the different hormones secreted by the anterior pituitary is made specific. The answer appears to be that the tissue is made up of several discrete cell types, only one of which synthesizes any one hormone. That is, one type of cell synthesizes ACTH, another synthesizes FSH, and so on. The situation is seen also in the endocrine pancreas, in which α-cells secrete glucagon and β-cells secrete insulin (section 9.2.6). In fact, it is probably generally true that any one cell type secretes only one type of protein in any amount. Each of the 10^6 or so different types of circulating immunoglobulin, for example, is synthesized by a different type of plasma cell (section 11.2.1.2).

The functions of the hormones secreted by the anterior pituitary are as follows. TSH, LH, FSH and ACTH are all trophic hormones and stimulate the growth and activity of their respective target organs (Fig. 9.3). Growth hormone (also called somatotrophin) stimulates most cell types to increased cell division, with growth of the resulting tissues and organs; hence its name. It is therefore a trophic hormone, though not so specific as the others. Since cartilage and bone-forming cells are a particularly important type of target cell, the outcome of a deficiency or an excess of growth hormone is most noticeable in terms of stature and skeletal size.

The secretions of the posterior lobe are under somewhat different control. In this instance the hormones (ADH and oxytocin) are actually synthesized in the hypothalamus, where they are packaged into granules bound to a protein called **neurophysin.** The neurophysin–hormone complex travels from hypothalamus to pituitary not by the blood-stream, as do the anterior pituitary releasing hormones, but by

axonal transport (section 12.2); that is, through the interior of axons that run from hypothalamus to posterior pituitary.

The functions of the posterior pituitary hormones are as follows. Oxytocin stimulates muscular contraction of the uterus during childbirth; during lactation, it stimulates milk ejection from the mammary glands. ADH (vasopressin) stimulates the active reabsorption of water by the renal tubules (section 7.2.2). It also has slight 'pressor' activity; that is, it constricts blood-vessels and raises the blood-pressure; like adrenalin, it promotes glycogen breakdown in liver.

Between the anterior and posterior lobe of the pituitary gland lies a small area known as the pars intermedia. From here melanophore stimulating hormone (MSH) is secreted. Melanophores are structures that are responsible for the darkening of skin; the mechanism is through the dispersion of a black granular pigment known as melanin. This function of MSH is only one of several, the others of which may be physiologically more important.

9.2.2 Adrenal

The adrenals are two small organs (each approximately 10 g in weight) situated above the kidneys. Each organ consists of two parts: an outer cortex and an inner medulla (Fig. 9.5). The **cortex** secretes several steroid hormones, which are divided into two classes by virtue of their function: the **glucocorticoids** and the

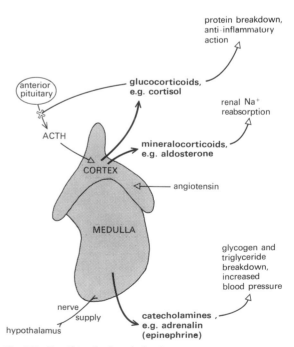

Fig. 9.5 Function of adrenal glands

mineralocorticoids. The **medulla** secretes the **catecholamines** adrenalin (also called epinephrine) and noradrenalin (also called norepinephrine).

The glucocorticoids, of which **cortisol** is one of the most important, act on many tissues and organs. The general effect is to cause a degradation of protein, with utilization of the glycogenic amino acids so produced to form carbohydrate (Fig. 2.22). Cortisol has a powerful anti-inflammatory action, and synthetic analogues of cortisol are used in the treatment of rheumatoid arthritis.

As mentioned above, the function of the adrenal cortex is stimulated by ACTH, secreted by the anterior pituitary. The plasma level of corticosteroids is controlled by feedback regulation of ACTH release. This is achieved by inhibition of the action of ACTH-releasing hormone from the hypothalmus, as well as by direct inhibition of ACTH release from the anterior pituitary.

The mineralocorticoids, of which **aldosterone** is the most important, regulate the Na^+ and K^+ content of body fluids. The main function of aldosterone is to stimulate the active reabsorption of Na^+ and the excretion of K^+ by the renal tubule (section 7.2.2.1).

The secretion of the catecholamines by the medulla is under control of the nervous system. Sympathetic nerve fibres controlled by the hypothalamus and by other centres of the brain stimulate the release of adrenalin and noradrenalin. Although the fibres are part of the sympathetic nervous system, transmission across the adrenal synapses is by acetylcholine, not noradrenalin (section 12.3.2).

Of the two catecholamines released by the adrenal medulla, **adrenalin** is the most important. It acts on many cell types, including those of liver, muscle, fat, and heart. Its main actions are:

i. to increase the heart rate and blood-flow to vital organs and

ii. to promote the degradation of carbohydrate and fat; the result is to make more glucose and fatty acids available to muscle for oxidation.

In short, to provide more muscular energy. Adrenalin release is triggered in just those situations that require an increase in energy consumption: the need to run away from, or to pursue, another individual or animal. In fact most stressful situations result in an increase of adrenalin.

The adrenal medulla is not an essential organ. It can be removed without overt harm. Presumably noradrenalin synthesized at other sites (see section 12.3.2) suffices to keep normal functions going.

The fact that the adrenal cortex, producing steroid hormones, is adjacent to the adrenal medulla, producing catecholamines, is not fortuitous. For steroids such as the glucocorticoids are inducers of the enzymes leading to catecholamine synthesis (section 9.3.2). Since cortex and medulla are directly connected by blood-vessels, stimulation of the cortex by ACTH leads directly to the secretion of catecholamines from the medulla. During stress, the combined action of glucocorticords and catecholamines ensures that a greatly increased supply of energy-yielding intermediates is made available (Fig. 9.11).

9.2.3 Gonads

The **gonads** comprise the testes in the male and the ovaries in the female. Their function is two-fold: to produce sperm or eggs (ova) respectively, and to secrete steroid hormones. Both processes are under endocrine control from the anterior pituitary gland.

The **testes** in an adult male are situated in a bag called the scrotum. A series of tubes, through which sperm passes during ejaculation, leads to the penis (Fig. 9.6). Testicular form and function is controlled by the gonadotrophic hormones, FSH and LH, secreted by the anterior pituitary. These hormones are the same in males as in females – hence their name. FSH controls spermatogenesis; LH controls steroid secretion. The testis secretes several steroid hormones, jointly called **androgens**, of which **testosterone** is the most important. Testosterone secretion maintains what are known as the secondary male characteristics: facial and body hair, deep voice, skeletal size, and so

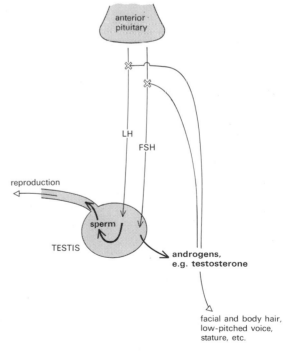

Fig. 9.6   Function of testes

forth. Testosterone regulates the secretion of gonado-
trophins from the anterior pituitary by negative feedback. This is exercised largely through inhibition of
FSH- and LH-releasing hormones from the
hypothalamus.

The **ovaries** in an adult female are situated on both
sides of the uterus, the cavity in which the foetus or
embryo develops (Fig. 9.7). The function of the
ovaries and related organs is controlled by the
gonadotrophic hormones FSH and LH, secreted by
the anterior pituitary. The ovaries secrete a number of
steroid hormones, jointly called **oestrogens**; the most
important are **oestradiol** and **oestrone.** Oestrogen secretion is stimulated by LH. Oestrogen secretion itself
controls the menstrual cycle and maintains the secondary female characteristics: enlarged breasts, body fat,
skeletal size, and so forth. Oestrogens also maintain
the mucosal lining of the vagina. Like testosterone,
oestrogens regulate the secretion of FSH and LH by
negative feedback.

During the menstrual cycle, a follicle containing an
ovum develops under stimulation by FSH. This produces oestrogens, which stimulate the growth and
thickening of the inner lining of the uterus, known as
the endometrium. The endocrine control in this instance is positive, not negative. For several days after
commencement of menstruation, the secretion of LH
by the anterior pituitary rises. This causes the follicle
to disintegrate, releasing the ovum; the remainder of
the follicle matures to become a corpus luteum. The
corpus luteum, stimulated by LH, produces a steroid
hormone called **progesterone** (which is also the precursor of the adrenal steroids, Fig. 2.17). Progesterone
aids in the futher development of the endometrium
and, should the ovum become fertilized at this time, in
the establishment and maturation of the foetus. If
fertilization does not take place, the corpus luteum
disintegrates, progesterone secretion ceases, and the
endometrium sloughs off and is eliminated through the
vagina in the form of the menstrual discharge.

The extent to which the secondary sex characteristics are expressed varies from person to person. In
extreme cases, as illustrated by certain disorders,
genetically constituted males–that is, with XY
chromosomal makeup – develop female characteristics
(feminizing syndrome); or genetically constituted
females – that is XX chromosomal makeup – develop
male characteristics (masculinizing syndrome). How is
this possible? The answer is that the secretion of
androgens or oestrogens is not an 'all-or-nothing'
phenomenon. Normal testes secrete some oestrogens
and normal ovaries some androgens. If androgen or
oestrogen synthesis is faulty, or if the synthesis of the
respective receptors (section 9.4) does not take place,
the balance between androgen and oestrogen function
is disturbed, and abnormal development results. Exactly how the relative amounts of androgen and oestrogen are normally set is not clear; presumably a gene
on the Y chromosome exerts some controlling effects.

There is an overlap not only between androgen and
oestrogen synthesis, but also between adrenal steroid
synthesis and gonadal steroid synthesis. That is, the
adrenals secrete a certain amount of androgen or
oestrogen. Since pregnenolone, the precursor of androgens and oestrogens, is also the precursor of
gluco- and mineralocorticosteroids (Fig. 2.17), this is
perhaps not surprising. The result is that there is an
interplay between gonadal hormones and metabolic
hormones, which may underlie the many non-sexual
features that differ between males and females
(Fig. 9.8).

9.2.4 Thyroid and parathyroid

The development and function of the thyroid gland is
under the control of TSH secreted by the anterior
pituitary (Fig. 9.4). The thyroid gland (about 20 g in
weight) secretes **thyroxine** and related compounds, of
which tri-iodothyronine is the most important,
(Fig. 9.9). These two thyroid hormones control basal

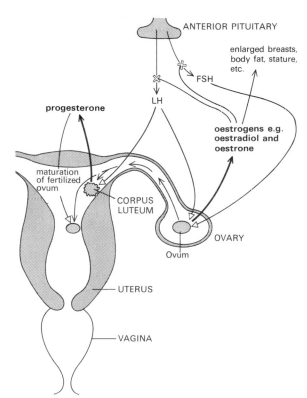

Fig. 9.7 Function of ovaries

See text for details of menstrual cycle

Fig. 9.8 Some metabolic differences between males and females

Property	Males	Females
Red cells in blood	5.0×10^{12}/l	4.5×10^{12}/l
Lipoproteins in plasma		
High density	0.95 g/l	1.24 g/l
Low density	2.00 g/l	2.10 g/l
Very low density	0.68 g/l	0.40 g/l
Vitamin C in plasma	4.8 mg/l	9.0 mg/l
Acid phosphatase in plasma	5.1 units/l	4.4 units/l
Binding of ribosomes to endoplasmic reticulum of liver	Stimulated by oestrogens	Stimulated by androgens

Some examples that are not obviously related to sexual function are quoted. (For differences in plasma levels of sex hormones, see Fig. 9.24.) Some of the values, which have been averaged for various age groups, have been taken from *Documenta Geigy: Scientific tables* (7th ed) J. R. Geigy S. A., Basle (1970), with permission.

metabolic rate (see section 8.2.2); that is, an increased concentration of thyroid hormones in plasma increases metabolic rate. The connection between metabolic rate and the concentration of thyroid hormones is so close that metabolic rate is generally assessed by determination of thyroxine level, which is easier to perform than assay of O_2 consumption and CO_2 output. The thyroid hormones affect brain function; lack of them can lead to cretinism. The thyroid hormones control TSH release from the anterior pituitary by negative feedback.

In certain instances the feedback system breaks down and hyperplasia of the thyroid results. This is seen in the case of a nutritional deficiency of iodine. Thyroxine cannot be synthesized (since it contains iodine); hence the plasma level of thyroxine falls below that necessary to prevent the release of TSH from the anterior pituitary. TSH secretion therefore continues unchecked, and since TSH is a trophic hormone, it stimulates the thyroid gland (albeit a nonfunctional gland) to hypertrophy and growth. The abnormally large thyroid gland that results is known as a goitre.

In addition to its major role of secreting thyroxine and tri-iodothyronine, the thyroid also secretes, from specialized cells, a hormone called **calcitonin.** The main function of calcitonin is to decrease Ca^{2+} levels in plasma by suppressing Ca^{2+} mobilization from bone. Such a fall in plasma Ca^{2+} stimulates the development and function of the parathyroid glands, which are four small glands (approximately 6 mm long) embedded in the thyroid.

The **parathyroids** secrete parathyroid hormone or **parathormone.** Parathormone acts antagonistically to calcitonin; that is, it increases the plasma concentration of Ca^{2+}. This is achieved in four ways. First, parathormone promotes the excretion of phosphate by the kidney (section 7.2.2); this lowers the plasma concentration of phosphate, which is restored by the resorption of bone; Ca^{2+} is released at the same time. Secondly, parathormone itself has a direct effect on bone, stimulating resorption and inhibiting deposition. Thirdly, parathormone reduces the secretion of Ca^{2+} in the digestive juices, thus retaining more in the circulation. Fourthly, parathormone inhibits calcitonin release from the thyroid gland. In addition to calcitonin and parathormone, the plasma concentration of Ca^{2+} is controlled by vitamin D (Fig. 8.5).

9.2.5 Pineal

The pineal is a small organ about 0·5 cm long, situated deep in the middle of the brain. It secretes a hormone

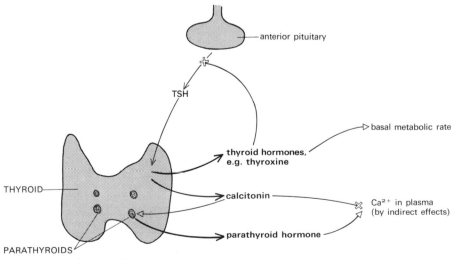

Fig. 9.9 Function of thyroid and parathyroid glands

called melatonin (not to be confused with melanin) which is thought to control reproduction by inhibiting the release of LH, at least in animals, from the anterior pituitary. It may have other functions also. Melatonin release is sensitive to light, and the pineal gland appears to play a part in the 'biological clock' rhythms associated with changes in the lightness and darkness of the environment.

9.2.6 Other organs

In addition to the specific endocrine glands discussed above, several organs contain hormone-secreting cells. The most important of these are pancreas, kidney, and gastrointestinal tract.

9.2.6.1 Pancreas

The major function of the pancreas is to secrete degradative enzymes into the alimentary tract, to aid the digestion of food. It has a second function, which is to secrete two hormones, **glucagon** and **insulin,** into the blood-stream. The two functions are structurally separated in that the enzyme-secreting cells, called glandular or exocrine cells, which make up the bulk of the pancreas, are distinct from the hormone-secreting or endocrine cells. The latter are organized into a structure referred to as islets of Langerhans. The endocrine cells are further divided into two types: α- and β-cells (Fig. 9.10). The α-cells secrete glucagon; the β-cells insulin.

The secretion of glucagon is stimulated by a fall in plasma glucose concentration. The effect of glucagon is to reverse this by promoting the breakdown of liver glycogen to glucose, and the formation of glucose by gluconeogenesis, followed by the release of glucose into the blood-stream. At the same time glucagon stimulates the β-cells to secrete insulin.

Insulin has an effect opposite to that of glucagon. That is, it stimulates the removal of glucose from the blood-stream into muscle and fat cells. Amino acid uptake into these tissues is also stimulated. In muscle, glucose is converted into glycogen as well as being degraded to supply energy. In fat cells, glucose is converted to glycerol and fatty acids, and stored as triglyceride. Insulin also stimulates the formation of glycogen in liver. Insulin secretion is stimulated not only via glucagon but also by a high plasma concentration of glucose and amino acids, especially if these arise from dietary intake of carbohydrate and protein. Insulin has in addition some more long-term effects, such as the stimulation of protein synthesis in muscle and liver (including, for example, the enzyme glucokinase, Fig. 2.67).

The antagonistic effects of glucagon and insulin on blood glucose levels may at first sight seem difficult to reconcile with the fact that glucagon stimulates insulin

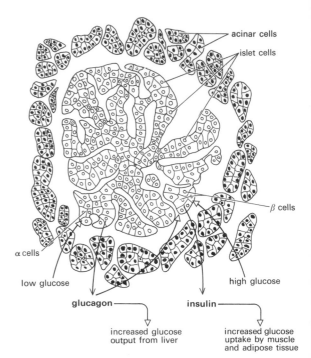

Fig. 9.10 Endocrine function of pancreas

A section through part of the pancreas is illustrated. The α and β islet cells together comprise the endocrine pancreas. The acinar cells, from which digestive enzymes and bicarbonate are secreted (Figs. 8.10 and 8.11), together comprise the exocrine pancreas.

release. However it must be remembered that the glucose released by glucogon is derived from liver, whereas the glucose removed by insulin goes to muscle and fat cells. Thus the combined action of glucagon and insulin is to cause a transfer of glucose from liver to muscle and fat. If one adds to this scheme the effect of cortisol in stimulating the conversion of protein to glucose (through an increased conversion of glucogenic amino acids), and the effect of adrenalin in stimulating the breakdown of glycogen and fat, a concerted action of these four hormones on intermediary metabolism becomes apparent (Fig. 9.11).

Returning to the examples of excess food intake, and of stress, cited at the beginning of this chapter, hormonal participation may be summarized as follows (Fig. 9.11). After a meal of fat or carbohydrate, the concentration of fatty acids and glucose in plasma rises. Insulin secretion is stimulated; glucagon secretion is depressed. The result is maximal storage of nutrients: glucose as glycogen in liver and muscle, and fatty acids as triglyceride in adipose tissue. During starvation these effects are largely reversed. Blood glucose is low, so glucagon secretion is stimulated, whereas insulin secretion is depressed. Glycogen stores in liver and muscle are broken down to glucose, and triglyceride is degraded to fatty acids and ketone

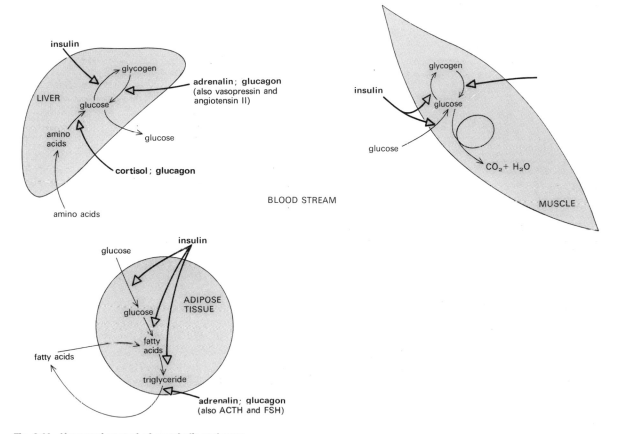

Fig. 9.11 Hormonal control of metabolic pathways

The hormones shown in brackets have been included for completeness; their action is in response to certain states of stress, rather than to dietary change. Note that a direct action of insulin or cortisol on liver cells has not been shown experimentally. See also Fig. 2.67.

bodies. Utilization of fatty acids and ketone bodies 'spares' blood glucose as much as possible.

After a meal of protein, the secretion of insulin and glucagon is stimulated. The result is maximal conversion of the amino acids derived from dietary protein into glucose (that is, gluconeogenesis stimulated by glucagon), followed by uptake and storage of glucose (stimulated by insulin). That is, the conversion of dietary amino acids into cellular carbohydrates, without change in the plasma glucose concentration.

During the need for extra muscular activity, as in stress, adrenalin is secreted. Glycogen is degraded to glucose in liver and muscle; triglyceride is degraded to fatty acids; fatty acids and glucose are used by muscle to provide extra energy. Cortisol is secreted at the same time and provides a further supply of glucose.

9.2.6.2 Kidney

The major function of the kidney is to regulate the fluid and ion balance (including pH) of plasma; it achieves this by the excretion of urine. This involves the reabsorption of some 99 per cent of plasma Na^+ from the glomerular filtrate (section 7.2.2.1). The reabsorptive process, which is dependent on the presence of aldosterone, is monitored by the kidney as follows (Fig. 9.12). A fall in plasma Na^+ concentration leads to the secretion of a hormone called **renin** from cells in the renal cortex. Renin acts to form another hormone, **angiotensin II,** which stimulates the formation of aldosterone by the adrenal cortex. The end-result is that aldosterone secretion is controlled by the level of plasma Na^+; in other words it is controlled by the target system on which it acts. In fact any change in plasma Na^+ concentration, such as that resulting from dietary lack or from excessive loss through sweating, results in a change in aldosterone release. This type of control, which operates also in the case of glucagon and insulin (Fig. 9.11) is to be contrasted with the other type of control, exemplified by the androgens, the oestrogens, the corticosteroids, and thyroxine, in which the hormone regulates its own release by a negative feedback mechanism *not* directly

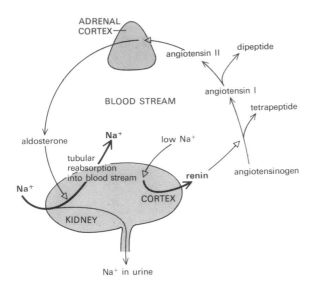

Fig. 9.12 Endocrine function of kidneys

A fall in plasma Na⁺, due to faulty reabsorption by the kidney, causes release of renin; renin indirectly stimulates the secretion of aldosterone from the adrenal cortex; aldosterone promotes the reabsorption of Na⁺ from the kindey.

involving its target systems (Fig. 9.2). The two systems also reflect the degree to which the nervous system participates in hormonal release (Fig. 9.14).

The release of angiotensin II occurs as follows (Fig. 9.12). Renin, which is an enzyme, degrades a plasma globulin (angiotensinogen) to form angiotensin I, which is further degraded by a serum enzyme to angiotensin II. In that the process involves cleavage of small peptides from circulating plasma proteins it is similar to the reactions of blood coagulation (Fig. 7.31). In a sense, the formation of a fibrin clot following damage to a blood-vessel can be thought of as a hormonal response, and the plasma proteins such as thrombin and autoprothrombin C that participate in the process, as hormones. Angiotensin II has an important property in addition to its stimulation of aldosterone release. It causes an increase in blood-pressure, due to constriction of the arterioles; like vasopressin, it promotes glycogen degradation in liver. In fact renin secretion is triggered by either a fall in plasma Na⁺ or by a decrease in blood-pressure (or by sympathetic nervous stimulation).

The renal cortex secretes another hormone, erythropoietin, which promotes the maturation of red blood cells by the bone marrow. Thus the renal cortex monitors blood in at least two ways: for its content of Na⁺, and for its content of red blood cells. Deficiencies are corrected for by the secretion of renin and erythropoietin respectively.

9.2.6.3 Gastrointestinal tract

The major function of the gastrointestinal tract is to digest and to absorb food (Chapter 8). Digestion is achieved through (i) the secretion of hydrochloric acid and pepsin in the stomach, and (ii) the secretion of bile, bicarbonate, amylase, lipase, trypsin, and chymo-trypsin into the duodenum. Bile is formed in the liver; bicarbonate and the digestive enzymes in the exocrine cells of the pancreas. Each process is under hormonal control (Fig. 9.13).

The presence of food in the **stomach** causes the pyloric cells to secrete **gastrin.** This peptide hormone stimulates the fundic cells to secrete hydrochloric acid; the stimulus is both direct and indirect. The indirect mechanism involves the formation of histamine, which itself stimulates HCl secretion. Histamine acts *locally* rather than *systemically*; that is, its concentration rises only in the cells of the stomach, and not in the blood-stream as a whole. Other hormones such as the catecholamines act in both a local and a systemic manner.

The acid juice leaving the stomach stimulates the **duodenal** cells to secrete two peptide hormones: **cholecystokinin** (also called pancreazymin) and **secretin.** Cholecystokinin has two main effects: to cause contraction of the gall bladder and hence increase the secretion of bile, and to stimulate the secretion of digestive enzymes from the pancreas. Secretin also stimulates biliary and pancreatic secretions; in this case the flow of water and ions, including bicarbonate, from liver and pancreas, is stimulated.

Gastrointestinal secretions are under feedback control: an excessive production of duodenal hormones (cholecystokinin, secretin, and others) leads to an inhibition of gastric secretion, which in turns decreases duodenal secretion.

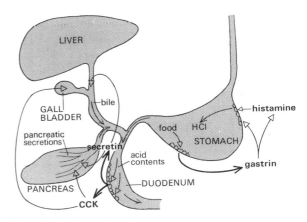

Fig. 9.13 Endocrine function of alimentary tract

The mucosal cells of the stomach and duodenum, from which the hormones are secreted, are indicated.
CCK = cholecystokinin

9.2.7 Functional classification of hormones·

Hormones, like other biological molecules, may be classified either according to structure or to function. A structural classification is given in section 9.3. In this section, classification according to function will be attempted. This is difficult for two reasons. First, all the functions of the various substances designated as hormones are not yet known. Secondly, many hormones have more than one function. Nevertheless the reader may find the allocation of hormones to different broad categories useful, provided that it is remembered that such a classification is only a rough approximation and that many exceptions and overlaps exist.

The scheme presented in Fig. 9.14 distinguishes two types of trigger that cause hormonal release: First, a change in the external environment, and secondly, a change in the internal environment. External changes may be subdivided into gradual changes, such as seasonal and diurnal variations, and more immediate changes, such as the onset of danger and other stress situations. Changes in the internal environment include nutritional changes such as an intake of food, and metabolic changes such as an increase in muscular activity; the actual trigger may be an alteration in the concentration of Na^+ or glucose in blood, of hydro-

chloric acid in the stomach, and so on. The receptors for external changes are nerve cells; the cells of the pineal gland may also function as receptors of external variation. The receptors for internal changes are specific cells, of mainly non-endocrine organs such as the α-and β-cells of the endocrine pancreas, the cortical cells of the kidney, or the mucosal cells of the gastrointestinal tract.

Interaction between trigger and receptor leads to the generation of a signal. This may be (a) electrical, as in the innervation of the hypothalamus from other centres in the brain, or in the innervation of the adrenal medulla from the hypothalamus, or (b) hormonal, as in the release of melatonin, insulin, glucagon, etc. Ultimately, all signals result in the release of a specific hormone.

The action of these hormones may be relatively slow acting and long lasting, or relatively fast acting and short lasting. The stimulation of growth by growth hormone, the elevation of basal metabolic rate by thyroxine, the development of secondary sex characteristics by androgens and oestrogens, the breakdown and conversion of protein into carbohydrate, and the resorption or deposition of bone are all examples of slow acting, long lasting hormonal responses. The increase in heart-rate and the degradation of fat and

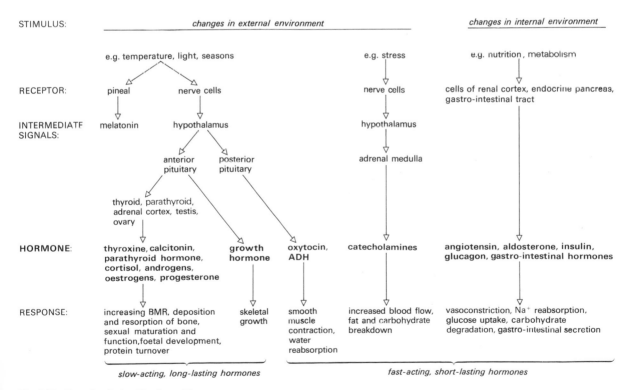

Fig. 9.14 Functional classification of hormones

The table summarizes the functions of some of the more important hormones.

carbohydrate by catecholamines, the degradation of carbohydrate by glucagon, the uptake of glucose by insulin, the constriction of blood-vessels by angiotensin II, the contraction of the uterus by oxytocin, the reabsorption of water by ADH, and the production of gastric, biliary, and pancreatic secretion by gastrointestinal hormones are all examples of fast acting, short lasting responses (Fig. 9.14). Some hormones, such as aldosterone, do not completely fit into the category under which they are listed; others, like insulin, have a secondary, slow effect as well as a fast effect.

9.3 Types of hormones and their turnover

Hormones fall into four broad categories with regard to structure: they are either

i. **polypeptides** (proteins and peptides);

ii. **amino acid derivatives;**

iii. **fatty acid derivatives;** or

iv. **steroids.**

In this section a brief outline of these four groups will be presented, together with a summary of the biosynthetic mechanisms involved. Most hormones are not secreted continuously from the cells in which they are formed. A separate storage and release mechanism exists, and this will be discussed. The lifetime of hormones, once they are released into the bloodstream, is rather short; it varies from a few minutes to several days for different hormones. The mechanism by which hormones are degraded will also be discussed.

9.3.1 Polypeptides

9.3.1.1 Structure

The distinction between a protein and a peptide is not clear-cut. Each is made up of a number of amino acids linked in linear sequence. Generally, compounds having a molecular weight greater than a few thousand are termed proteins, while compounds having a molecular weight less than this are termed peptides. But there is no dividing line, and this is well illustrated in the case of hormones (Fig. 9.15).

The molecular weights assigned to the polypeptide hormones listed in Fig. 9.15 refer to the basic units: either a single polypeptide chain, or two or more chains covalently linked by disulphide bonds, as in the case of insulin. Like other proteins, the basic units may associate with each other to form dimers, trimers, and so forth (quaternary structure).

The structure of some of the hormones listed in

Fig. 9.15 Polypeptide hormones

Hormone	Molecular weight (approx.)	Number of amino acid residues
Erythropoietin	50–60000	>200
FSH	34000	>200
TSH	28300	>200
LH	27400	>200
Prolactin	24000	198
Growth hormone	21500	190
Parathyroid hormone	9500	83
Insulin	5700	51
ACTH	4500	39
CCK	4000	33
Calcitonin	3600	32
Glucagon	3500	29
Secretin	3000	27
Gastrin	2000	17
Growth hormone-RH	1400	14
α MSH	1600	13
LH- and FSH-RH	1200	10
ADH (vasopressin)	1100	9
Oxytocin	1000	9
Angiotensin II	1000	8
TSH-RH	400	3

-RH = -releasing hormone, see Fig. 9.4.
Some of the hormones are glycoproteins; the molecular weights quoted include the carbohydrate portion. In the case of hormones that associate to form dimers, etc, the molecular weight of the monomeric unit has been quoted. In some instances, the data refer to hormones of animal, not human, origin.

Fig. 9.15 shows an interesting feature. This is that different hormones share parts of the same sequence. αMSH, for example, is virtually identical with the first 13 residues of ACTH; TSH, FSH, and LH are similar; and oxytocin and ADH differ at only two residues. In short, polypeptide hormones fall into a series of 'families' (Fig. 9.16). Such similarities imply a common mechanism of action, just as a similarity in structure between trypsin, chymotrypsin, elastase, and thrombin is reflected in the fact that each is a hydrolytic enzyme, having a serine residue at its active site. In fact one mechanism of action of these, and many other hormones, *is* the same: the activation of a membrane associated enzyme, adenyl cyclase.

9.3.1.2 Biosynthesis

Polypeptide hormones are synthesized, like other proteins, by ribosomes according to a mRNA template. In the case of a hormone such as insulin, which consists of two different subunits linked by disulphide bonds, one might imagine that two separate mRNA molecules are involved. In fact this is not so. There is only one insulin gene and this codes for a single mRNA. The mRNA is translated into a protein of molecular weight 9400. This protein has been isolated

(a) α **MSH and N-terminal end of ACTH**

α MSH: Ac-Ser-Try-Ser-Met-Glu-His-Phe-Arg-Try-Gly-Lys-Pro-Val

ACTH: Ser-Try-Ser-Met-Glu-His-Phe-Arg-Try-Gly-Lys-Pro-Val-Gly-(amino acid)$_{20}$

(b) **ADH (vasopressin) and oxytocin**

ADH: Cys-Tyr-Phe-Glu-Asn-Cys-Pro-Arg-Gly
             └─S─────S─┘

Oxytocin: Cys-Tyr-Ile-Glu-Asn-Cys-Pro-Leu-Gly
             └─S─────S─┘

Two examples are shown; several others exist.

Fig. 9.16 Structural similarity among polypeptide hormones

from pancreatic β-cells and is termed proinsulin. Proinsulin, which has no hormonal activity, contains the amino acid sequence of both the A- and the B-chain, and has the disulphide groups in exactly the same positions as have the separate A- and B-chains in the insulin molecule (Fig. 9.17). Conversion of proinsulin to insulin involves hydrolysis at two specific sites; insulin and an inactive peptide of some 30 amino acids are released. This manner of forming two separate polypeptide chains from a larger precursor is to be contrasted with that in which separate mRNA molecules are involved, and in which the subunits associate *after* they have been formed. The latter type

of mechanism is exemplified by proteins such as lactate dehydrogenase, haemoglobin and the immunoglobulins.

Post-translational modification of protein hormones involves not only the removal of peptides, but also the addition of carbohydrates. Erythropoietin, FSH, TSH, LH, and other hormones are glycoproteins and their biosynthesis is completed by the addition of sugar residues in the Golgi apparatus (section 2.5.1.2).

A glance at Fig. 9.15 shows that some of the polypeptide hormones are extremely small. TSH-releasing hormone, for example, consists of only three amino acid residues (glutamate, histidine, and proline).

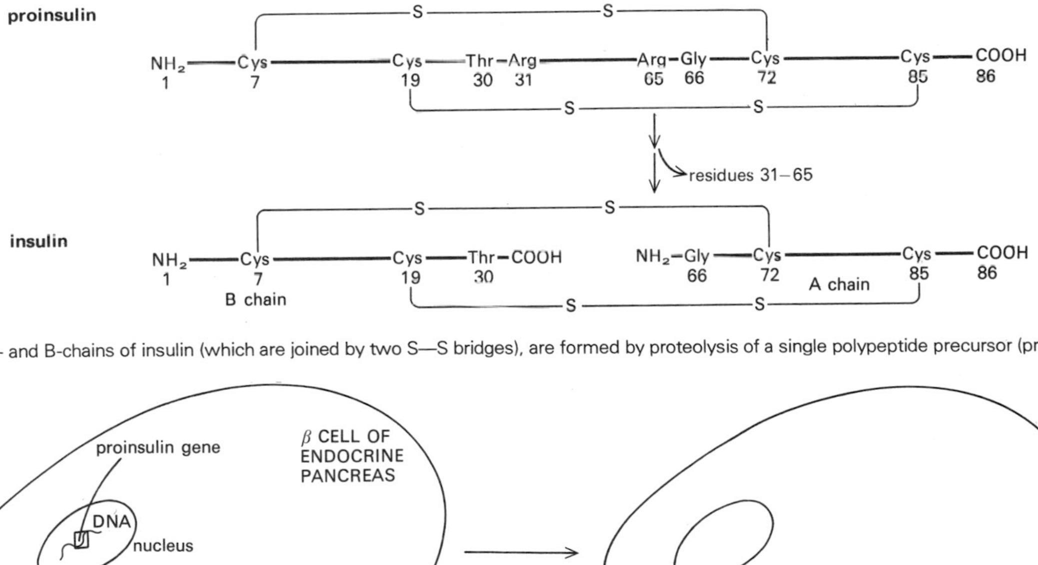

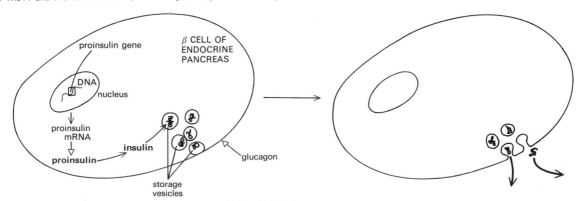

a The A- and B-chains of insulin (which are joined by two S—S bridges), are formed by proteolysis of a single polypeptide precursor (proinsulin).

b Glucagon triggers the release of insulin by exocytosis (see Fig. 6.16).

Fig. 9.17 Formation and release of insulin

Are such small peptides synthesized on an mRNA template, or are they synthesized by specific enzymes acting in a defined sequence? The answer appears to be that all hormones are synthesized on an mRNA template, specified by its respective gene. The initial product of mRNA translation is, as in the case of insulin, larger than the final product. In fact, hydrolytic degradation of a precursor protein or peptide to a smaller product is a typical feature of post-translational modification. It occurs in the case of enzymes such as trypsin, chymotrypsin, and pepsin, with blood clotting factors such as thrombin and fibrin, and with many, if not all, protein and peptide hormones (for example insulin, angiotensin II, and the hypothalamic releasing hormones).

In some instances, both products of hydrolysis have independent hormonal activity (Fig. 9.18). The precursor of αMSH, for example, is hydrolysed in the cells of the intermediate lobe of the pituitary (to yield αMSH, and a hormone known as 'corticotrophin-like intermediate lobe peptide' (CLIP), which appears to function as an insulin-releasing hormone). In addition to these post-translational modifications, there is pre-translational modification involving 'secretory peptides', that commit a protein to being secreted by a cell (section 5.2.2).

The synthesis of peptides by way of a single mRNA template may be contrasted with the synthesis of peptides by successive enzymes. In the case of the tripeptide glutathione (Fig. 7.17), for example, it is known that synthesis is through the action of two successive condensing enzymes. Synthesis by way of a single mRNA template is genetically a more efficient process than synthesis by way of successive condensing enzymes, each specified by a separate gene. This is especially true if the enzymes that degrade the precursor peptide are non-specific proteases, as in the case of

the proinsulin → insulin conversion. If they are not, as in the angiotensinogen → angiotensin I conversion (Fig. 9.12), an extra gene is required (in this instance for renin synthesis). Just how non-specific proteases recognize precursor peptides and proteins at the right positions is not yet fully understood.

9.3.1.3 Release

Most of the protein and peptide hormones are stored in the cells in which they are synthesized in the form of vesicles. Insulin, for example, is stored in the β-cells of the endocrine pancreas as granules surrounded by a vesicular membrane. When β-cells are stimulated, by glucagon, by nervous stimulation (vagus), or by an increase in glucose concentration, the contents of the vesicles are released into the blood-stream by the mechanism of exocytosis; that is, by fusion of the vesicles with plasma membrane (Fig. 9.17). In other words it is secretion of insulin, not its synthesis, that is the immediate response of the β-cells to stimulation. The β-cells are therefore specialized in two quite distinct ways. First, they possess a mechanism for storing insulin. This involves transcription of the proinsulin gene (which is kept permanently repressed in all other cells of the body) and conversion of proinsulin to insulin. Secondly, they possess the mechanism for releasing insulin. This involves a receptor on the cell surface that is specifically sensitive to glucagon and other agents. Interaction of glucagon with the cell surface leads to an increased concentration of Ca^{2+} and this somehow stimulates exocytosis. Exactly how the rate of proinsulin synthesis and its conversion to insulin is controlled in relation to the demand for insulin release is not clear.

The type of storage and release mechanism just described operates also in the case of many other hormones. It is a way of ensuring that the response to

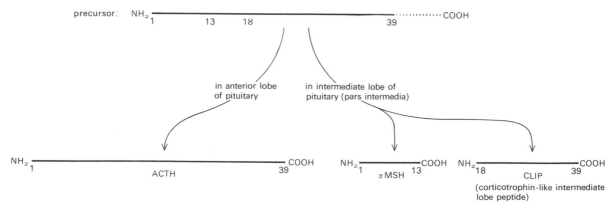

Fig. 9.18   Common precursors of polypeptide hormones

The anterior and intermediate lobes of the pituitary possess specific enzymes that hydrolyse a common precursor as shown; the exact site of synthesis of the precursor is not known. This type of mechanism may operate for other hormones also.

stimulation is both rapid and extensive. It also implies, of course, that cells may undergo a refractive period immediately after stimulation, during which time replenishment of the storage granules takes place. However, the capacity of hormone-secreting cells is enormous: the posterior pituitary, for example, contains enough ADH (vasopression) for 40 days' supply.

9.3.1.4 Degradation

Most of the protein and peptide hormones are degraded by hydrolytic enzymes. Insulin is degraded by an insulinase, angiotensin II by an angiotensinase, and so forth. Whether each hormone is degraded by a specific enzyme, or whether some degradative enzymes are relatively non-specific, has not been established. As is the case with other proteins, and with mRNA itself, however, the relative longevity of different molecules is controlled more by the rate of their synthesis than by the rate of their degradation (see section 5.3).

9.3.2 Amino acid derivatives

Four types of hormone that have been discussed in section 9.2 fall into this category: the catecholamines, the thyroid hormones, melatonin, and histamine (Fig. 9.19). In each case biosynthesis involves specific enzymes. It is the genes coding for the enzymes, not for some precursor protein (section 9.3.1.2), that are specifically transcribed in the hormone-producing cells.

The catecholamines adrenalin and noradrenalin are derived from phenylalanine or tyrosine. The pathway, which operates in the cells of the adrenal medulla, is depicted in Fig. 9.19. The same pathway, at least as far as noradrenalin, operates also in all sympathetic nerve endings in which noradrenalin is the transmitter substance (section 12.3.2).

Adrenalin and noradrenalin are stored within the adrenal cells in the form of granules, the chromaffin granules. ATP is also present, perhaps to neutralize the positive charge on the catecholamine molecule by the negative charge on the phosphate groups. The ATP–catecholamine complex is surrounded by a membrane to form a vesicle. When the adrenal cells are stimulated by sympathetic nerve transmission, the contents of the vesicle are discharged by exocytosis, exactly as in the case of insulin release (Fig. 9.17). The neurotransmitter that mediates the stimulation is acetylcholine. Interaction of acetylcholine with its receptor at the adrenal cell surface results in a transient influx of Ca^{2+}, which – as in insulin release – stimulates exocytosis.

The action of adrenalin and noradrenalin, which is at the cell surface, is terminated largely by uptake into the receptor cell. In the liver, catecholamines are also 'detoxified' by methylation. A specific enzyme transfers a methyl group from S-adenosyl methionine (SAM) to the 3-hydroxyl of the benzene ring, yielding an inactive product. The 3-methoxy derivatives are further metabolized by oxidation and conjugation with sulphate or glucuronide, and eventually excreted in the urine.

The thyroid hormones thyroxine (T_4) and tri-iodothyronine (T_3) are formed in two stages (Fig. 9.20). First, a protein called thyroglobulin is made; this is the form in which the thyroid hormones are stored. Thyroglobulin is particularly rich in tyrosine residues (120 per molecule) and these become iodinated under the influence of TSH. Iodine, which is present as iodide in salt and other foods, is oxidized within the thyroid cells to free iodine, which displaces one or two of the aromatic hydrogens of the tyrosine residues to form mono- and di-iodotyrosine derivatives. Thyroglobulin is stored extracellularly in the lumen of the thyroid follicles as a colloidal suspension. The second stage, also stimulated by TSH, is the uptake of thyroglobulin into the thyroid cells by pinocytosis (section 6.4.2). Ingested thyroglobulin is degraded by lysosomes to form T_3 and thyroxine (T_4). These become vesiculated prior to release by exocytosis. In short, thyroid release involves both pinocytosis (of thyroglobulin) and exocytosis (of T_3 and thyroxine).

The thyroid gland is the only organ in the body capable of incorporating iodide (by the mechanism depicted in Fig. 9.20). Thus if radioactive iodine is administered, it becomes concentrated in the thyroid gland. This has found a use in the analysis of thyroid disorders by X-ray analysis, and in the treatment of thyroid tumours (Graves' disease). In the latter case an excess of radioactive iodide is given so that the radioactive emission actually destroys the tissue.

In the blood-stream, thyroxine, and to a lesser extent T_3, become bound to a specific globulin which transports thyroxine to its target tissues. Lack of the binding protein is as serious a defect as lack of iodine. The action of thyroxine may be controlled by degradation of the binding protein, rather than by degradation of thyroxine itself.

Melatonin is formed from tryptophan, by way of 5-hydroxytryptamine, serotonin (Fig. 9.19). This conversion is not unique to the pineal gland, as serotonin is formed in many other tissues, particularly in other parts of the brain (Fig. 12.10); it has a variety of pharmacological actions. Two specific enzymes convert serotonin to melatonin, which is released into the blood-stream. Melatonin action is terminated by further metabolism involving hydroxylation, followed by conjugation with sulphate or glucuronide.

Histamine is formed from histidine by decarboxylation (Fig. 9.19). Its action is terminated by a specific

a catecholamines

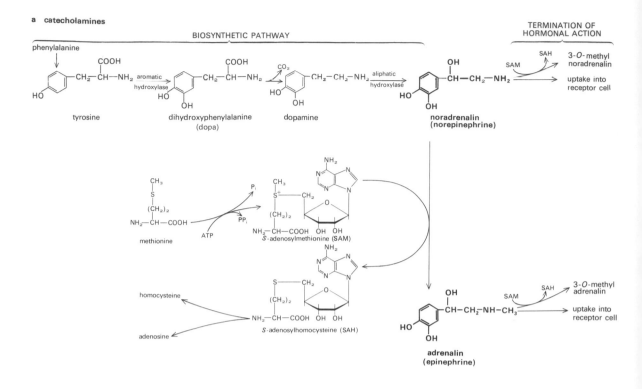

b thyroid hormones

c melatonin

d histamine

Fig. 9.19 Biosynthesis of amino acid-derived hormones

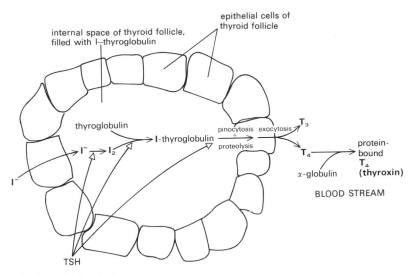

Fig. 9.20 Formation and release of thyroid hormones

histaminase enzyme. It should be pointed out that the major function of histamine is as a mediator of humoral sensitivity (section 11.2.2); in addition, it may act as a neurotransmitter in the brain. Whether neurotransmitters such as histamine or serotonin should be regarded as 'local' hormones is a matter of definition. What is important is that many amino acid derivatives, including the catecholamines, have both a local, largely transmitter, role as well as a systemic, largely hormonal, role.

The metabolic changes which generate the four hormones just discussed show certain common features. First, in each case, a carboxyl group of the parent amino acid is removed, making the products more basic. Secondly, the number of enzymic steps involved is rather small; this may be contrasted with the biosynthesis of steroid hormones from cholesterol, in which the metabolic sequence is considerably longer. Finally, in several cases the action of the hormone is terminated by specific metabolic changes, which result in the formation of compounds that are more readily excreted in the urine.

9.3.3 Fatty acid derivatives

A class of compounds known as **prostaglandins** exhibits biological function that is best described as hormonal. The reason why prostaglandins were not mentioned in section 9.2 is that they are not formed in any one specific type of cell. Nor do they act on any one type of cell. Rather they act within the tissue in which they are formed. They are, in short, truly 'local' hormones.

Prostaglandins have a variety of actions. A major one is to control the contraction of smooth muscle

and other functions of the autonomic nervous system (see section 12.1). Thus intestinal motility and the contraction of the uterus are increased; gastric secretion is decreased. Various reproductive functions are also affected. Prostaglandins appear to play a role in the processes of inflammation and the generation of pain, in that aspirin, a known anti-inflammatory and analgesic drug, inhibits the synthesis of prostaglandins.

The structure of prostaglandins, of which there are several different types, is that of unsaturated, hydroxylated fatty acids (Fig. 9.21). The biosynthetic pathway begins with the essential, polyunsaturated fatty acids (section 8.2.3.3). From linolenic acid, for example, PGE_1 and $PGF_{1\alpha}$ are synthesized; from arachidonic acid, PGE_2 and $PGF_{2\alpha}$ are synthesized.

Prostaglandins are not stored in the form of vesicles, but are secreted directly from the cells in which they are synthesized. The mechanism by which their action is terminated involves reduction of the double bonds and hydroxylation at carbon atom 15. Eventually they are oxidized to dicarboxylic acids and excreted. Like other hormones, prostaglandins have relatively short life-times.

9.3.4 Steroids

The steroid hormones fall into two main categories: gonadal steroids and adrenal steroids. The major gonadal steroids are the androgens, of which testosterone is the most important, and the oestrogens, of which oestrone and oestradiol are the most important; progesterone may be included in this category since its action is largely connected with gonadal function. The major adrenal steroids are the glucocorticoids, of which cortisol is the most important, and the mineralocorticoid, aldosterone (Fig. 9.22).

$CH_3(CH_2)_4[CH{=}CH{-}CH_2]_3\ CH{=}CH(CH_2)_3\ COOH$
arachidonic acid

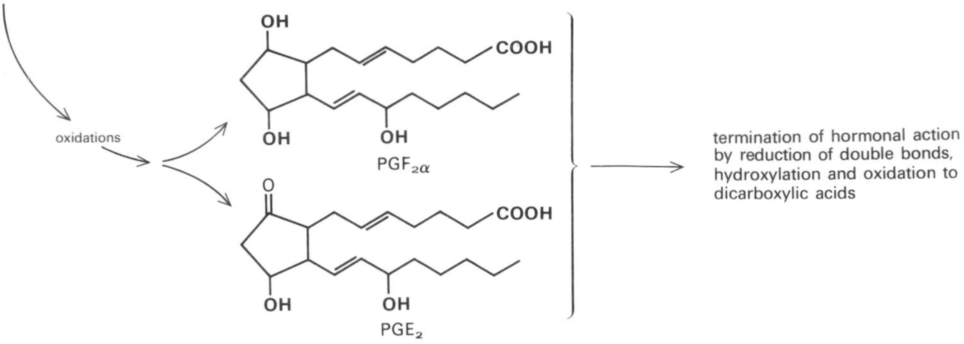
oxidations
OH
COOH
OH
OH
PGF2α
O
COOH
OH
OH
PGE2
termination of hormonal action by reduction of double bonds, hydroxylation and oxidation to dicarboxylic acids

Fig. 9.21 Biosynthesis of prostaglandins

The corresponding prostaglandins having only one double bond in the molecule are known as $PGF_{1\alpha}$ and PGE_1.

category	hormone	formula	site of synthesis
[oestrogen]	progesterone	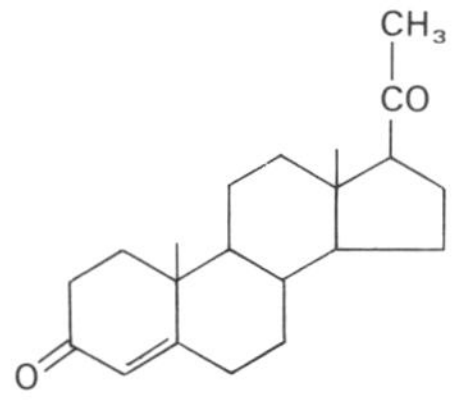 CH3 CO O	ovary, corpus luteum; also testis and adrenal cortex
androgen	testosterone	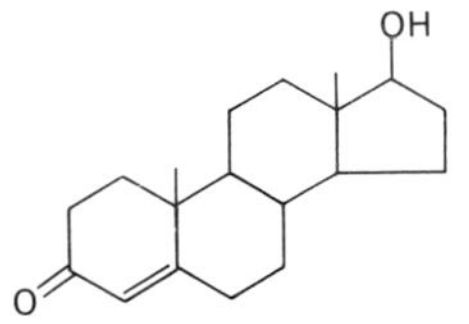 OH O	testis; also ovary
oestrogen	oestradiol	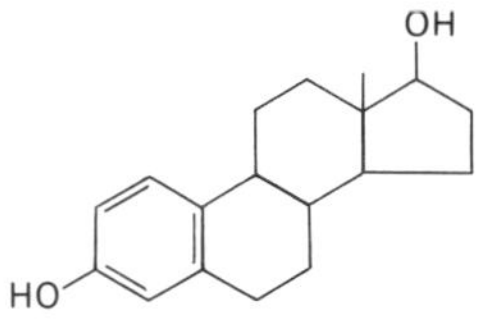 OH HO	ovary; also testis
glucocorticoid	cortisol	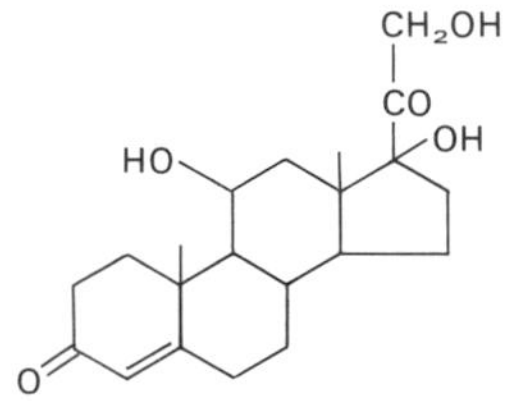 CH2OH CO OH HO O	adrenal cortex
mineralocorticoid	aldosterone	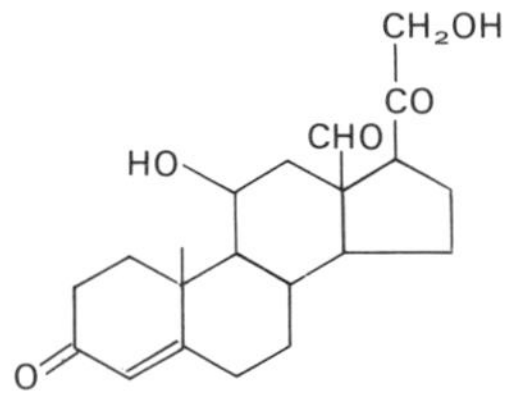 CH2OH CO CHO HO O	adrenal cortex

Fig. 9.22 The major steroid hormones

The biosynthesis of all steroid hormones starts from cholesterol. This is converted, by a number of oxidative steps in which the sidechain is removed, to pregnenolone, which is a common precursor of the other steroid hormones (Fig. 2.17). The reactions, which include hydroxylations, isomerizations, and decarboxylations, occur partly within the mitochondria and partly on the endoplasmic reticulum of the hormone-producing cell. Note that testosterone is not only an intermediate in oestrogen synthesis, it is actually secreted in small amounts by the ovary. Equally, testicular cells secrete a limited amount of oestrogens. Testosterone is not, in fact, the form in which its biological activity is exerted. In hormone-responsive cells, such as those of prostate, seminal vesicle, and other secondary male organ, testosterone is reduced to dihydrotestosterone (DHT) at the endoplasmic reticulum, and it is DHT that acts on the nucleus of such cells (Fig. 9.23). Feedback control at the anterior pituitary, however, is exerted by testosterone itself. Unlike testosterone, DHT cannot be converted to oestrogens and it therefore exerts a purely 'androgenic' action if introduced into the blood-stream.

Secretion of steroids takes place directly from the endoplasmic reticulum (or Golgi apparatus) into the blood-stream. Storage, and subsequent release, of hormone-containing vesicles does not occur. All steroid hormones are rather insoluble in water, and are transported in the blood-stream bound to globulins. These have considerable specificity towards different hormones, which are bound in hydrophobic regions of the globulin molecule.

The action of steroids, which have half lives measured in hours, is terminated in the liver. In the case of the corticoid steroids and testosterone, an inactive metabolite is produced by reduction; the reduced metabolite is then conjugated with sulphate or glucuronate to increase its solubility, and excreted in the urine. Oestrogens are conjugated without prior reduction.

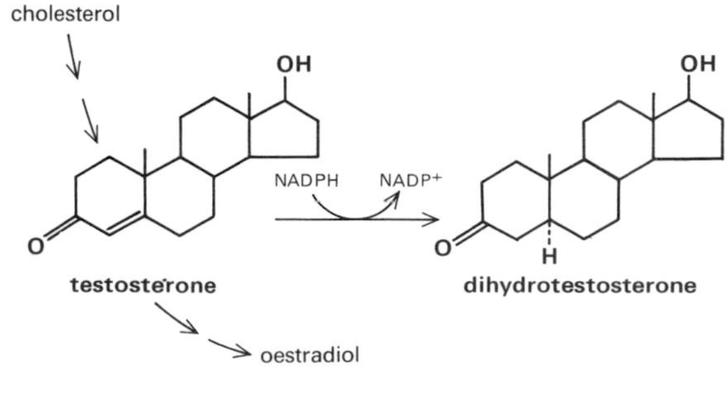

Fig. 9.23   Testosterone metabolism

9.3.5 Measurement of hormones in plasma

The concentration of most hormones in the bloodstream is rather low (Fig. 9.24). This is not surprising, since hormones act as cofactors of metabolism (section 9.4), not as substrates.

Measurement of such small quantities, which is important for the clinical evaluation of endocrine function, is generally carried out by **radioimmunoassay.** The principle of the method is as follows (Fig. 9.25). An antibody is raised against the hormone to be measured. In the case of hormones that are not antigenic, such as the amino acid derivatives or the steroids, an antibody is raised against the hormone coupled to an inert protein (that is, as hapten). The antibody is now used to react with radioactively labelled hormone, that is, the antigen, to produce a complex which can be separated from excess free hormone and measured. If a sample of blood containing non-radioactive hormone is added to the antigen–antibody mixture, the amount of radioactive hormone that is complexed is reduced. By comparing the degree of reduction with that brought about by a known amount of non-radioactive hormone, the concentration of hormone in the sample can be assessed. The principle, namely that of saturation analysis, is the same as that involved in RNA–DNA hybridization (Fig. 4.19). The method is both sensitive and specific. It can be used to measure hormone binding proteins also, and has been

Fig. 9.24   Plasma levels of major hormones

Hormone		Concentration (approx.)	
		(mg/l)	(nmol/l)
Cortisol		150	400
Thyroxine		80	100
Testosterone	(men)	6	20
	(women)	0.3	1
Progesterone	(men)	0.3	1
	(women)	1–11*	3–33*
Insulin		8	1
Noradrenalin		0.2	1
Oestradiol	(men)	0.05	0.2
	(women)	0.06–0.2*	0.3–0.8*
Aldosterone		0.1	0.3
Adrenalin		0.05	0.3
Growth hormone		3	0.15
Glucagon		0.1	0.03
ACTH		0.04	0.01
Angiotensin II		0.01	0.01

These values, which are approximate averages taken from a number of sources, refer to normal adults. The values fluctuate markedly accordingly to nutritional status, time of day, stress situations, etc. Note that values quoted for progesterone and oestradiol refer to premenopausal women.

* According to phase of menstrual cycle.

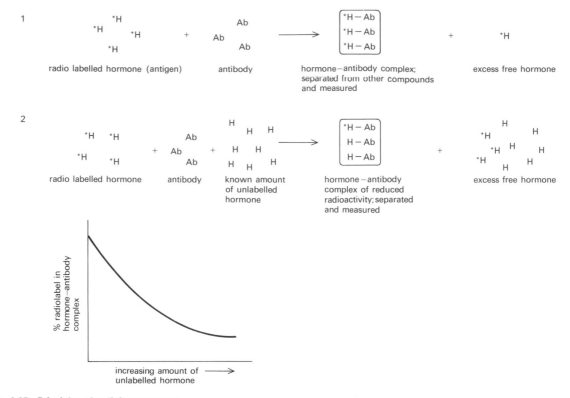

Fig. 9.25 **Principles of radioimmunoassay**

extended to measure other substances, such as drugs, against which an antibody can be raised.

9.4 Mechanism of hormone action

It would be convenient to be able to relate the physiological function of hormones, as outlined in section 9.2, to their biochemical action; in other words, to explain their physiological effect in molecular terms. Unfortunately this is as yet not possible for a number of reasons. First, the biochemistry of some of the physiological responses is not known. The molecular basis of virility, of foetal development, of the control of organ and body size, to cite but a few examples, is not understood. Surprisingly enough, not even the control of basal metabolic rate can yet be described in molecular terms. Secondly, the biochemical action of many hormones is not yet fully clarified. Testosterone and the oestrogens, for example, stimulate protein synthesis in target organs, but how is this made specific? Thirdly, more than one biochemical action has been attributed to several hormones. Insulin, for example, stimulates glucose uptake and glycolysis by muscle, glycogenesis and protein synthesis in liver, and lipogenesis in fat cells, while it

inhibits gluconeogenesis in liver and glycolysis in fat cells. Which is the action primarily related to its hypoglycaemic (reduction in blood glucose) function?

At present, then, it is not profitable to do more than describe what biochemical actions have been elucidated and to relate these, where possible, to known physiological functions. All hormones have one property in common. They interact with a specific receptor at the target cell, and as a result of that interaction, the cell responds in one or more ways. The interactions may conveniently be divided into two types: interaction with a receptor at the cell surface and interaction with a receptor inside the cell.

9.4.1 At the cell surface

Many hormones are polypeptides, which do not cross the plasma membrane. This suggests that they interact with a receptor at the cell surface. It is true that proteins, and even larger molecules, are taken up by cells by pinocytosis or phagocytosis. But the entry of a substance by pinocytosis or phagocytosis is not the same as entry across the plasma membrane. The contents of a pinocytotic or phagocytotic vesicle become hydrolysed by lysosomal action as soon as they enter a cell. In other words, molecules ingested in this manner cannot exert any specific action within the cell. (Viruses are an exception. Their nucleocapsids (protein–RNA or protein–DNA complexes), which become exposed once they are inside a cell, are so arranged as to make them relatively resistant to lysosomal hydrolysis.)

In fact not only polypeptides, but also amino acid-derived hormones such as the catecholamines and fatty acid-derived hormones such as the prostaglandins, act at the cell surface. As a generalization, nonsteroid hormones act at the cell surface, whereas steroid hormones (and possibly thyroid hormones) act intracellularly. Three types of surface response have been described.

9.4.1.1 Adenyl cyclase

The enzyme adenyl cyclase catalyses the conversion of ATP to **cyclic AMP** (cAMP). It is situated on the plasma membrane, 'facing inwards': in other words reactant and product remain in the cytoplasm (Fig. 6.14). Many hormones stimulate the enzyme (Fig. 9.26), as a result of which cAMP has been termed a **'second messenger'**, the first messenger being the hormone itself. (Cyclic GMP, which tends to be elevated in situations that depress cyclic AMP, and vice versa, may be another type of second messenger.) Binding of hormone is not directly to the enzyme, but rather to a protein receptor on the plasma membrane that 'faces outwards'. Receptor and enzyme then interact, as a result of which enzymic activity is increased.

The cAMP that is formed has several actions. The main one is to stimulate the **phosphorylation** of **proteins.** Structural proteins such as histones, and enzymes such as glycogen phosphorylase, triglyceride lipase, and glycogen synthetase, exist in two forms: phosphorylated and unphosphorylated. Phosphorylation, which is catalysed by a specific **protein kinase,** generally occurs on the hydroxyl group of a serine

residue. The two forms have differing activities. Histones, for example, bind DNA less effectively in the phosphorylated form, and thus in some way 'free' DNA to enable replication and transcription to take place. Glycogen phosphorylase and triglyceride lipase are more active in the phosphorylated form, whereas glycogen synthetase is less active.

Thus activation of adenyl cyclase can lead to DNA replication and transcription, to glycogen and triglyceride breakdown, and to inhibition of glycogen synthesis. Exactly how cAMP that is produced near the plasma membrane reaches the nucleus to affect histone phosphorylation is not clear. There may be a second type of adenyl cyclase situated at the nuclear membrane, which actually controls nuclear events. How protein hormones that do not enter cells stimulate DNA replication and transcription is also not clear.

It may be wondered how stimulation of a *general* metabolic event such as phosphorylation of proteins can be the mechanism by which a *specific* response -- a characteristic of hormone action – is achieved. The answer is that the specificity lies at the cell surface. ACTH stimulates adenyl cyclase in cells of the adrenal cortex, not liver or muscle, while glucagon and adrenalin stimulate adenyl cyclase in cells of liver and muscle but not of adrenal cortex; the reason is that the receptor proteins for ACTH, glucagon, and adrenalin are different. The metabolic potential of the target cell is also specialized. The main cAMP-sensitive proteins of the adrenal cortex are those associated with steroid synthesis; the main cAMP-sensitive protein of liver and muscle is glycogen phosphorylase.

Of course cells possess other cAMP-sensitive proteins, but then cells respond to hormones in more than one way. ACTH, for instance, stimulates gene activity and protein synthesis in adrenal cortex cells as well as steroidogenesis (it is a trophic hormone), and glucagon stimulates glucogenesis as well as glycogen breakdown

Fig. 9.26 Hormonal stimulation of adenyl cyclase

Cell type	Hormone	Major response
Liver	adrenalin and other β adrenergic stimuli*; glucagon	glycogen breakdown
Muscle	adrenalin and other β adrenergic stimuli*	glycogen breakdown
Adipose tissue	adrenalin and other β adrenergic stimuli*; glucagon	triglyceride breakdown
Adrenal cortex	ACTH	corticosteroid secretion
Pancreatic β cell	glucagon	insulin secretion
Gonads	FSH and LH	androgen and oestrogen secretion
Kidney	ADH (vasopressin)	water reabsorption
Brain	catecholamines, dopamine (see Figure 12.7)	neurotransmission
Thyroid, ovary, anterior pituitary, etc.	prostaglandins†	various

* See section 12.3.2.   † Note that in other cell types (e.g. adipose tissue), prostaglandins *inhibit* adenyl cyclase.

in liver. In short, hormones do not stimulate one or other metabolic event inside a cell. **Hormones stimulate cells.** The response depends on the cell type. An endocrine cell, such as an adrenal cortex, thyroid, or gonad cell, which is programmed to grow and divide as well as to synthesize specific hormones, will respond by increasing all these activities. A non-endocrine cell, such as a liver cell, will respond by mobilizing the pathways leading to increased glucose output: glycogen degradation, gluconeogenesis, conversion of glucose 6-phosphate to glucose, inhibition of glycogen synthesis, and so forth.

The difference between endocrine cells, which are stimulated by the trophic hormones (Fig. 9.3), and non-endocrine cells, which are stimulated by the other hormones, is that non-endocrine cells are programmed to respond in one of two (rarely more) ways. Muscle can either absorb and store glucose, or it can degrade it when energy is required elsewhere in the body. Liver is a little more complicated, but in general it either produces and excretes glucose, or it absorbs and utilizes it: an increase in cAMP leads to the degradative reactions; a decrease in cAMP leads to the synthetic reactions. Non-endocrine cells therefore have two separate receptor systems: receptors for hormones such as glucagon and adrenalin that increase cAMP, and receptors for hormones that decrease cAMP. The situation is analogous to the innervation of involuntary muscle or secretory gland cells by both sympathetic and parasympathetic nerves (section 12.1). Insulin is the main hormone responsible for a decrease of cAMP, but how this is achieved is not clear; it is not by inhibition of adenyl cyclase, since the insulin receptor does not interact with adenyl cyclase.

The effect of hormones such as adrenalin in causing an increase in intracellular cAMP may not be as direct as is implied in Fig. 2.4. That is, the increase may be caused by factors other than the activity of adenyl cyclase. Ca^{2+} is such a factor, and there is good evidence that adrenalin and other hormones stimulate cells through an increased uptake of Ca^{2+}.

9.4.1.2 Calcium entry

Many metabolic processes are modulated by Ca^{2+} ions. The degradation of glycogen, for example, is stimulated by Ca^{2+}. Other enzymes, such as the phosphodiesterase that degrades cAMP to AMP, are inhibited by Ca^{2+}. An increase in intracellular Ca^{2+} therefore potentiates the effect of adenyl cyclase on glycogen breakdown in two ways: by direct stimulation of the respective enzymes, and by increasing the concentration of cAMP through inhibition of its degradation (Fig. 9.27). Several hormones that were originally though to act entirely by stimulating adenyl cyclase have now been shown to stimulate the **entry** of Ca^{2+} into cells (Fig. 9.28). Exactly how much of the hormonal effect is due to Ca^{2+} entry and how much to activation of adenyl cyclase is at present unclear.

Another intracellular event that is stimulated by Ca^{2+} is membrane fusion (Fig. 6.17). Since many hormones, especially the trophic hormones, stimulate their target cells to release a second hormone by exocytosis, and since the key event in exocytosis is membrane fusion, it is not surprising that such hormones cause uptake of Ca^{2+}.

The mechanism by which Ca^{2+} enters cells is not known. Since the cytoplasmic concentration of Ca^{2+} is several orders of magnitude lower than the plasma

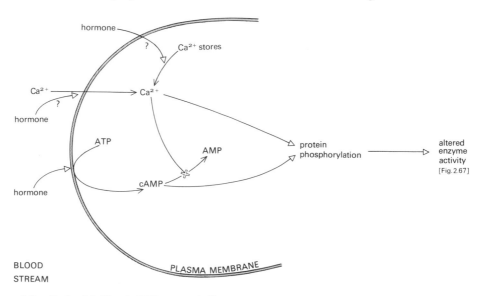

Fig. 9.27   Synergistic effects of Ca^{2+} and cAMP on metabolic processes

Fig. 9.28 Hormonal stimulation of Ca²⁺ uptake

Cell type	Hormone	Major response
Liver	adrenalin and other α-adrenergic stimuli*; ADH and angiotensin II	glycogen breakdown
Smooth muscle	adrenalin and other α-adrenergic stimuli*	muscle contraction
Salivary gland	adrenalin and other α-adrenergic stimuli*	secretion of saliva

* See section 12.3.2.

concentration of Ca^{2+}, it is likely that the entry is passive. But whether entry is caused by opening of a Ca^{2+} 'channel', or by inhibition of the normal Ca^{2+}-exit mechanism, is not clear. Indeed, it is not even clear that Ca^{2+} *necessarily* has to enter cells: a redistribution of internal Ca^{2+}, which happens to result in Ca^{2+} entry, may be the primary response. Nor have the hormone receptors – presumably proteins – on the cell surface been characterized. The receptors are probably the same as the α-adrenergic receptors on the surface of efferent cells of the autonomic nervous system (section 12.3.2).

9.4.1.3 Other mechanisms

There is a third type of mechanism, as yet unclear, by which interaction between hormone and receptor at the cell surface leads to an intracellular change. The change is brought about by a stimulation of **nutrient uptake** (Fig. 9.29). One hormone that acts in this manner is insulin, Muscle and fat cells are relatively impermeable to glucose and amino acids. Entry of these nutrients is by facilitated diffusion; that is, specific transport proteins catalyse the entry process (section 6.3.1). The activity of the proteins is stimulated by the binding of insulin to a cell surface receptor. Whether the transport proteins are themselves the receptor, whether there is a separate receptor that interacts with the transport protein in the way that hormonal receptors interact with adenyl cyclase (Fig. 6.14), or whether the uptake is stimulated from within the cell by some type of 'second messenger', is not known.

Fig. 9.29 Hormonal stimulation of nutrient uptake

Cell type	Nutrient	Hormone	Major response
Muscle	glucose	insulin	glycogen synthesis
	amino acids	insulin	protein synthesis
Adipose tissue	glucose	insulin	triglyceride synthesis
Liver	amino acids	glucagon	protein synthesis

The entry of glucose into muscle results in the synthesis of glycogen and in an increased rate of glucose degradation. An increased uptake of amino acids leads to an increase in the synthesis of protein (Fig. 9.29). Although proteins do not serve a storage function for amino acids in the way that glycogen serves to store glucose, or triglyceride to store fatty acids, the amount of protein in tissues nevertheless decreases during starvation or insulin deficiency (diabetes mellitus), and is increased when the nutritional or endocrine status is returned to normal. Since some 40 per cent of tissue mass is in the form of muscle, this represents a major site of protein turnover. Whether the increase in protein synthesis mediated by insulin is entirely due to an increased uptake of amino acids is not clear; a second mechanism involving an increase in translational activity (Fig. 9.32) seems likely, though how this is achieved is not known.

In fat cells, an increased uptake of glucose results in an increased synthesis of triglyceride. This is because the predominant pathway of glucose metabolism in fat cells is to generate glycerophosphate and fatty acids. In addition, free fatty acids of plasma are utilized for triglyceride synthesis.

The situation with regard to liver is unclear. On the one hand insulin undoubtedly stimulates glycogen and protein synthesis in liver. On the other hand the uptake of glucose and amino acids is so rapid in the absence of insulin that a stimulatory effect is difficult to observe. Moreover the increase in glycogen synthetase is due partly to an increase in the amount of active (glucose 6-phosphate-independent, Fig. 2.56) enzyme present. Another intracellular enzyme that is increased is glucokinase. In short, there is considerable evidence that insulin stimulates the formation of specific proteins. As with muscle and fat cells, it is not clear how this is brought about. Whatever the mechanism, it is clear that insulin has a generally anabolic (that is, leading to the synthesis of carbohydrate, fat, and protein) effect on cells, whereas glucagon, adrenalin, and cortisol, have the opposite or catabolic (that is, leading to the breakdown of carbohydrate, fat, and protein) effect (Fig. 9.11).

9.4.2 Intracellular

The theory of hormonal action at the cell surface accounts fairly satisfactorily for the mechanism of action of most of the polypeptide, amino acid-derived and fatty acid-derived hormones. On the other hand steroid hormones, which readily cross the plasma membrane, act within cells. The mechanism of entry is as follows.

Like the steroid cholesterol from which they are derived, steroid hormones are rather hydrophobic (apolar) and bind preferentially to lipids and proteins

containing hydrophobic regions. The degree of hyd-
rophobicity of steroid hormones is less than that of
cholesterol. Steroid hormones lack the long aliphatic
side-chain; they also have additional keto and hydroxyl groups, which introduce a certain degree of
hydrophilic (polar) character (Fig. 2.17). As a result,
steroid hormones bind less to the aliphatic sidechains
of triglyceride and phospholipids, and more to the
hydrophobic regions of certain proteins. Thus cholesterol is transported in plasma bound largely to triglyceride and phospholipid in lipoproteins, whereas
steroid hormones are transported bound largely to
hydrophobic areas on proteins (chiefly α- and β-
globulins).

At the cell surface, cholesterol readily changes its
binding to the aliphatic side-chains of plasma membrane phospholipids; that is, it enters the plasma
membrane. Steroid hormones however, preferentially
bind to hydrophobic proteins that are present in the
cytoplasm; that is, they enter cells. The specificity of
steroid hormone action is therefore achieved through
the presence of specific receptor proteins in the cytoplasm of target cells. Because there is little barrier to
their permeability, steroid hormones enter all cells. In
the absence of an intracellular receptor, however, the
equilibrium between intracellular and extracellular
hormone is in favour of extracellular hormone (where
it is bound by plasma globulins). In target cells, on the
other hand, the equillibrium is in favour of intracellular binding to the receptor proteins.

Steroid hormone receptors have not been extensively purified. What is known is that the receptors
readily pass into the nucleus (Fig. 9.30). During passage, the size of the receptor changes. Initially its
sedimentation coefficient (section 4.6.1) is approximately $9s$; in the nucleus it is approximately $4s$. When
steroid hormones enter the nucleus, they become
rather tightly bound to the **chromatin;** binding is most
likely to the acidic proteins of chromatin. The receptor
protein returns to the cytoplasm (Fig. 9.30). Therefore
if a radioactive steroid hormone, such as oestradiol, is
injected into the blood-stream, the net result is that
the radioactivity becomes concentrated in the nucleus
of target cells.

In the case of testosterone, an additional metabolic
step is involved. Testosterone is bound by an
NADPH-linked reductase on the endoplasmic reticulum, which converts testosterone to the dyhydro-
derivative (Fig. 9.23). Dihydrotestosterone then enters the nucleus and binds to chomatin.

9.4.2.1 Transcription

The fact that steroid hormones become concentrated
within the nucleus of target cells has led to the concept
that the mechanism of their action is to **stimulate
transcription** of DNA to specific species of mRNA

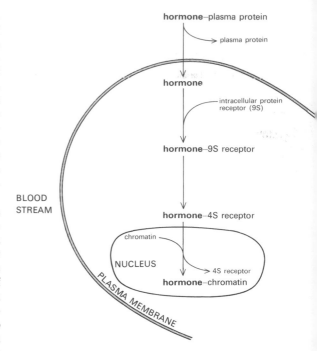

Fig. 9.30   Intracellular transport of steroid hormones

(Fig. 9.31). The concept is supported by three lines of
evidence. First, the hormones bind preferentially to
the non-histone proteins of chromatin, and it is non-
histone proteins that are believed to regulate the
expression of specific genes; that is, to regulate transcription (section 4.6.2). Secondly, the activity of isolated chromatin in synthesizing mRNA is stimulated
by steroid hormones. In several instances, the mRNA
has been shown to synthesize a specific protein.

Fig. 9.31   Hormonal stimulation of transcription

Cell type	Hormone	Major response
Liver	glucocorticoids	gluconeogenesis, due to increased synthesis of transaminases and other amino acid-metabolizing enzymes
Renal tubule	aldosterone	Na⁺ reabsorption, due to increased synthesis of specific protein(s)
Prostate	dihydrotesto-sterone	tissue growth and other 'trophic' effects
Oviduct of chicks	oestradiol	synthesis of egg white proteins

The experimental evidence on which stimulation of transcription is
based is: firstly, that the hormones becomes bound to chromatin (see
Fig. 9-30) and secondly, that they increase the concentration of
specific mRNA species in the target tissues.
  Note that some of the hormones listed in Fig. 2.67 have not been
studied in such detail, and hence do not appear in this Figure (or in
Fig. 9.32).

Thirdly, the responsiveness of target organs is inhibited by actinomycin D. This implies that RNA synthesis is required to elicit the response. It is true that the use of inhibitors in biochemical experiments is never a totally satisfactory approach, but nevertheless, such results provide confirmatory evidence.

If one accepts the transcriptional machinery as a major site of steroid hormone action, the effect of corticosteroids in stimulating protein breakdown and gluconeogenesis can be explained. For stimulation is due to the synthesis, and not the activation, of specific enzymes of that pathway: transaminases, pyruvate carboxylase (pyruvate → oxaloacetate), PEP carboxykinase (oxaloacetate → PEP), fructose diphosphatase (fructose 1,6-diP → fructose 6-P) and glucose 6-phosphatase (glucose 6P → glucose). Similarly aldosterone stimulates the synthesis of a Na^+ transport protein.

9.4.2.2 Translation

Several hormones **stimulate** the **translational** stage of protein synthesis (Fig. 9.32). Changes include an alteration in the ratios of free to bound ribosomes, of the turnover of phospholipids and proteins associated with the endoplasmic reticulum, and of the overall translational efficiency of the endoplasmic reticulum. Whether the effect is a primary one, or whether it is a reflection of increased transcription of ribosomal and other genes, is not clear.

In addition to the hormones listed in Fig. 9.32, thyroxine appears to stimulate the translational machinery and this may be its main site of action; an increase in the efficiency of translating particular proteins can certainly account for its action in raising the basal metabolic rate. (The reason that thyroxine is not listed in Fig. 9.32 is that it has not been shown to satisfy the criteria mentioned in the legend.) Although a low molecular weight compound such as thyroxine has little difficulty in crossing the plasma membrane, the major site of thyroxine binding has not been established. One hypothesis for the action of thyroxine is based on another action of this hormone, the ability to uncouple oxidative phosphorylation in mitochondria. Uncoupled mitochondria (section 3.4) require more oxygen to produce a given amount of ATP and this results in an increased metabolic rate. There are however two drawbacks to a hypothesis ascribing the physiological action of thyroxine to this effect: (i) it is liver mitochondria (which constitute less than 10 per cent of the total mitochondrial population of the body), not muscle mitochondria, that are sensitive to uncoupling by thyroxine, (ii) rather high concentrations of thyroxine are required to uncouple oxidative phosphorylation, and (iii) the mitochondrial response in an intact animal occurs *after* the basal metabolic rate has been raised.

Increasing the translational efficiency of protein synthesis is a rather unspecific response. But this is also a fair description of the action of some of the hormones listed in Fig. 9.32. Growth hormone stimulates a general increase in biosynthetic activity, and so does insulin, in *its receptive target organs*. As has already been stressed, differentiated cells, such as those of muscle, uterus, cartilage, or bone, are programmed, through the de-repression of certain genes, to synthesize particular proteins. What hormones – especially 'developmental' hormones such as thyroxine, oestrogens, or growth hormone — do, is to accelerate the rate at which that pre-set pattern of proteins is produced.

Fig. 9.32 Hormonal stimulation of translation

Cell type	Hormone	Major response
Liver	glucagon	gluconeogenesis, due to increased synthesis of transaminases
Muscle, cartilage, bone, and other tissues	growth hormone	growth, due to increase in general protein synthesis
Muscle and heart	insulin	stimulation of polysome formation, leading to increased synthesis of proteins (structural and enzymes); increased synthesis of glycogen resulting from enzyme induction (Fig. 2.67).

The experimental evidence on which control at the translational level is based is firstly that inhibitors of transcription such as actinomycin D are without effect and secondly that rapid changes in the activities of individual components of the protein-synthesizing machinery (ribosomes, initiation factors, etc., Fig. 5.7) occur.

Note that thyroxin and some of the hormones listed in Fig. 2.67 have not been studied in such detail and hence do not appear in this Figure (or in Fig. 9.31).

9.4.3 Unresolved problems

It will by now have become clear that there are several unresolved problems regarding the molecular basis of endocrine action. One of the major ones concerns the site of action of protein hormones. How do ACTH, TSH, and similar hormones exert their 'trophic' effect and how do hormones such as insulin and growth hormone stimulate protein synthesis without entry into cells? In the case of the trophic hormones, an increase in cytoplasmic cAMP alone is unlikely to explain the effect; likewise, in the case of insulin and growth hormone, a change in intracellular cyclic AMP (which is actually decreased in the case of insulin) is unlikely to account for the observed response.

9.5 Disease

The function of hormones is to regulate the metabolic activities of the body. If the production of a particular hormone becomes deficient or excessive, the resulting **imbalance** can lead to a variety of disorders (Fig. 9.33). Examples of diseases in which hormonal secretion is deficient are diabetes mellitus (lack of insulin), Addison's disease (lack of adrenal steroids), myxoedema (lack of thyroid hormones), and hypogonadism (lack of gonadal steroids, leading to sterility and loss of secondary sex characteristics).

Examples of diseases in which hormonal secretion is excessive are Cushing's syndrome (excess of adrenal steroids) and thyrotoxicosis (excess of thyroid hormones). Over-production of hormone often results from endocrine-derived tumours, such as pituitary tumours (excess of any one of the pituitary hormones), parathyroid tumours (excess of parathyroid hormone), or adrenal tumours (a malignant form of Cushing's syndrome). Some examples of endocrine deficiency or excess are illustrated in Fig. 9.34.

Endocrine disorders result not only from an aberrant rate of hormone release, but also from faulty receptors at the target organs, or from faulty carrier proteins. Many disorders that are as yet poorly understood may prove to fall into one or other of these categories.

Fig. 9.33   Some diseases of the endocrine system

Hormone	Alteration	Cause	Outcome	Clinical description
Insulin	Decrease	Unknown	Increased blood sugar (hyperglycaemia)	Diabetes mellitus
	Increase	Neoplasia of Islet cells	Decreased blood sugar (hypoglycaemia)	Hypoglycaemic coma
Growth hormone	Decrease	Damage to hypothalamus or pituitary	Decreased skeletal growth	Dwarfism
	Increase	Pituitary tumour	Increased skeletal growth	Gigantism; acromegaly
ACTH	Decrease	Damage to hypothalamus or pituitary	Atrophy of adrenal cortex	Pituitary–adrenal insufficiency
	Increase	Pituitary tumour	Increased adrenal cortical function	Cushing's syndrome
Parathyroid hormone	Decrease	Post-thyroidectomy	Hyocalcaemia; tetany	Hypoparathyroidism
	Increase	Neoplasia of parathyroid gland	Demineralization of bone; hypercalcaemia	Hyperparathyroidism
TSH	Decrease	Damage to hypothalamus or pituitary	Atrophy of thyroid and decreased metabolic activity	Pituitary hypothyroidism
FSH and LH	Decrease	Damage to hypothalamus or pituitary	Atrophy of gonads; loss of gonadal secretion	Hypogonadism
Thyroxine	Decrease	I_2 deficiency; auto-immune thyroiditis	Decreased metabolic activity	Endemic goitre; primary hypothyroidism (myxoedema)
	Increase	Increased thyroid stimulation (not due to excess TSH; autoimmunity, Fig. 11.16)	Increased metabolic activity and increased size of thyroid	Thyrotoxicosis
Catecholamines	Increase	Tumour of adrenal medulla	Hypertension and other pharmacological effects	Pheochromocytoma
Glucocorticoids and mineralocorticoids	Decrease	Auto-immunity-induced atrophy or tuberculosis of adrenal cortex	Hypertension; increased ACTH (leading to hyperpigmentation)	Addison's disease
Glucorticoids	Increase	ACTH excess; tumour of adrenal cortex	Obesity and other metabolic effects	Cushing's syndrome
Mineralocorticoids	Increase	Tumour of adrenal cortex	Hypertension; increased plasma Na^+; low K^+	Aldosteronoma (Conn's syndrome)
Oestrogens	Decrease	Pituitary or ovarian damage	Failure of sexual development in females; amenorrhea	Female hypogonadism
Androgens	Decrease	Pituitary or testicular damage	Failure of sexual development in males	Male hypogonadism

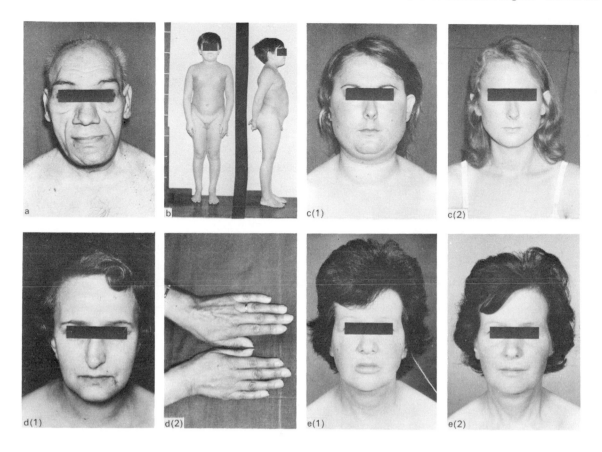

a Acromegaly. Note the effect of excessive growth hormone on the facial bones.
b Dwarfism. The effect of growth hormone deficiency in a 14 year-old boy is illustrated.
c (1) Cushing's syndrome. Note the effect of excessive cortisol in 'mooning' and reddening of face. (2) Same patient after restoration of normal cortisol levels by adrenal surgery.
d Addison's disease. Note the effect of cortisol deficiency in increased pigmentation of face (1) and hands (2).
e (1) Myxoedema. Note the effect of thyroxine deficiency on facial characteristics. (2) Same patient after treatment with thyroxine.

Photographs by courtesy of Professor J. S. Jenkins.

Fig. 9.34 Illustration of some endocrine diseases

Treatment of an endocrine disease is often straightforward: administration of the missing hormone in the case of a deficiency or removal of part of an endocrine gland in the case of excess. The relative ease with which accurate determinations of hormone concentration in plasma can be made (Fig. 9.25) is an important factor in the detection and treatment of endocrine disease.

9.6 Summary

Nature of endocrine function

The main function of the endocrine system is to **control metabolism** in the body. Control is achieved through the action of **hormones,** which are substances secreted by specific endocrine organs, in response to changes in the internal or external environment. The secretion of hormones is also regulated by **negative feedback;** that is, an increase in the concentration of a hormone in plasma results in a descreased secretion of that hormone.

Endocrine organs

1. The most important endocrine organ is the **pituitary** gland (or hypophysis). The gland is divided into three functionally distinct parts: an anterior lobe, a posterior lobe, and an intermediate part. The **anterior** lobe secretes **growth hormone** (GH, or somatotrophin), **thyroid stimulating hormone** (TSH),

adrenocorticotrophic hormone (ACTH), **luteinizing hormone** (LH), **folicle stimulating hormone** (FSH), and (in females) **prolactin.** The function of these hormones (known as **trophic** hormones) is to stimulate the growth and function of other endocrine organs (thyroid gland by TSH, adrenal cortex by ACTH, gonads by LH and FSH. GH stimulates the growth of cells in general, especially those of bone and other types of connective tissue. The secretion of trophic hormones is controlled by 'releasing hormones' (and by 'release-inhibiting' hormones), secreted by the **hypothalamus.**

The **posterior** lobe secretes **antidiuretic hormone** (ADH or vasopressin) and **oxytocin.** These hormones are synthesized in the hypothalamus, and pass directly to the posterior lobe of the pituitary gland by **axonal transport.** ADH stimulates the reabsorption of water in the kidney; oxytocin stimulates the contraction of certain smooth muscles.

The intermediate part of the pituitary gland secretes melanophore stimulating hormone (MSH), which functions in the darkening skin.

2. The **adrenal** glands are each divided into two parts: a cortex which secretes **glucocorticoids** and **mineralocorticoids,** and a medulla which secretes **catecholamines.** Glucocorticoids (such as **cortisol**) stimulate the breakdown of protein into carbohydrate, and have an anti-inflammatory action; mineralocorticoids (such as **aldosterone**) stimulate the reabsorption of Na^+ in the kidney, and thus control the ionic composition of plasma. The catecholamines (such as **adrenalin,** also called epinephrine) stimulate the breakdown of glycogen and triglyceride in their respective stores and increase blood-flow; by this means extra glucose and fatty acids are made available (for ATP production) to muscle and other cells.

3. The male and female **gonads** secrete sex hormones, which stimulate primary and secondary sexual characteristics. **Testes** secrete **androgens** (such as testosteronc); **ovaries** secrete **oestrogens** (such as oestradiol and oestrone).

4. The **thyroid** gland secretes thyroid hormones (such as **thyroxine**) which stimulate the overall metabolic rate (that is, the oxidation of foodstuffs throughout the body). The thyroid gland also secretes **calcitonin,** which lowers the plasma concentration of Ca^{2+} (by inhibiting Ca^{2+} release from bone); a neighbouring gland (the parathyroid) secretes **parathormone,** which raises Ca^{2+} levels (by stimulating Ca^{2+} release from bone).

5. Hormones are also secreted by specific cells in organs that have a predominantly non-endocrine function. The **pancreas** secretes **insulin** and **glucagon;** insulin stimulates an uptake of glucose from the plasma (into muscle and other cells), while glucagon stimulates release of glucose into the blood-stream

(from liver glycogen). The **kidney** secretes **renin,** which through another hormone stimulates the secretion of aldosterone from the adrenal cortex. The **gastrointestinal tract** secretes a number of hormones that stimulate secretion of hydrochloric acid and digestive enzymes.

Types of hormones

1. GH, TSH, ACTH, LH, FSH, prolactin, insulin, glucagon, calcitonin, parathormone, renin, MSH, ADH, and oxytocin are **polypeptides.** Catecholamines and thyroid hormones are **amino acid derivatives.** All these hormones are stored in the cells in which they are synthesized in the form of small vesicles; release of hormones, by stimulation of the cells in which they are stored, is by **exocytosis** of the vesicular contents.

2. **Prostaglandins,** which are a class of 'local' hormones not produced by specific cells, are derivatives of **polyunsaturated fatty acids;** they control smooth muscle contraction. Glucocorticoids, mineralocorticoids, androgens, and oestrogens are steroids (derived from cholesterol).

Mechanism of action

Hormones act either at the **cell surface** (most polypeptide hormones and catecholamines) or in the **interior** of cells (most steroid hormones). Action at the cell surface is through an activation of **adenyl cyclase** (leading to an increase in intracellular cyclic AMP), through increased Ca^{2+} concentration, or through increased **nutrient** (glucose and amino acids) entry. Action within cells is on **transcription** of specific genes (within the nucleus) or on **translation** of mRNA in the cytoplasm.

Disease

A decrease or increase in the synthesis of a particular hormone leads to an imbalance of metabolism (for example diabetes mellitus, Addison's disease, myxoedema, Cushing's syndrome, or thyrotoxicosis). An increase is often due to an endocrine tumour.

FURTHER READING

Reviews:
G.Litwack (ed.) In *Biochemical actions of hormones* (Vol. 1, 1970; Vol. 2, 1972; Vol. 3, 1975; Vol. 4, 1977). Academic Press, New York.

General effects of hormones:
M.C.Perry (1974). The hormonal control of metabolism. In *Companion to biochemistry* (eds A.T.Bull, J.R.Lagnado,

J.O.Thomas, and K.F.Tipton), p. 587, Longman, Harlow.

M.Rodbell, M.Lin, and C.Londos (1977). Hormonal regulation of adenylate cyclase systems. In *Receptors and recognition* (eds P.Cuatrecasas and M.F.Greaves), Series A, Vol. 3. Chapman and Hall, London.

Specific effects of hormones

D.A.Hems (1977). Short-term hormonal control of liver glycogen metabolism. *Trends biochem. Sci.* **2,** 241.

R.H.C.Strang (1977). The pineal gland: an example of a biological rhythm. *Trends biochem. Sci.,* **2,** 135.

Polypeptide hormones:

S.Reichlin *et al.* (1976). Hypothalamic hormones. *Ann. Rev. Physiol.* **38.**

S.J.Chan and D.F.Steiner (1977). Preproinsulin, a new precursor in insulin biosynthesis. *Trends biochem. Sci.,* **2,** 254.

Steroid hormones:

J.Gorski and F.Gannon (1976). Current models of steroid hormone action: a critique. *Ann. Rev. Physiol.* **38.**

B.W.O'Malley and W.T.Schrader (1976). The receptors of steroid hormones. *Scient. Am.* **234,** 32.

D.Schulster, S.Burstein, and B.A.Cooke (1976). *Molecular endocrinology of the steroid hormones.* Wiley, London.

Endocrine diseases:

Chapters on diabetes mellitus and hypoglycaemia, diseases of the adrenal gland, disorders of iodine metabolism, and the pancreas, in R.H.S.Thompson and I.D.P.Wootton (eds) (1970). *Biochemical disorders in human disease* (3rd edition). J. and A. Churchill, London.

10
Musculoskeletal system

10.1 Introduction

Approximately 40 per cent of the body weight is skeletal muscle; 10 per cent is bone, with another 10 per cent in the form of cartilage, tendon, and skin. Fat depots and the internal organs such as liver, spleen, lung, heart, kidney, gastrointestinal tract, and endocrine organs make up the remainder. Of the internal organs, muscles are found associated chiefly with heart (cardiac muscle) and the gastrointestinal tract, bladder, and uterus (smooth muscle).

The musculoskeletal system may be divided into **muscle** and **connective tissue.** The structural difference between these is that muscles are cells, albeit of abnormal size and shape, whereas connective tissue is made up of extracellular material. Components of connective tissue predominate in tissues such as tendon, cartilage, and bone; they are present in considerable amounts in tissues such as blood-vessels, trachea, and skin; and they are present in relatively low amounts in organs such as kidney, spleen, or liver. No organ is without some connective tissue, since it is connective tissue that holds cells together.

The function of muscle and connective tissue is different also. Muscle is designed to undergo **movement,** connective tissue is designed to confer **rigidity.** The difference is reflected in the constituent molecules. Muscle contains the proteins actin and myosin; connective tissue contains the protein collagen, various proteoglycans, and the inorganic salt calcium phosphate, in varying amounts depending on the type of tissue.

Movement is not restricted to muscles. All cells are capable of a certain amount of movement, whether it be mitosis and cytokinesis, exocytosis and endocytosis, or flagellar movement and locomotion by spreading. It is therefore not surprising to find that actin and myosin (or molecules very similar to actin and myosin) are present in all cells. The difference between for example a liver cell and a muscle cell is that the liver cell contains actin and myosin in very minor amounts whereas in a muscle cell actin and myosin constitute over 50 per cent of the proteins. Moreover, in muscles, movement is concerted. In liver, one cell may be undergoing mitosis, another may be in the S phase of the cell cycle synthesizing DNA, and another may be in G_0 actively secreting protein or bile. As a result, significant movement on the macroscopic scale occurs only in groups of muscle cells. It is such movements that are important in terms of the energy required (Fig. 8.1) and the work achieved.

10.2 Muscle

10.2.1 Types

There are three types of muscle (Fig. 10.1). The most abundant is **skeletal muscle.** This is the type that is

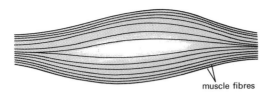

all muscles are composed of fibres. The structure of a fibre depends on the function of the muscle : there are three types

a   skeletal muscle fibre
(striated, voluntary)

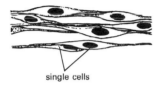

giant multinucleated cell

b   smooth muscle fibre
(non-striated, involuntary)

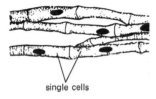

single cells

c   cardiac muscle fibre
(striated, involuntary)

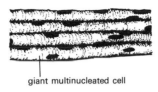

single cells

Fig. 10.1   Types of muscle

responsible for moving the bones that make up the skeleton; in other words for moving arms, legs, fingers, toes, trunk, head, and so forth. This movement is generally under conscious control, and skeletal muscle is therefore known also as voluntary muscle, to distinguish it from the other two types in which movement is involuntary, that is, not under conscious control. Another name for skeletal muscle is striated, since the appearance under the electron microscope is one of regular striations (Fig. 10.1).

The next most abundant type of muscle is **smooth muscle.** This is the type that is responsible for the peristaltic movement in the intestinal tract, for the release of water from the bladder, and for ejection of the foetus from the uterus. In each case the muscles are arranged in a ring around a cavity or tube (gastrointestinal tract, bladder, or uterus) in such a way that contractions lead to the propulsion of the contents along the tube. The name smooth muscle is based on the fact that there are no regular striations in the structure. Movement of smooth muscle is involuntary.

The third type of muscle is **cardiac muscle.** It is responsible for pumping the blood round the vascular system. Cardiac muscle is intermediate between smooth and skeletal muscle in that it is striated – though the pattern is not as regular as that of skeletal muscle – as well as involuntary (Fig. 10.1). Movement is continuous at around 80 constrictions or beats per minute, from a few months before birth until death.

The difference between the three types of muscle is not quite as clear-cut as implied in Fig. 10.1. Thus some types of muscle are striated and involuntary, while others are smooth and voluntary.

It will be noticed from Fig. 10.1 that, in the case of skeletal muscle, there is no plasma membrane separating the fibres. Each cell is several millimetres long. It may be wondered how such a cell is formed. Does it grow out from a smaller precursor cell, in the way that an axon grows? The answer is that it does not. Instead, skeletal muscle cells are formed by fusion of several thousand of the precursor myoblast cells. During development of muscle tissue, myoblasts line up in long and regular rows. They then undergo simultaneous cell–cell fusion to form a long, multinucleated myotube. Only after this has been formed do the actin and myosin fibres begin to take shape inside it. Once fusion has taken place, mitosis no longer occurs.

In the case of smooth muscle and cardiac muscle, cell fusion does not take place. The fibres run from cell to cell, each of which is mononucleated, and of normal size.

10.2.2 Function

All types of muscle have the following properties in common. Active movement is achieved by a shortening of the muscle spindle; after each contraction, the muscle returns to its original length by a relaxation process that is unable to generate force. In other words, muscles can pull, but cannot push. The ability to move a limb in one of two directions is achieved by pulling on one or other of two muscles that are attached to the joints at separate points (Fig. 10.2).

The contraction of muscle is brought about by a sliding past each other of two filaments that make up a muscle fibril. The molecular mechanism by which this is achieved is discussed in section 10.2.3.

The **trigger** for **muscle contraction,** whether voluntary or involuntary, is a **nerve impulse.** Nerves that stimulate smooth muscle and cardiac muscle are part of the autonomic nervous system. The contraction of smooth muscle is controlled both by stimulatory impulses and by inhibitory impulses. Contraction of cardiac muscle is stimulated not only by the autonomic nervous system but also by a centre known as the pacemaker, within the heart itself. During foetal development the heart begins to beat *before* the nerve

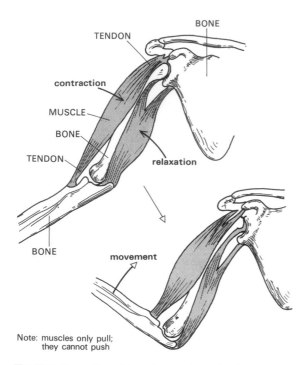

Note: muscles only pull;
they cannot push

Fig. 10.2   Generation of force by muscle contraction

supply is established; this is because the pacemaker is already functioning. A nerve impulse releases a neurotransmitter (section 12.3.2), that binds to a specific receptor at the muscle cell plasma membrane. As a result of binding, the permeability of the plasma membrane is momentarily increased, so that Na^+ enters and K^+ leaves. this depolarization is the stimulus for muscle contraction.

The **energy** for muscle contraction is supplied in the form of **ATP.** ATP is generated by the oxidation of glucose, fatty acid, or ketone bodies. In the case of skeletal and smooth muscle, a certain amount of ATP can be generated by anaerobic glycolysis also. The oxygen that diffuses into muscle cells is temporarily stored in the form of a haem-containing protein called myoglobin. The oxygen-binding characteristics of myoglobin are such (Fig. 7.19) as to favour a displacement of oxygen from arterial blood-supply into muscle cell.

The uptake of glucose is stimulated by insulin; inside the cell glucose is either stored as glycogen (during times of inactivity in the case of skeletal

muscle; smooth and cardiac muscle store little glycogen), or degraded to yield ATP. Fatty acids or ketone bodies are not stored, but are used as they enter cells.

The ATP that is formed in skeletal muscle is either used directly to drive contraction, or is temporarily stored in the form of phosphocreatine. **Phosphocreatine** is a 'high energy' compound (Figs 3.5 and 3.6; $\Delta G^{\ominus}$ of hydrolysis $\simeq -43$ kJ/mol). It is generated from ATP during periods of inactivity, and is converted back to ATP when muscular activity commences (Fig. 10.3). The total amount of creatine in muscle, and hence the amount of phosphocreatine that accumulates during inactivity, is insufficient to maintain contraction for any length of time. In other words, creatine phosphate ensures a rapid start to muscle contraction; it does not maintain it. During prolonged contractions ATP generated directly by metabolism is used instead.

An additional way of generating ATP is by the action of adenylate kinase (also known as myokinase). Adenylate kinase is an enzyme that catalyses the conversion of two molecules of ADP into ATP and AMP (Fig. 10.4). The equilibrium is slightly in favour of ADP formation; during muscle contraction, however, ATP is used faster than it is generated by metabolism, so that its concentration falls relative to that of ADP, and adenylate kinase operates to generate more ATP from ADP (Fig. 10.4). An additional factor helps to shift the equilibrium in favour of ATP formation. This is the presence of the enzyme adenylate deaminase (section 2.2.3), which hydrolyses the free amino group of AMP to yield IMP. AMP is eventually reformed from IMP; the reaction requires energy in the form of GTP (Fig. 2.43) and cannot therefore occur until more ATP is generated by metabolism. The reconversion of AMP back to ADP, by reversal of the adenylate kinase reaction, requires ATP also. In other words, like phosphocreatine, adenylate kinase serves to supply only a limited amount of ATP in the absence of metabolism, equivalent at most to half the cellular concentration of ADP.

10.2.3 Constituents

The generation of the contractile force of muscle depends on interactions between its constituent molecules. Actin, myosin, other proteins, ATP, and

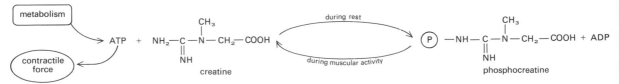

Fig. 10.3   Function of phosphocreatine

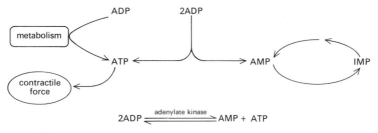

Fig. 10.4 Function of adenylate kinase

Ca^{2+} ions are the chief components involved. The mechanism to be described relates to skeletal muscle, which is the type that has been most extensively studied. The general principles apply also to smooth and cardiac muscle, though the details may differ. For example the myosin of smooth muscle is different from the myosin of skeletal muscle.

The myofibrils that constitute a muscle fibre are made up of two types of filament: a **thick filament** composed of **myosin** and a **thin filament** composed of **actin.** When viewed under the electron microscope, the filaments and the spaces between them appear as a series of light and dark bands. During contraction the thin filaments **slide** past the thick filaments (Fig. 10.5). Contractile movement does not therefore involve a shortening or lengthening of actual molecules. The molecules remain intact; what changes is the position of one molecule relative to the other. Of course the filaments are not single molecules; each filament is made up of many hundreds of actin, or myosin, molecules in polymerized form.

The relaxed state of myofibrils is the lengthened form, and this is maintained until an impulse is received. How is the impulse transmitted from the cell surface to the myofibril, and how is sliding of the filaments achieved? The mechanism is not completely understood, but in outline is as follows.

Sliding is caused by the making and breaking of cross-bridges between the myosin and actin filaments. ATP is involved in the process, during which it becomes hydrolysed to ADP and P_i (Fig. 10.5). Successive rounds of hydrolysis, removal of ADP, and rebinding of ATP, cause the filaments to slide further and further past each other. When one fibre has become fully contracted, it is pulled back to the extended, relaxed form by contraction of an adjacent fibre. In other words ATP is involved only in the contraction, not in the relaxation, of muscle.

The enzyme responsible for ATP hydrolysis during contraction is myosin itself. It requires Ca^{2+} for activity; as for other enzymes that react with ATP, magnesium is also required to keep ATP in the required configuration for binding to the active site of the enzyme. The shortening of muscle fibres can be demonstrated in the test-tube. The proteins of muscle are first extracted by solutions of high ionic strength. If such solutions are squirted into water in the presence of magnesium and ATP, actin–myosin filaments in the relaxed state are regenerated. Such filaments contract when Ca^{2+} is added.

The Ca^{2+} binds not to myosin or actin itself but to a protein called troponin that is part of the actin-containing filament. Troponin has an extremely high affinity for Ca^{2+} as a result of chelation by certain amino acids that form a lattice around the Ca^{2+} atom. Exactly how the binding of Ca^{2+} to troponin causes the myosin ATPase to act is not clear. Yet another molecule called tropomyosin is required and it is an effect of the Ca^{2+}–troponin–actin–tropomyosin–myosin–ATP complex that causes rupture of the cross-bridges and hydrolysis of ATP. In smooth muscle, it is likely that Ca^{2+} binds directly to an actin–myosin–ATP complex, since troponin and tropomyosin, (which constitute about 1 per cent of the fibrillar proteins of skeletal muscle), are present in much smaller amounts. One of the differences between skeletal and smooth muscle is in their rates of contraction, and the rapidity of contraction of skeletal muscle is probably due to the participation of troponin and tropomyosin.

Ca^{2+}, then, is the **internal trigger** for muscle contraction (Fig. 10.5). It is stored in the endoplasmic reticulum (called sarcoplasmic reticulum) of muscle cells. The sarcoplasmic reticulum forms a continuous network across the muscle cell, from plasma membrane to actin–myosin filaments. When muscle cell plasma membrane is depolarized, the entry of Na^+ somehow causes the release of Ca^{2+} at the site of the myofilaments. In order to prepare for a second round of contraction, Ca^{2+} ions have to be removed from the troponin complex at the site of the cross bridges. This is achieved by a specific Ca^{2+}–ATPase (section 6.3.3.2) which pumps Ca^{2+} back into the sarcoplasmic reticulum. In the case of smooth and cardiac muscle cells, the Ca^{2+} is stored not within the cell, but enters from the extracellular fluid. That is, stimulation of the cell leads directly to an influx of Ca^{2+}; following contraction of the muscle fibril, a plasma membrane Ca^{2+}–ATPase pumps the Ca^{2+} out again.

The energy required to achieve muscle contraction is therefore supplied by the hydrolysis of ATP at two distinct sites:

i. at the cross bridges holding the actin and myosin filaments apart and

ii. in the membrane of the sarcoplasmic reticulum.

Since the products of ATP hydrolysis, namely ADP and inorganic phosphate, need to be recondensed rapidly to regenerate fresh ATP, the mitochondria, in which this reaction occurs, are situated at regular intervals close to the actin–myosin filaments. The muscle cell is thus one of the most highly organized of cells. It provides an excellent example of the close relationship between biological structure and function.

a *sliding filaments*

1 thin section electron micrograph of skeletal muscle fibre undergoing contraction (courtesy of Dr. H.E. Huxley)

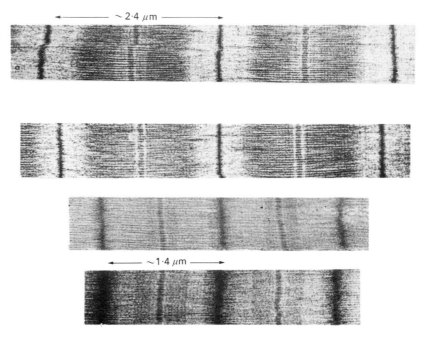

2 structural interpretation: actin filaments slide past myosin filaments

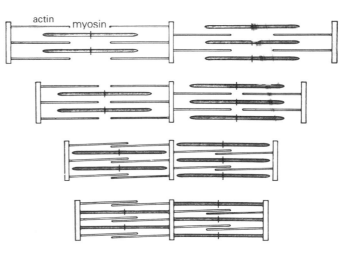

Fig. 10.5 Molecular basis of muscle contraction

3 molecular basis: sliding
  is achieved by successive
  making and breaking
  of 'cross-bridges' between
  actin and myosin

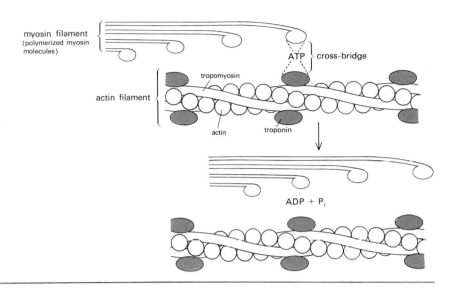

b *role of Ca^{2+}*

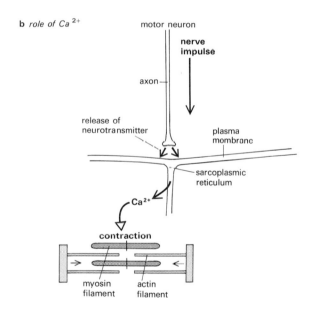

10.2.4 Disease

Some diseases that affect muscle are listed in Fig. 10.6. Metabolic diseases, such as diabetes mellitus or the glycogen storage diseases are the result of a failure in the supply of glucose to muscle cells. In diabetes, glucose uptake from the blood-stream is impaired, in glycogen disorders, glucose production from glycogen is impaired. The muscle fibres are not primarily affected in either case, but are unable to function properly because of lack of ATP. In other diseases, such as muscular dystrophy or myesthenia gravis, the muscle fibre itself is affected. In the first, the whole muscle fibre is gradually destroyed; leakage of creatine kinase into the blood-stream is a useful indicator of muscle degradation. In the second disease, which is probably an autoimmune disease, the receptor for the nerve impulse (acetyl choline) is affected. Inflammatory diseases such as rheumatoid arthritis (section 10.3.3) also affect muscle fibres.

Failure of cardiac muscle is obviously more dangerous than failure of skeletal muscle. Lack of oxygen, due to a blockage of the blood-supply, as in coronary thrombosis, (section 7.5), is one of the most common causes of death. Inflammation of heart muscle (myocarditis) and non-specific degeneration of heart muscle (cardiomyopathy) are less common diseases of the heart.

Fig. 10.6   Some diseases of muscle

Muscle type	Cause	Outcome	Clinical description
Skeletal muscle	Decreased utilization of nutrients	Muscle weakness	Diabetes mellitus
	Genetic defect of certain enzymes	Muscle weakness	Glycogen storage diseases (Fig. 2.57) (rare)
	Unknown	Degeneration of muscle fibres	Muscular dystrophy
	Auto-immunity (Fig. 11.16)	Degeneration of acetyl choline receptors, leading to defective muscle contraction.	Myasthenia gravis
Heart muscle	Obstruction of blood-supply	Myocardial infarction	Coronary artery thrombosis
	Inflammation of heart muscle due to infection	Heart failure	Myocarditis
	Degeneration of heart muscle due (?) to auto-immunity	Heart failure	Cardiomyopathy

10.3 Connective tissue

10.3.1 Types

Tissues in which extracellular materials play an important role are bone, cartilage, tendon, and skin, as well as the walls of the blood-vessels, of trachea, and of intestinal tract. In each case collagen (section 10.3.2.1) is a key constituent; because of its high content these tissues are tougher than organs such as liver, pancreas, spleen, or kidney.

10.3.1.1 Bone

Bone is the chief constituent of the skeleton. Approximately 10 per cent of the body is skeleton; the musculoskeletal system as a whole makes up some 60 per cent of the body weight. In addition to being tough, bone is also hard. The hardness of bone is the basis of its two functions: to protect softer tissues and to form inelastic structures that can be used for movement. The skull and the rib cage are examples of protective bones, preventing external damage to brain and to heart and lungs; the bones of the limbs are examples of bones used for movement.

The **hardness** of bone is due to the presence of solid crystals of hydrated calcium phosphate known as **hydroxyapatite**, $3Ca_3(PO_4)_2.Ca(OH)_2$. Approximately 80 per cent of bone is inorganic matter. The remainder is protein and some proteoglycan (see Fig. 10.10). Most of the protein in bone is collagen. The fibres of collagen form a regular network into which calcium phosphate becomes deposited during the formation of bone. Before calcium phosphate is deposited, the tissue is cartilaginous. In fact cartilage, which is made up largely of collagen and proteoglycan is the precursor of bone; in young children the bones contain relatively more collagen and less calcium phosphate, and as a result are softer and less brittle than adult bones.

Although bones are made up almost entirely of extracellular material some cells are nevertheless present, since the deposition of calcium phosphate is a cellular event. The cells, which are in contact with the blood-stream, are situated largely on the inside of bone (Fig. 10.7). Bone is metabolically not as static a structure as might be thought, and a certain amount of degradation accompanies synthesis. In other words, bone is turned over. The cells that degrade bone are different from the cells that lay down bone. Degradative cells are known as osteoclasts; synthetic cells as

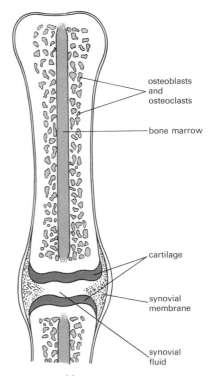

osteoblasts and osteoclasts

bone marrow

cartilage

synovial membrane

synovial fluid

Fig. 10.7   Structure of bone

osteoblasts. Together the two cells cause bones to be remoulded throughout life. Remoulding by turnover is, of course, restricted to those areas of bone that are in contact with cells and the vascular system.

The turnover of bone is in a sense secondary to the maintainance of plasma Ca^{2+} levels. That is, it is a kind of store for plasma Ca^{2+}. If Ca^{2+} levels fall as a result of a nutritional deficiency, for example, degradation of bone exceeds synthesis. If this continues, bones become soft and liable to fracture.

The cavities of bones, through which the blood flows, are filled with a mass of cells known as bone marrow (Fig. 10.7). The cells are the precursors of the blood cells. It is in the **bone marrow** that red cells, white cells and platelets are formed. The process is known as **haemopoiesis** (Fig. 10.8). In order to maintain the extensive supply of these cells, all of which have a relatively short life span, the precursor or stem cells are in a rapid state of cell division. Thus for every cell that matures into a blood cell, at least one stem cell has to be produced in order to maintain the stem cell population. Whether red cells and white cells in fact originate from the same stem cell is not at present clear.

The production of blood cells should not be regarded as a function that is specifically related to the structure of bone. It just happens that bones form cavities suitable for this purpose. Prior to the formation of bones in the foetus, for example, the production of blood cells takes place in the liver and other organs. Since haemopoiesis does, however, take place

in the bone marrow from birth onwards, defects that occur in bone can influence the formation of blood. An example is the effect of radioactive fall-out following the explosion of an atomic bomb. Fall-out contains radioactive strontium, ^{90}Sr. Sr is an analogue of Ca and is absorbed and deposited like Ca^{2+}; it therefore becomes concentrated in bone. The radioactive emission from ^{90}Sr has little effect on the structure of bone. It is, however, highly toxic to the adjacent bone marrow cells. Toxicity is due to the fact that radioactive radiations are mutagenic, and that mutations are generally deleterious rather than beneficial. In haemopoietic cells, mutations can lead to the emergence of leukaemic cells (section 7.4), and a high incidence of leukaemia in survivors of atomic bombs has been related to their ingestion of ^{90}Sr.

10.3.1.2 Cartilage, tendon, and skin

Cartilage occurs in two forms: at the ends of bones, where it forms a pad of softer tissue, and as a separate structure such as the septum of the nose or the wall of the trachea. As mentioned above, cartilage is a precursor of bone: in the foetus the skeleton is largely cartilaginous. The function of cartilage is to form a tissue that is tough and at the same time slightly elastic. The constituents that contribute to this function are collagen and proteoglycans.

Tendon is composed almost exclusively of collagen. The function of tendon is to join skeletal muscle to bone, that is, to transmit the force of muscle contraction to the movement in bones (Fig. 10.2).

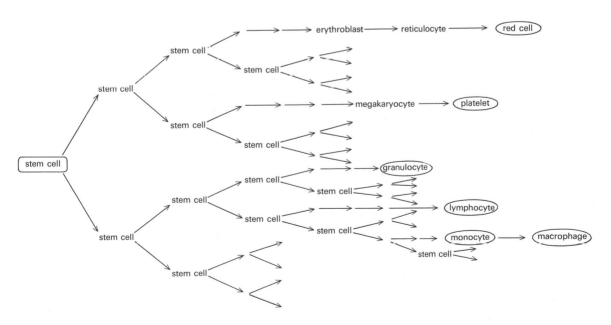

Fig. 10.8   Function of bone marrow: haemopoiesis

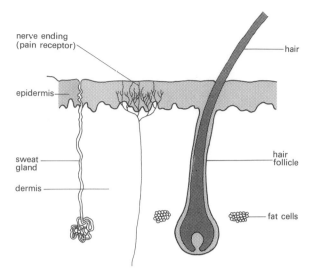

Fig. 10.9 Structure and function of skin

Skin consists of an outer layer, the epidermis, and an inner layer, the dermis. The epidermis is largely cellular, though most of the cells are keratinized and dead; the dermis is largely extracellular material containing much collagen and dermatan sulphate. Skin also contains a blood-supply, lymphatic drainage, nervous supply, sweat glands, fat cells, and other cells, some of which are illustrated.

Skin is a much more complex tissue. It contains glands that secrete sweat and cells that produce hair and nails, and is much more highly vascularized than cartilage, tendon, or bone (Fig. 10.9). It does, however, contain much extracellular material in the form of collagen, elastin, proteoglycans and keratin, and may therefore be regarded as connective tissue. Its main function is to protect the body against external damage and invasion by infectious microbes. The cells at the outermost layer, known as the epidermis, are dead in the sense that they are no longer able to divide. They are continuously being sloughed off and replaced by new cells rather like the epithelial cells of the gastrointestinal tract.

The cells that form cartilage, tendons, and skin, namely chondroblasts and fibroblasts, are a major site of action of growth hormone (somatotrophin).

10.3.1.3 Other types

The walls of the blood-vessels are, like the walls of trachea, rich in connective tissue. Collagen, elastin, and proteoglycans are the main constituents. The layer underlying the epithelial cells of the intestinal tract and kidney tubule also contain much collagen and proteoglycans; this part of the tissue is known as basement membrane.

An entirely fluid type of connective tissue, known as synovial fluid, is found in the spaces between the ends of two adjacent bones. Its function is to lubricate the bones as they move against each other, to prevent erosion. A major constituent of synovial fluid is the proteoglycan hyaluronic acid.

10.3.2 Constituents

10.3.2.1 Collagen

The major constituent of connective tissue is **collagen,** which makes up about a third of all body proteins (Fig. 10.10). Collagen is a protein with a particularly high content of proline and glycine; some of the proline residues, as well as some of the lysine residues, are hydroxylated and subsequently become glycosylated with galactose and glucose residues. Because of the high content of proline, collagen does not form regions of α-helix (Fig. 1.6); instead it forms β turns through the glycine residues; approximately every third residue is glycine. The tertiary and quaternary structure of collagen is unique also. Collagen has three subunits that are wound round each other in the form of a triple helix that forms long continuous fibres (Fig. 10.11).

The composition of collagen varies from tissue to tissue. The collagen of cartilage is not the same as that of tendon, which is again different from the collagen of basement membrane. At least four different types of collagen (types I, II, III, and IV) exist, each coded for by different genes (Fig. 10.11). Moreover, the three subunits are not necessarily identical. In tendon and bone, for example, two of the subunits are the same, the third one different. Again each subunit is specified by a separate gene. Note that tissues such as skin contain more than one type of collagen. The situation is therefore analogous to the existence of cell-specific isoenzymes such as LDH (Fig. 2.10).

Unlike the formation of actin and myosin filaments within muscle cells, the assembly of collagen fibres takes place outside cells. The cells that secrete collagen – chondroblasts in the case of cartilage, and

Fig. 10.10 Constituents of connective tissue (approximate percentage of dry weight)

Type of tissue	Ca phosphate	Collagen	Elastin	Proteoglycan
Bone	80	18	—	1
Cartilage	—	50	—	50 (chondroitin sulphates)
Tendon	—	90	10	1
Skin	—	45	5	45 (dermatan sulphate)

The figures are very approximate values derived from a number of sources; there is much variation between different sorts of cartilage, tendon, or skin. Note that the water content of bone is exceptionally low (10–15 per cent), compared with that of other tissues (70–80 per cent).

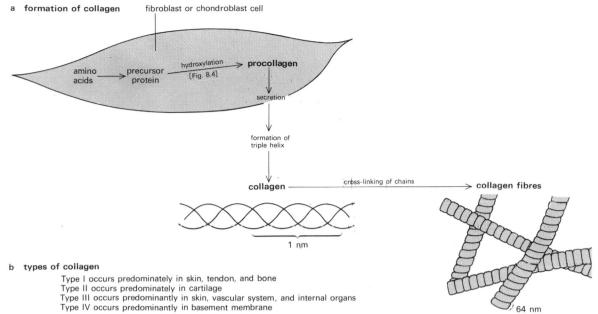

a   formation of collagen

fibroblast or chondroblast cell

amino acids → precursor protein → hydroxylation [Fig. 8.4] → **procollagen**

secretion

formation of triple helix

collagen ——— cross-linking of chains ——→ collagen fibres

1 nm

b   types of collagen

Type I occurs predominately in skin, tendon, and bone
Type II occurs predominately in cartilage
Type III occurs predominantly in skin, vascular system, and internal organs
Type IV occurs predominantly in basement membrane

64 nm

Fig. 10.11   Formation and structure of collagen

fibroblasts in the case of skin and other tissues – synthesize a soluble precursor of collagen. This is extruded by a mechanism that is not yet clear. As more and more collagen is produced it is modified by covalent cross-linking to form the fibrous, insoluble strands typical of connective tissue (Fig. 10.11).

The function of collagen is to provide **tensile strength.** It is easy to see how a network of fibres of the type seen in Fig. 10.11 can achieve this. Collagen also plays a role in wound healing. The function of collagen, secreted by neighbouring fibroblasts, is to 'plug' any hole in blood-vessels or skin by interacting with the blood clot that forms in the first instance (section 7.5.2). The blood clot gradually retracts, but the meshwork of collagen fibres, to which proteoglycan is added, remains to keep the wound sealed until a new layer of cells has grown over the site of injury.

Another function of collagen is in the formation of bone. Calcium phosphate is deposited within the framework of a matrix formed by the fibres of collagen. During bone formation, the predominant type of collagen changes from type II to type I.

A protein that has properties similar to collagen, and that is generally found associated with it, is elastin. Like collagen, it is rich in glycine and proline. Unlike collagen its proline and lysine residues are not hydroxylated.

10.3.2.2 Proteoglycans

The predominant types of proteoglycan in connective tissue are hyaluronic acid, chondroitin sulphate and dermatan sulphate. Hyaluronic acid is strictly speaking not a true proteoglycan since it is only loosely linked to protein. Chondroitin sulphate and dermatan sulphate on the other hand are covalently linked to protein.

Hyaluronic acid is made up of a repeating unit of N-acetyl-D glucosamine linked to D-glucuronic acid (Fig. 10.12). The molecule binds much water and forms a rather large and asymmetric structure. The resulting compound is jelly-like in its properties, and has a high viscosity. This makes it ideally suitable for lubricating the surfaces of bones, and hyaluronic acid is a major constituent of synovial fluid (Fig. 10.7). Hyaluronic acid is found in high concentration also in the jelly-like layer that surrounds an ovum. Hyaluronic acid occurs not only in all types of connective tissue, but also in the cementing or 'ground' substance that holds cells together.

The carbohydrate part of **chondroitin sulphate** and **dermatan sulphate** is made up of a repeating unit of N-acetyl-D-galactosamine linked to a uronic acid. In chondroitin sulphate the uronic acid is D-glucuronic acid; in dermatan sulphate it is L-iduronic acid. The linkage region between protein and carbohydrate consists of a serine residue joined through its hydroxyl group to xylose (Fig. 2.58). Both proteoglycans contain sulphate groups, linked by ester bonds to the carbohydrate portion. The position of the sulphate groups varies (Fig. 10.12). Chondroitin sulphate is, like hyaluronic acid, a constituent of all types of connective tissue as well as of the ground substance that holds cells together. Chondroitin sulphate is

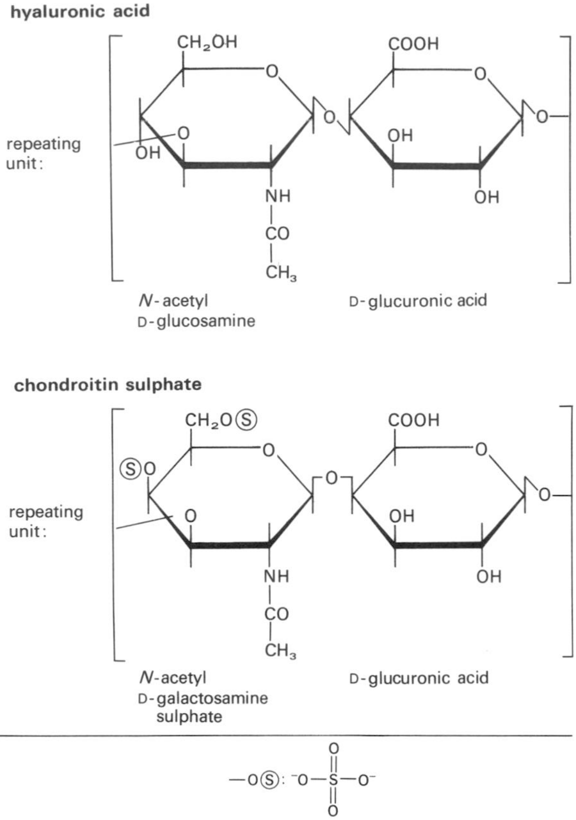

hyaluronic acid

repeating unit:

N-acetyl
D-glucosamine   D-glucuronic acid

chondroitin sulphate

repeating unit:

N-acetyl
D-galactosamine
sulphate   D-glucuronic acid

Fig. 10.12 Structure of proteoglycans

In chondroitin 4-sulphate, and in dermatan sulphate, the sulphate group is on the 4-position; in chondroitin 6-sulphate it is on the 6-position. In dermatan sulphate, D-glucuronic acid is replaced by L-iduronic acid.

There are ~2500 repeating units in each carbohydrate chain of hyaluronic acid, ~40 repeating units in each carbohydrate chain of chondroitin sulphate and ~20 repeating units in each carbohydrate chain of dermatan sulphate. Note that in addition each proteoglycan has some other sugars in the 'linkage region' between the repeating units and the protein 'core' (Fig. 2.58).

found chiefly in cartilage; dermatan sulphate (also called chondroitin sulphate B) in skin (Fig. 10.10). The ratio of chondroitin-4 sulphate to -6 sulphate increases with age.

One of the main functions of chondroitin sulphate appears to be to interact with collagen. The molecule is laid down in a regular pattern in sequence with the banding of the collagen fibre. The sulphate groups bind Ca^{2+}, so that the result is a type of organic Ca^{2+} matrix, which may be contrasted with the inorganic Ca^{2+} matrix of bone.

10.3.3 Disease

Diseases of connective tissue, unlike those of muscle (Fig. 10.6), are structural in the sense that cells are not involved; diseases of the bone marrow are, of course, an exception (Fig. 10.13). Connective tissue diseases may be further divided into nutritional and non-nutritional. The most important nutritional diseases

are rickets and scurvy. Rickets, a disease in which the bones are soft and more easily fractured, is due to deficiency of Ca^{2+}, which is itself usually a consequence of a deficiency in vitamin D. Scurvy affects bone as well as other connective tissues. The main symptoms are pain in the joints and a tendency for haemorrhage to occur in mucous membranes, skin, and muscles. Wound healing is impaired. The defect is a lack of mature collagen. The contributing factor is a lack of ascorbic acid, vitamin C, with a resulting failure to hydroxylate the proline and lysine residues of immature collagen (Fig. 8.4).

Hormonal disease in which the balance of circulating Ca^{2+} is upset, also affects the structure of bone. In hyperparathyroidism for example, excessive loss of Ca^{2+} occurs. The resulting decalcification leads to cystic areas in bone. The excess Ca^{2+} in the blood-stream may lead to the formation of renal stones.

An important structural disease of connective tissue is rheumatoid arthritis. This affects most of the types of

Fig. 10.13   Some diseases of connective tissue

Cause	Outcome	Clinical description
Over-production of immature white cells by bone marrow	Anaemia; haemorrhage; infection	Leukaemia
Over-production of red cells by bone marrow	Thrombosis	Polycythemia
Vitamin D deficiency	Failure of bone mineralization	Rickets; osteomalacia
Vitamin C deficiency	Failure of collagen formation	Scurvy
Over-production of parathyroid hormone	Demineralization of bone	Hyperparathyroidism
Auto-immunity (see Fig. 11.16)	Degeneration of collagen	Rheumatoid arthritis.

connective tissue, especially cartilage, tendons, and bone. The disease, which may result from autoimmunity or from some other cause, is due to the destruction of collagen (and possibly hyaluronic acid also). Excessive degradation may be caused by an increase in the activity of collagenase and hyaluronidase, as well as by immune reaction against collagen. Exactly which is the precipitating cause is not at present clear. Corticosteroid analogues are used in therapy, and so is aspirin, but as yet no really effective cure exists.

10.4 Summary

Muscle

1. The function of muscle cells is to cause **movement. Skeletal** muscles move parts of the skeleton; **smooth** muscles cause contraction of internal organs such as the gastrointestinal tract, bladder, and uterus; **cardiac** (heart) muscle causes the beating of the heart and hence the flow of blood through the vascular system.

2. Movement is obtained through the **contraction** of **fibres.** The fibres consist of two types of filament: a thick filament composed of the protein **myosin** and a thin filament composed of the protein **actin.** Contraction is achieved by a **sliding** past each other of the two types of filament. Movement is coupled to the hydrolysis of **ATP**; the trigger for hydrolysis is an increase in cytoplasmic concentration of **Ca^{2+}** (itself brought about by nervous stimulation).

3. If the structure of muscle fibres is faulty, as in muscular dystrophy, movement is impaired. Faulty metabolism within muscle cells, as in diabetes mellitus and other metabolic disorders, likewise leads to malfunctioning of muscle fibres. If heart muscle is prevented from functioning (for example by failure of the oxygen supply through a blood clot) the consequence (coronary thrombosis) is often fatal.

Connective tissue

1. The function of connective tissue is to impart **tensile strength** to certain parts of the body such as **bone, cartilage, tendon,** and **skin.** Most of the components of connective tissue lie outside cells.

2. The main constituent of bone is **hydroxyapatite** ($3Ca_3(PO_4)_2.Ca(OH)_2$). The main constituent of cartilage, tendon, and skin is collagen. **Collagen** is a protein that is particularly rich in the amino acids glycine, proline, and hydroxyproline; collagen exists in the form of three polypeptide chains wound around each other in a **triple helix.** Other constituents of cartilage, tendon, and skin are another protein, **elastin,** and the proteoglycans **hyaluronic acid, chondroitin sulphate** and **dermatan sulphate.**

3. The spaces inside bone that are in contact with the vascular system are filled with a mixture of cells known as **bone marrow.** These cells, which divide rapidly, are the precursors of the **blood cells** (red cells, white cells, and platelets).

4. A decrease in the deposition of hydroxyapatite in bone, caused by nutritional deficiency of Ca^{2+} or by hormonal imbalance of Ca^{2+} metabolism, leads to a weakening of bone. Faulty deposition of collagen, as occurs in deficiency of vitamin C (scurvy) leads to defects in all types of connective tissue. Inflammations, as in rheumatoid arthritis, likewise affect most sorts of connective tissue.

FURTHER READING

Muscle contraction

A.J.Buller (1975). *Contractile behaviour of muscle* (Oxford Biology Reader no. 36). Oxford University Press.
G.Offer (1974). The molecular basis of muscular contraction. In: *Companion to biochemistry* (eds A.T.Bull, J.R.Lagnado, J.O.Thomas, and K.F.Tipton), p. 673. Longman, Harlow.

Bone

J.J.Pritchard (1974). *Bones* (Oxford Biology Reader no. 47). Oxford University Press.

Bone marrow

J.Breton-Gorins and F.Reyes (1976). Ultrastructure of human bone marrow cell maturation. *Int. Rev. Cytol.* **46,** 252.

Chemistry of proteoglycans

C.F.Phelps (1972). *Polysaccharides* (Oxford Biology Reader no. 27). Oxford University Press.

Diseases of the musculo-skeletal system

Chapters on Diseases of muscle, Connective tissue disorders, Disorders of bone and calcium metabolism. In R.H.S.Thompson and I.D.P.Wootton (eds) (1970). *Biochemical disorders in human disease* (3rd edition). J. and A.Churchill, London.

11

The immune system

11.1 Nature of the immune response

The human body is protected against infectious agents in three ways. First, those parts of the body that are in contact with the environment, namely skin and the respiratory and alimentary tracts, are covered by a layer of cells, one function of which is to exclude invasive organisms. This is achieved by the secretion of keratin and fat (skin) or mucus (respiratory and alimentary tract) onto the exposed surface. In addition, secretion of lysozyme (in sweat), hydrochloric acid (in stomach), and other bactericidal agents helps to prevent infection.

Should an agent succeed in penetrating this first barrier against infection, it comes up against a second line of defence within the body. This consists of the ability of macrophages and other phagocytic cells to ingest and degrade foreign agents. Bacterial toxins stimulate the complement system, and this helps to eliminate invasive microbes (section 11.2.1.3). Cell that have become infected with a virus begin to secrete interferon, a substance that prevents the transmission of virus to uninfected cells by protecting them against viral attack.

Should this second barrier also fail to eliminate an infectious agent, a third line of defence, namely that involving lymphocytes, comes into play. The first two barriers are sometimes referred to as 'innate' or 'non-specific' immunity; the third barrier is referred to as 'acquired' or 'specific' immunity. This chapter deals predominantly with the third line of defence, that of acquired immunity.

11.1.1 Protection

The main function of the **immune system,** then, is to **protect** the body against **infection.** The cells that participate in the process are the white blood cells: lymphocytes, monocytes, granulocytes and macrophages. It has been known for well over 100 years that a person who has suffered from an infectious disease is less likely to succumb to a second bout of infection than one who has not previously been exposed to the disease: exposure renders a person immune or 'safe'. It is now clear that exposure does not need to elicit the actual disease. The infective agent – be it virus, bacterium, or other organism – can be administered in such small doses, or in a heat-killed or 'attenuated' form, that only the mildest symptoms of the disease ensue. The route of administration can be through the surface of the skin, as in vaccination against smallpox; by intramuscular injection, as in immunization against diphtheria, typhoid, or measles; or orally, as in immunization against poliomyelitis. The reason why oral vaccines against poliomyelitis are effective is because the virus appears to be sufficiently resistant to digestion in the alimentary tract, and is therefore able to penetrate to the blood-stream.

The mechanism by which protection is achieved is as follows. The infectious agent, which is known as an antigen, elicits the build-up of a specific population of lymphocytes. Not only infectious agents, but any cell or macromolecule that is 'foreign' to the host acts as an antigen. If the antigen is reintroduced into the circulation, the **lymphocytes,** some of which are very long-lived, are recalled or restimulated to neutralize and eliminate the antigen from the body (Fig. 11.1). Neutralization involves one of two systems: the humoral system or the cellular system.

The humoral system is responsible for immunity against most types of bacterial disease. Administration of an antigen causes the lymphocytes to secrete a specific protein, known as an antibody, into the blood-stream. Antibody reacts with antigen to form an antibody–antigen complex. The complex is much more sensitive to destruction by phagocytic cells (neut-rophils, monocytes, and macrophages) than is antigen alone. The structure and biosynthesis of antibodies and the formation and destruction of antibody–antigen complexes is described in section 11.2.1.

The cellular system is responsible for immunity against many viral infections. It is also responsible for the rejection of foreign cells if these are introduced into the body in the form of a skin graft or organ transplant. The physiological advantage of possessing this type of rejection mechanism is not obvious. It may be related to the fact that the system probably serves also to reject potentially malignant cells that arise within the body. The cellular response system is described in section 11.3.1.

11.1.2 Sensitivity

Many people are particularly sensitive to the presence of foreign compounds in the blood-stream. The inhalation of dust or pollen, the absorption of certain foods

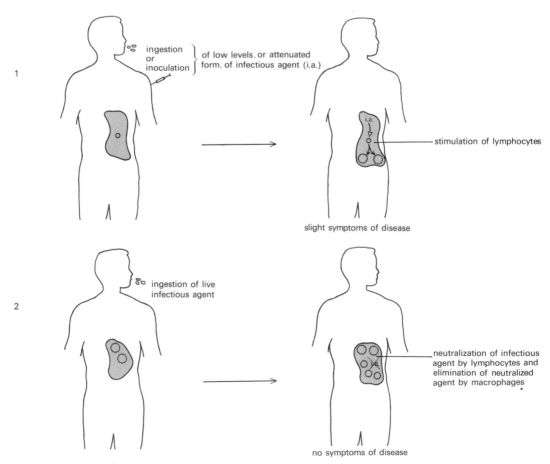

Fig. 11.1   Role of the immune system in protection

or drugs such as penicillin, or the effects of wasp and bee stings, lead to a response varying from a slight rash to asthma, anaphylactic shock, and death. The underlying cause behind such **hypersensitivity** or **allergy** is the same: an immune response leading to the release of histamine and other pharmacological mediators of anaphylaxis.

Histamine, which is released from tissue mast cells and from basophilic granulocytes (Fig. 7.29), causes a dilation of capillary blood-vessels and a contraction of smooth muscle. More seriously, it also constricts the air passages in the lungs, which restricts oxygen uptake and results in heart failure (systemic anaphylaxis) (Fig. 11.2).

Surprisingly, the effects of exposure to an allergic stimulus become progressively more pronounced at each encounter. That is, instead of becoming immune to the foreign agent, the body becomes increasingly more sensitive. In other words the effects of the aller-

gic stimulus are potentiated, not neutralized, by the immune system.

The immune system is involved in allergy, as in protection, in one of two ways: through circulating antibodies (humoral system) and through circulating lymphocytes (cellular system). The two responses are described in sections 11.2.2 and 11.3.2 respectively.

11.2 Humoral system

11.2.1 Protection

11.2.1.1 Structure and function of antibodies

Antibodies are proteins that are present in the blood-stream. Most of the γ-globulin fraction of serum consists of antibodies; an alternative name for γ-globulins is **immunoglobulins** (Ig). There are at least

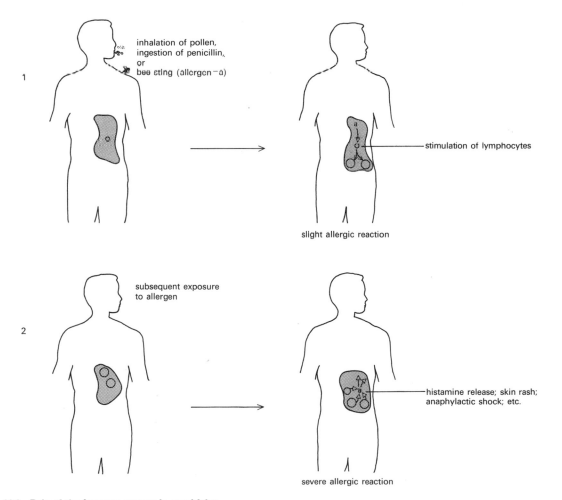

1

inhalation of pollen,
ingestion of penicillin,
or
bee sting (allergen – a)

stimulation of lymphocytes

slight allergic reaction

2

subsequent exposure
to allergen

histamine release; skin rash;
anaphylactic shock; etc.

severe allergic reaction

Fig. 11.2   Role of the immune system in sensitivity

Fig. 11.3   Types of Immunoglobulin

Name	Sub-units	Mol. weight	% Carbohydrate	Conc. in plasma	Function
IgG	$\gamma_2\kappa_2$ or $\gamma_2\lambda_2$	~150 000	3	8–16 g/l	main immunoglobulin; responsible for humoral immunity
IgA	$(\alpha_2\kappa_2)_{1-3}$ or $(\alpha_2\lambda_2)_{1-3}$	~$(160\,000)_{1-3}$	8	1.4–4 g/l	present in secretions; 'defends' body surfaces
IgM	$(\mu_2\kappa_2)_5$ or $(\mu_2\lambda_2)_5$	~900 000	12	0.5–2 g/l	B lymphocyte receptor; establishment of humoral immunity
IgD	$\delta_2\kappa_2$ or $\delta_2\lambda_2$	~185 000	13	<0.5 g/l	lymphocyte receptor;?
IgE	$\varepsilon_2\kappa_2$ or $\varepsilon_2\lambda_2$	~200 000	12	<0.001 g/l	histamine release; humoral sensitivity. Mainly bound to basophils and mast cells

Adapted from I. Roitt (1977). *Essential Immunology* (3rd edition). Blackwell Scientific Publications, Oxford, with permission.

five different types of immunoglobulin: IgG, IgM, IgA, IgE, and IgD, (Fig. 11.3).

The basic structure of all antibodies is the same: a combination of light (short) and heavy (long) polypeptide chains, joined by disulphide bonds. Antibodies are very heterogenous molecules. A person is capable of producing some 10^6 different immunoglobulin molecules in his blood-stream. The differences between molecules are slight and involve no more than a few amino acids. The amino acids that differ are all situated within one region of the antibody molecule, known as the variable region. It is at this site that combination with antigen takes place. The rest of the molecule is known as the constant region. Each molecule contains two identical light chains and two identical heavy chains (Fig. 11.4).

An individual synthesizes two types of immunoglobulin, that differ in the constant region of the light chain. Approximately half the immunoglobulin molecules contain κ (kappa) light chains; the other half contain λ (lambda) light chains. There are also variations in the heavy chains among the different types of immunoglobulin: IgG contains γ-chains, IgM contains μ-chains, and so forth (Fig. 11.3). As a result, the different types of immunoglobulin have different molecular weights. Another difference is in the number of subunits that make up the immunoglobulin molecule: IgM is made up of five identical units of the type depicted in Fig. 11.4; IgA is made up of one to three such units. All immunoglobulins have some carbohydrate attached to the constant part of the heavy chains; the amount varies among the different types of immunoglobulin.

The function of antibodies is to bind antigens. **Antigens** have two properties. First they elicit the **synthesis** of an **antibody.** Secondly, they **bind** specifically to that antibody. All compounds 'foreign' to an individual are potential antigens. That is to say, all compounds other than those normally present in the blood-stream or on the surface of cells. The immune system recognizes as foreign some compounds better than others. Proteins and polysaccharides, for example, are strong antigens; so are bacteria, bacterial toxins, and viruses, since they have proteins and polysaccharides exposed at their surface. Lipids and nucleic acids are weak antigens; most low molecular weight compounds are not antigenic at all (Fig. 11.5).

Low molecular weight compounds that are foreign can be made antigenic by covalently linking them to macromolecules that are not in themselves antigenic. 2,4-Dinitrophenyl (DNP) glycine, for example, is not antigenic nor is serum albumin. DNP glycine linked to serum albumin, however, is antigenic (Fig. 11.6). The antibody that is formed in response to DNP glycine–albumin binds either to DNP glycine–albumin or to DNP glycine; it does not bind to albumin. In other words the antibody elicited by DNP glycine–albumin is

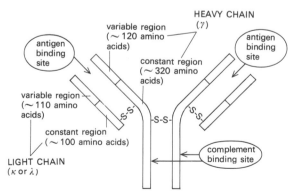

Fig. 11.4   Structure of immunoglobin

Note that the two light chains, and the two heavy chains, are identical.

Fig. 11.5 Antigenicity of different compounds

Strong antigens:	Proteins
	Polysaccharides
	Glycolipids
	Viruses
	Bacteria
Weak antigens:	Nucleic acids
	Phospholipids
Not antigenic:	Sugars
	Amino acids
	Steroids
	Drugs

Antigenicity means the capacity to elicit synthesis of a specific antibody.
[Note that non-antigenic compounds can be made antigenic ('haptens') by coupling to high molecular weight compounds such as proteins (Fig. 11.6).]

directed against DNP glycine. DNP glycine is known as a **hapten.** The usefulness of being able to elicit antibodies against haptens that are drugs, such as codeine or phenobarbitone, is that the drugs can then be specifically assayed by the method of radioimmunoassay (Fig. 9.25).

The reader may wonder why, if any foreign protein or polysaccharide is antigenic, the body is capable of neutralizing any such molecule by the formation of only some 10^6 different antibody molecules. The number of different proteins present in nature is considerably greater than this number. The answer is that the number of antigenic determinants is finite; that is, some antigens are so similar that an antibody does not distinguish between them, and binds to either. Antibody raised (in rabbits) against human globulins, for example, binds also to monkey globulins. In other words, an antibody can 'cross react' with several antigens, provided they are of sufficiently similar structure. Antibodies are like enzymes in this respect: hexokinase, for example, phosphorylates glucose, fructose, mannose, and glucosamine.

There is another similarity between antibodies and enzymes. Different enzymes recognize, and attack, different regions of a single protein molecule. Chymotrypsin hydrolyses peptide bonds adjacent to an

aromatic residue, trypsin hydrolyses peptide bonds adjacent to a basic residue, and so forth (Fig. 8.11). The same is true of antibodies. That is, one foreign protein elicits the formation of several different antibodies. Any polypeptide region that is sufficiently different from a region of host polypeptide chain is antigenic. And just as some nuclease enzymes do not distinguish between different regions of nucleic acids because these are too much alike, so antibodies do not distinguish between foreign and host nucleic acids. Hence nucleic acids are weak antigens. Fortunately this does not matter so far as infectious disease is concerned, since no virus or microbe consists only of nucleic acid, or has its nucleic acid exposed on its surface.

The binding between antibody and antigen, like that between enzyme and substrate, does not involve covalent bonds. Hydrogen bonds, hydrophobic bonds, and ionic interactions hold antigen and antibody together. The most important factor is the three-dimensional arrangement of the amino acids in the variable region of the antibody; this forms a complementary 'fit' around the antigen. Since each antibody has at least two antigen-binding sites, and some antigens have several antibody-binding sites, the resulting complex may involve several antigen and antibody molecules.

The function of antigen–antibody interactions is to increase the extent to which macrophages and other phagocytic cells destroy antigens; for the bigger the molecule, the better it is taken up by phagocytosis (Fig. 11.7). Moreover, these cells have receptors for antibody–complement complexes, which help to localize antigen at their surface. It must also be remembered that most foreign substances like viruses and bacteria are engulfed by macrophages anyway, especially in the presence of complement (see below). What antibodies do is to increase the efficiency of that destruction. The increase may make the difference between successful elimination, or survival, of a pathogenic microbe. It is to this end that the humoral part of the immune system has evolved.

11.2.1.2 Synthesis of antibodies

The presence of an antigen in the blood-stream triggers the synthesis of antibody molecules directed against it. The mechanism involves the class of lymphocytes known as B lymphocytes. Lymphocytes may be divided into two groups, B and T (Fig. 11.8). B lymphocytes are formed in the bone marrow and other lymphoid organs; the letter B is derived from the fact that in birds, these lymphocytes are formed at an anatomically distinct site site known as the Bursa of Fabricius. T lymphocytes are formed in the thymus; the stem cells of T lymphocytes are derived from the bone marrow also. The thymus is a small organ

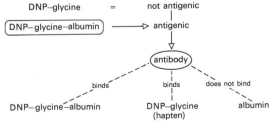

Fig. 11.6 Function of haptens

The function of a hapten such as 2,4-dinitrophenyl (DNP) glycine is illustrated. Note that if the albumin that is presented to the immune system is foreign to it (e.g. human albumin to a rabbit), it acts as an antigen in its own right and elicits antibody that does bind to it.

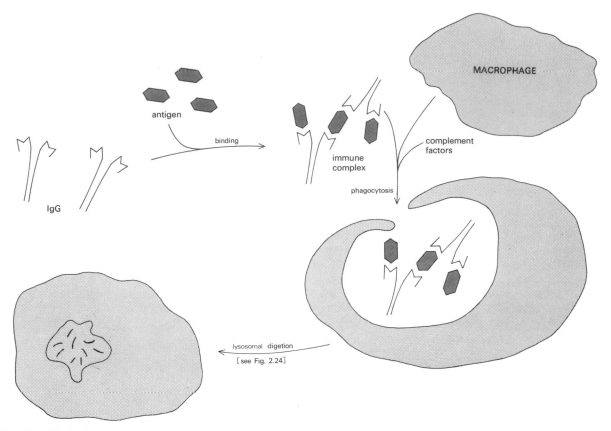

Fig. 11.7   Elimination of antigens

situated over the heart. It weighs approx. 10 g. in young children; thereafter it decreases in size until it is no more than a vestigial organ in the adult. The difference between B and T cells is illustrated by the fact that if the thymus is removed from a young

Fig. 11.8   Types of lymphocytes

Name	Origin	Function
B cells	thymus-independent	humoral synthesis of antibodies
T$_H$ cells	thymus-dependent	helper cells in humoral synthesis of antibodies
T$_S$ cells	thymus-dependent	suppression of various components of immune system
T$_C$ cells	thymus-dependent	cytotoxic to virally infected and foreign cells
T$_{DH}$ cells	thymus-dependent	causes delayed hyper-sensitivy to certain allergens

The distinction between the different types of lymphocyte is not clear-cut. Note that there exists a class of *non-lymphocytic* cells, referred to as K or 'Killer' cells, that is also cytotoxic to foreign and otherwise-altered (cancer?) cells.

animal, it renders it incapable of performing T cell functions, whereas B cell function is unimpaired.

B lymphocytes have on their cell surface a **receptor** that recognizes antigen. The receptor is **IgM.** There are thus some 10^6 different B lymphocytes in the circulation, each carrying a different IgM. When an antigen enters the circulation, it binds to its respective IgM-bearing B lymphocyte. That is, it specifically selects out one of the 10^6 different cells and binds to it (Fig. 11.9). Binding results in the stimulation of the B lymphocyte. A subpopulation of T cells, known as T 'helper' or T$_H$ cells, participates in the binding process so as to increase the effectiveness of stimulation. The mechanism of this potentiation is not clear. As stated above, it is not an absolutely necessary part of stimulation, since animals deprived of T cells undergo the humoral response.

Prior to binding, B lymphocytes are in a quiescent, G$_0$ state. Stimulation triggers cell cycle activity, leading to cell division. At the same time the size of the stimulated cell, now a blast cell, is increased. During the divisions that ensue, differentiation into plasma cells takes place. Plasma cells have the capacity to secrete IgG molecules of the same specificity as the

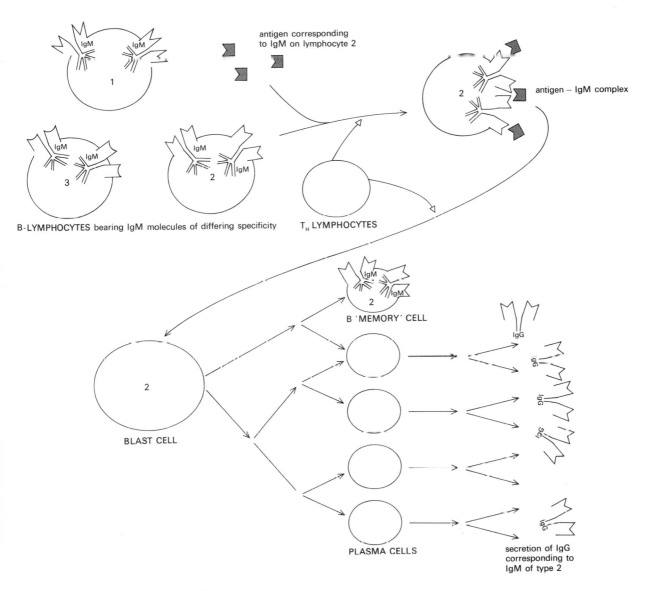

Fig. 11.9   Antigenic stimulation of B cells

Antigen cross-links receptor IgM molecules, causing them to cluster on the cell surface; this is part of the trigger for subsequent events.

IgM receptor on the parent B lymphocyte. In short, the effect of **antigenic stimulation** is to produce a population, or clone, of **IgG-secreting cells** from a single B lymphocyte. The IgG molecules in the circulation are now able to neutralize the remaining antigen molecules so as to complete their elimination by macrophages.

During the formation of plasma cells, some of the original B lymphocytes are regenerated in increased numbers. These are referred to as memory cells (Fig. 11.9), for the reason that if antigen is introduced a second time, the process of plasma cell formation and

IgG secretion is greatly amplified. Note that the secreted IgG molecules themselves do not survive for any length of time.

The first exposure to antigen thus corresponds to the immunization of an individual; the second exposure corresponds to successful elimination of an infectious agent in an immunized person. If the first exposure involves too much antigen, as would occur during an infectious attack of a non-immunized person, the capacity of the immune system is exceeded, and effective neutralization does not occur.

The life span of B memory cells, though longer than

that of the average white cell, is not infinite. Immunization against typhoid, for example, should be repeated every year; against smallpox every 3 years; against diphtheria or tetanus, every 5–7 years.

The production of IgG by plasma cells involves the normal process of protein synthesis (Chapter 5). Several genes are involved (Fig. 11.10). First, there are separate genes (each located on a different chromosome) for κ chains, for λ chains, and for the different heavy chains; the chains are linked by disulphide bridges *after* their synthesis. In that regard, IgG synthesis differs from, for example, synthesis of the disulphide-linked A and B chains of insulin, which are

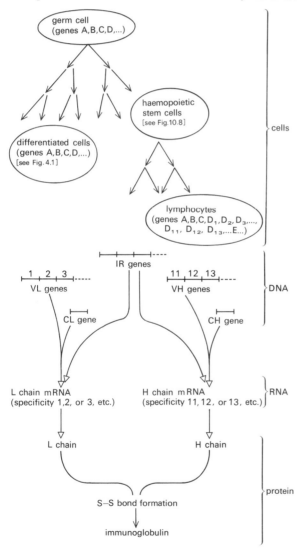

VL = variable part of a light chain   CH = constant part of a heavy chain
CL = constant part of a light chain   IR = immune response
VH = variable part of a heavy chain

Fig. 11.10   Synthesis of immunoglobulins

derived from a single chain (Fig. 9.17). Secondly, the variable and constant parts of each light or heavy IgG chain are specified by separate genes. Synthesis of the chains, however, occurs straight through from the NH_2-terminal in the variable region, to the COOH-terminal in the constant region. That is, the two genes coding for an immunoglobulin chain are joined in some manner *prior* to the translational event.

There is some evidence that the 10^6 or so genes coding for the variable parts of immunoglobulin chains are not present in the germ cells. In other words they arise, perhaps by mutation of a much smaller number of 'parental' genes, during development. Their formation occurs *after* the synthesis of most of the host proteins and other macromolecules, to which circulating lymphocytes have access, has taken place. Genes coding for antibodies that would bind to such non-foreign molecules are somehow eliminated, or never arise, or are prevented from being expressed; the existence of a class of 'suppressor' T lymphocytes (T_S cells) lends support to the latter possibility. Only genes coding for antibodies against potentially 'foreign' compounds are retained in an active form.

The elaboration and maturation of the immune system occurs around birth. If for some reason a host protein becomes accessible to the immune system *after* this has become established (or the 'suppressor' function of T_S lymphocytes breaks down), the protein will be recognized as foreign by one of the 10^6 circulating B (or T) lymphocytes and be eliminated. This type of immune response is referred to as **autoimmunity.**

The emergence of genes coding for the variable region of antibodies is therefore an exception to the rule that development is the result of a change in the *expression* of specific genes (Fig. 4.1), not of a change in their *presence* (Fig. 11.10).

A third set of genes is involved in the immune system (Fig. 11.10). These genes control the degree to which an individual responds to an antigenic stimulus. They have accordingly been termed immune response (IR) genes. Lack of the correct IR genes renders an individual less able to combat particular infectious or other diseases. The IR genes are located on that part of the chromosome that also codes for the histocompatibility genes (section 11.3.1.2). There is thus an overlap between possession of a particular type of histocompatibility gene and predisposition to certain disease (Fig. 11.15).

The whole process of antigenic stimulation of B cells and secretion of IgG takes place not in the circulation, but at the sites at which lymphocytes accumulate: tonsils, lymph nodes, and spleen.

The mechanism just described is the same for the synthesis of IgM on the surface of B lymphocytes, and indeed for the synthesis of all other immunoglobulin types (IgA, IgD, IgE) also.

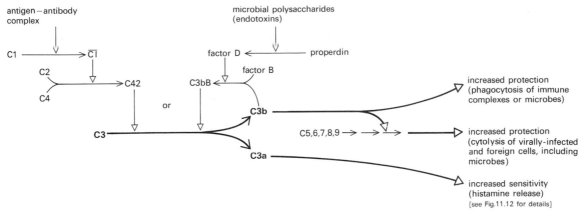

Fig. 11.11 Activation of complement

The major components of the complement system are shown. Each activation step (designated by —▷) involves a proteolytic degradation, as in blood coagulation (Fig. 7.31). The result is an amplification of the initiating signal; component clq (part of C1), for example, is present in the blood stream at ~100 μg/ml, whereas component C3 is present at ~1200 μg/ml.

11.2.1.3 Structure and function of complement

There is a group of plasma proteins that is collectively known as **complement.** The proteins do not function in the form in which they circulate in the blood-stream, but are first activated by a series of enzymic degradations (Fig. 11.11). The process is similar to the activation steps leading to the formation of a fibrin clot.

The proteins of complement are specified by numbers as shown in Fig. 11.11. One of the proteins, C1q, is particularly sensitive to heating and is destroyed by conditions (for example 56 °C for 30 minutes) that leave proteins such as immunoglobulins unaffected. Component C1q, which is the first component involved in the activation of complement by antigen–antibody complexes, has a structure similar to that of collagen. It is rich in proline and lysine residues, many of which are hydroxylated. Deficiency of vitamin C therefore affects the immune system as well as the function of connective tissue. Because of the complexity of the complement system, it is usual to refer simply to 'complement' without specifying which factors are involved in any particular situation.

One of the main features of complement is the ability to bind to antibodies that have formed an antigen–antibody complex. Binding, which is known as complement fixation, is to the constant region of immunoglobulin molecules (Fig. 11.4). As a result of binding, the activity of antibodies is increased (Fig. 11.12). Thus an antigen–antibody-C3b complex is more effectively engulfed by macrophages than is an antigen–antibody complex alone. Complement factor C3a fixed to an antigen–IgE complex on the surface of mast cells increases the release of histamine by such cells.

Attachment to complement of an antibody that is directed against a cell surface component makes the complex able to lyse such cells. Group A red blood cells, for example, are lysed by anti-A serum (which contains antibody against A group glycolipid, Fig. 7.23) in the presence of activated complement. Cells in general are lysed by anti-cell antibodies bound to activated complement. Such lysis is the humoral counterpart of cell destruction brought about by cytotoxic T cells in the absence of complement (section 11.3.1 below).

Complement is able to afford protection against infectious agents in the absence of antibodies. It does this in two ways (Fig. 11.12). First, complement alone is able to stimulate macrophages to destroy foreign agents. This is achieved by the properdin sequence of reactions (Fig. 11.11); attachment of factor C3b to the surface of macrophages makes the macrophages 'angry'. Destruction of infectious and other toxic agents by angry macrophages is a system that probably evolved before the immune system. It is likely to play at least as important a role in combating infections as does the immune system.

The second way in which complement affords protection is by direct lysis of infectious microbes. That is,

Fig. 11.12 Function of complement

Factor	Function
C3a	Histamine release from mast cells and basophils (partly by acting as chemotactic attractant for basophils and other granulocytes)
C3b	Increased phagocytosis of immune complexes by binding to immunoglobulins and macrophages (making macrophages 'angry')
	Increased phagocytosis of microbes by binding to macrophages (making macrophages 'angry').
	Activation of C5 and subsequent components, leading to cytolysis of target cells (including microbes)
	Increased hypersensitivity reactions

241

certain bacterial cells are lysed by activation of factor C5 (Fig. 11.11). The presence of lysozyme, an enzyme that destroys bacterial cell walls, aids in the lytic process. Lysozyme is present in several body secretions, and is also formed by macrophages, especially by macrophages made 'angry' by activation of complement as described above.

11.2.2 Sensitivity

The **allergic response** to agents such as pollen, insect or snake venom, penicillin or other drugs, involves an interaction between the agent (antigen) and an IgE molecule (antibody). As in the mechanism leading to protection, the antigen performs two roles. First, it elicits the synthesis of a specific antibody, and secondly, it binds to that antibody. The result of antigen–antibody binding, however, is not to prevent a pathological response, but to cause one. The response is a release of **histamine** and other pharmacologically active compounds at the site of injury and in the circulation. This results in an inflammation of the skin (known as a wheal) and a sequence of events collectively referred to as anaphylactic shock.

The secretion of IgE is elicited by antigens in the same way as is the secretion of IgG. That is, the antigen binds to a B lymphocyte bearing a receptor IgE on its surface. Stimulation of B lymphocytes, aided by helper T cells, leads to blast formation and the production of a clone of IgE-secreting plasma cells. In contrast to IgG, the secreted IgE becomes bound to the surface of mast cells. These are either circulating basophils, or 'fixed' mast cells in skin, blood-vessels, or other organ.

When an antigen is introduced topically or into the circulation, it binds to its receptor IgE; the presence of activated complement potentiates binding. Binding results in the release of histamine from vesicles within the mast cells (Fig. 11.13). The vesicles also contain ATP and heparin, each of which helps to neutralize the positively charged histamine molecules. The contents of the vesicles are released by exocytosis (Fig.

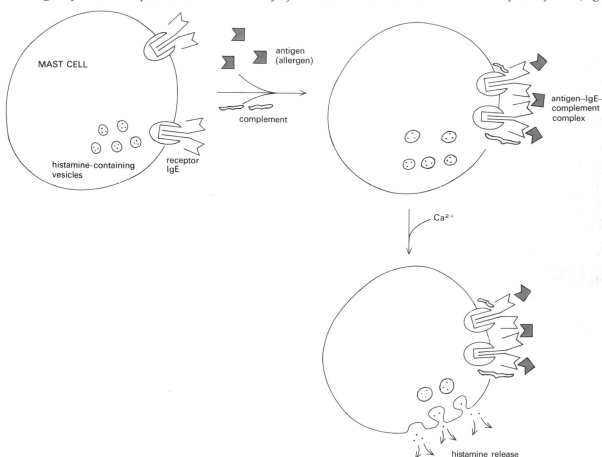

Fig. 11.13   Antigenic stimulation of mast cells

Antigen cross-links receptor IgE molecules, causing them to cluster on the cell surface; this is part of the trigger for subsequent events.

6.16), in exactly the same way as are the contents of hormone-containing vesicles (Fig. 9.17) or of transmitter-containing vesicles (Fig. 12.8) from their respective cells. As in other forms of exocytosis, the internal stimulus is an increase in the concentration of Ca^{2+} ions brought about by an influx of Ca^{2+}. The increase of Ca^{2+} permeability is the direct result of the IgE–antigen binding reaction.

Platelets are also involved in the allergic response. The release of serotonin and other compounds from platelets contributes to local inflammation and anaphylactic shock. Whether platelets contain IgE on their surface or whether they release their contents in response to a primary release of histamine from mast cells is not known.

The allergic response is generally not serious and most individuals recover quickly. Indeed, the response has a beneficial basis. Some 10 per cent of the population, however, is particularly sensitive (hypersensitive) to stimulation by such antigens. In this case, the resulting anaphylactic shock can be so intense as to lead to death. Hypersensitivity is controlled by the immune response genes; a person hypersensitive to one antigen, such as penicillin, tends to be hypersensitive to another, such as bee venom, even though different IgE molecules are involved.

11.3 Cellular system

The cellular system operates not by way of circulating immunoglobulins, but by way of circulating lymphocytes. The lymphocytes are thymus-derived or T cells (Fig. 11.8). Certain antigens elicit the multiplication of T cells that bear a specific receptor, in the same way as B cells are stimulated by other antigens (Fig. 11.9). The resulting population of T cells then eliminates (resulting in protection) or potentiates (resulting in sensitivity) the antigen. The structure of the antigen receptor on T cells is not yet known. It is not an intact immunoglobulin, but may be a molecule possessing sequences similar to the variable region of immunoglobulins.

The role of the various classes of T lymphocyte is as follows. T_c lymphocytes are involved in cell-mediated **protection**, T_{DH} lymphocytes in cell-mediated **sensitivity**. The distinction is not at all clear-cut, however, and there is much interplay between the two types of lymphocyte. Two other types of T lymphocyte have already been mentioned: T_H cells help to transform B lymphocytes into antibody-secreting plasma cells; T_S cells have the opposite effect: they suppress transformation. Failure of T_S cell function underlies the development of autoimmunity. In fact T_S cells probably suppress the function of lymphocytes in general. That is, T_S cells suppress not only the immune responses

leading to protection, but also the immune responses leading to sensitivity. In short, T_S cells regulate the extent to which the humoral and cellular immune systems respond to an antigenic stimulus. The entire immune system is thus balanced between a state of stimulation and one of suppression. Moreover the humoral and cellular system should not be viewed as alternatives, but rather as interrelated systems.

11.3.1 Protection

11.3.1.1 Against viral infection

The main protective function of the cellular system is to provide immunity against viral infections. The cells that are involved, namely T_C cells, act by lysing other cells. How does such lysis of *cells* achieve the elimination of *viruses*? The answer is that it is not intact viruses that are eliminated by the cellular system. What elimination of intact virus does occur is through non-specific destruction by macrophages or through the humoral system. The cellular system acts only against viruses that have successfully infected a cell.

Viral infection involves entry into a cell, replication of the viral genome (DNA or RNA), and synthesis of the proteins that are associated with the intact virus. The release of virus is in most cases preceded by the appearance of viral proteins on the cell surface (Fig. 11.14). It is these proteins that are recognized as foreign by T_C lymphocytes.

The first appearance of viral protein on the cell surface leads to the proliferation of that type of T_C lymphocyte (approximately 1 out of some 10^6 different types) that is complementary to the antigen. That is, a clone, or sub-population, of T_C lymphocytes directed against the viral antigen becomes established; the process is exactly the same as in the establishment of B 'memory' cells. At the second appearance of viral antigen on infected cells, sufficient antiviral T_C lymphocytes are formed to kill all virus-bearing cells (Fig. 11.14).

The mechanism by which T_C lymphocytes lyse cells, and eliminate intracellular virus particles, is not clear. Neither circulating antibodies nor complement factors are involved. Macrophages, on the other hand, do play a role. They somehow become 'recruited' to the site of lysis, perhaps through the action of lymphokines (section 11.3.2); lymphokines may also play a direct role in T_C cell killing. The neutralization of viral components, through phagocytosis and lysosomal degradation, is completed by macrophages.

11.3.1.2 Against foreign cells

The cells of one individual resemble those of another individual in almost every regard. The composition of liver, muscle, skin, or other cell, is virtually the same

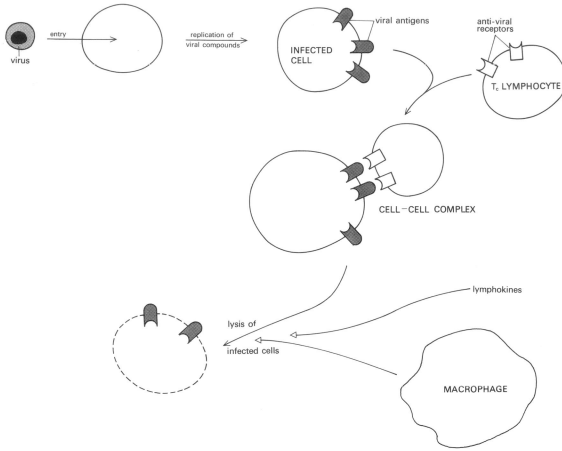

Fig. 11.14   Destruction of virally infected cells

in all human beings. But one difference is in a group of surface glycoproteins known as **histocompatibility antigens.** These glycoproteins, which are present on the surface of all cell types, vary from one person to the next. The variation is in the protein, not the carbohydrate part of the molecule. Only in identical twins are the histocompatibility antigens likely to be exactly the same.

As a result of variation, if an organ such as a kidney, or a piece of tissue such as skin, is grafted from one person to another, it will be recognized as foreign by circulating T_C lymphocytes and be eliminated. The humoral system also plays a part in the rejection of grafted organs. Only by suppressing the immune response through anti-proliferative and anti-inflammatory drugs, and by matching the histocompatibility antigens as much as possible (some individuals are more alike in their histocompatibility antigens than others), can organs be successfully grafted from one person to another.

The genes that specify the structure of the histocom-

patibility antigens, known as HLA genes, map in the same region of the chromosome as the immune response genes. Other functions that map in this region are some of the complement factors, and another surface glycoprotein known as HLA-IA (IA standing for immune associated antigen); unlike the histocompatibility antigens, IA glycoproteins are present only on B and T lymphocytes and on macrophages.

It is unlikely that histocompatibility antigens are the actual proteins specified by the immune response genes. On the other hand the products of the immune response genes may somehow be related to the histocompatibility antigens, since there is a significant correlation between different subgroups of histocompatibility antigen and susceptibility to certain types of disease. This is summarized in Fig. 11.15.

A possible function of the histocompatibility antigens is in development. The processes of cell recognition and cell adhesion, both of which play a part in development, are likely to be mediated through the carbohydrate portion of cell surface glycoproteins such

Fig. 11.15   Diseases associated with HLA genes

Antigen specified by gene	Phenotype (allelic variant)	Disease	Relative risk
B	7	multiple sclerosis	1.5
B	8	diabetes (insulin-dependent)	1.9
B	8	thyrotoxicosis	2.5
B	8	myasthenia gravis	5.0
B	8	Addison's disease	6.4
B	27	Reiter's disease	48
B	27	ankylosing spondylitis	81
D	W2	multiple sclerosis	5.0

The relative risk is defined as the chance of a person of particular phenotype contracting the disease, compared with persons not of that phenotype.

Note that different phenotypes of each gene represent 'normal' alternatives, like the blood group antigens (Figs 7.23 and 7.24); this is in contrast to the situation of one 'normal' phenotype and several 'abnormal' phenotypes, that applies in the case of haemoglobins (Fig. 7.22).

Adapted from I. Roitt (1977). *Essential immunology*, 3rd edition. Blackwell Scientific Publications, Oxford, with permission.

as those having histocompatibility function. This implies that the immune function of the histocompatibility antigens is 'by the way'. Certainly immune function cannot underly the basis of development. First, the development of cell specificity takes place several days after fertilization, whereas immune competence does not occur until after birth. Secondly, all metazoan organisms are made up of specific cell types, whereas only higher animals appear to possess an immune system of the type described in this chapter. If the immune system is involved in any way in cell specialization, it is likely to be in the *maintenance* of predetermined cell types, and not in the *initiation* of their specificity.

The function of histocompatibility antigens, then, might be due to their carbohydrate portion. If the structure of the protein portion is not important for development, mutations in the relevant gene will not be deleterious, and will occur at high frequency. This is exactly what is found. Because the immune system recognizes an altered surface protein as foreign, however, such mutant glycoproteins are highly antigenic. This explains cell rejection between different individuals. The variation in protein structure, which is at least as great as that of the variable region of immunoglobulins, is unlikely to result from random mutation alone, however, and probably does have some function in regard to the individuality of different persons. Histocompatibility antigens are the only type of molecule so far discovered that vary from person to person, and it is clear, from external appearances alone, that individuals (other than identical twins) do vary in a great many respects.

Whatever the function of histocompatibility antigens proves to be, it is likely that the carbohydrate portion plays an important role. In that sense, these glycoproteins may be contrasted with the other class of highly variable molecules, the immunoglobulins. In histocompatibility antigens, the functional portion of the molecule, that is, the carbohydrate part, is relatively constant (its variation is between different cell types but not between different individuals), while the nonfunctional part, which makes the molecule antigenic, is extremely variable. In immunoglobulins, the functional portion of the molecule is extremely variable (it is the region that binds antigen), whereas the nonfunctional part (which happens to contain carbohydrate) is relatively constant.

A possible reason for the evolution of histocompatibility antigens is that they provide protection against the emergence of cancer. If the surface of cancer cells is sufficiently different from that of normal cells, then any circulating cancer cell will, like a virally-infected cell, be recognized as foreign by the immune system, and be eliminated. The persistence, and growth, of cancer cells would result only if there were a failure of such an immune surveillance system. As yet, a difference between the histocompatability antigens of cancer cells and normal cells has not been detected. On the other hand cancer cells do possess antigenic markers on their surface that distinguish them from other cells. Moreover it is known that persons who have received immunosuppressive drugs are at greater risk in developing leukaemia (though not in developing other cancers), than is the population at large.

The histocompatibility antigens may also protect against viral infection. For there is evidence that the elimination of virally infected cells by T_c lymphocytes involves an interaction with the histocompatibility antigens on the surface of infected cells, as well as with viral proteins.

11.3.2 Sensitivity

The cellular immune system plays a role in sensitivity. A special class of T lymphocytes known as T_{DH} cells is involved: the reason for the notation is that delayed hypersensitivity (DH) is the main manifestation of this part of the immune system.

When an antigen such as wasp or bee sting venom enters the circulation, the response occurs within a matter of minutes. With other antigens that result in allergy, the response does not occur for several hours. The latter type of response has been termed delayed hypersensitivity, in order to distinguish it from the more immediate type of response. The fast response is generally mediated through the humoral system; the slower response through the cellular system. Exactly how the delayed response is mediated is not clear. It is likely that the antigen reacts first with T_{DH} lymphocytes. These are stimulated to divide and then to secrete, rather like plasma cells, a special class of compounds. The compounds are known as **lymphokines;** their structures have not been elucidated. Lymphokines perform a variety of functions, including the initiation of an inflammatory response, by reacting with other cells such as platelets and macrophages. It is the inflammatory response that leads to delayed hypersensitivity.

In addition to T_{DH} cells, T_S cells may play a role in delayed hypersensitivity, and in immediate hypersensitivity, also. Normally T_S cells limit the extent to which humoral sensitivity (that is, IgE-mediated histamine release) and cellular sensitivity (that is, lymphokine-mediated inflammation) occur. A failure in T_S cell function is a contributing factor to the establishment of immediate or delayed hypersensitivity.

11.4 Disease

Diseases of the immune system fall into two types. Those due to an insufficient response, and those due an excessive response (Fig. 11.16).

An insufficient response underlies the establishment of infectious diseases. Failure to eliminate an invading virus, microbe, or other parasite results in the replication of the infective agent within the body. The failure may be due to a deficiency of one of the white cells, a condition known as leucopenia: that is, to an insufficiency of B or T lymphocytes, or to an insufficiency of granulocytes, monocytes, or other phagocytic cell. Alternatively the failure may be due to an insufficiency of one of the complement factors, which itself may be the result of a defect in the immune response genes, as illustrated by the diseases listed in Fig. 11.15.

If one accepts the hypothesis that cancer results

Fig. 11.16   Diseases of the immune system

Defect	Outcome
Insufficiency of response	
General	All infectious diseases, i.e. viral (e.g. influenza, measles, poliomyelitis, smallpox, etc.), bacterial (e.g. streptococcal, staphylococcal, *Salmonella*), and protozoal (e.g. malarial) infections.
	? Cancer
Specific: e.g. C3 deficiency	Certain bacterial infections
Excessive response	
General	Allergies; e.g. hay fever, asthma, susceptibility to bee and wasp stings, to penicillin, etc. Delayed hypersensitivity; e.g. tuberculosis.
Specific: circulating auto-antibodies: e.g. against red cells	Autoimmune diseases: autoimmune haemolytic anaemia
platelets	idiopathic thrombocytopenic purpura
intrinsic factor	pernicious anaemia
thyroid gland surface antigen	Hashimoto's disease
thyroid gland TSH receptor	thyrotoxicosis
glomerular membrane	Goodpasture's syndrome
muscle cell acetyl choline receptor	myasthenia gravis
myelin components	? multiple sclerosis
DNA*	systemic lupus erythematosus
IgG*	? rheumatoid arthritis

* It is the presence of immune complexes (DNA-anti DNA; IgG-anti IgG) at certain sites in the body that sets off secondary reactions which culminate in the disease.

Some other diseases that have a relation to the immune system are listed in Fig. 11.15.

from a deficiency of immune surveillance, then neo-plastic diseases fall into the category of immune disease. A malfunctioning of the cellular immune system may be responsible also for the establishment of intracellular bacterial infections such as lepromatous leprosy.

An excessive immune response underlies allergic reactions to substances such as pollen, certain foods, wasp and bee stings, drugs such as penicillin, and so forth. The sensitivity is due to an abnormally high amount of circulating IgE, most of which is bound to the surface of basophils and mast cells. In the same category may be listed diseases of delayed hypersen-sitivity, such as tuberculosis. Hypersensitivity may also be viewed as an insufficiency of the immune response, in the sense that a failure of T_S cell function probably plays a part.

Autoimmunity, which may underline several diseases, is obviously an immune disease. If the mechanism by which non-foreign molecules (that is, host proteins and polysaccharides) are prevented from acting as antigens is by T_S lympho-cytes, then the emergence of autoimmunity is clearly the result of a defect in the immune response. If autoimmunity arises by the accidental exposure (which may be the result of a viral infection) of non-foreign molecules to the immune system, then autoimmune diseases fall into a category of their own, involving an excessive, rather than a deficient, response. Both mechanisms are likely to play a role in the emergence of autoimmunity. The diseases associated with it may conveniently be listed in the category of excessive immune response, in the sense that they represent an allergic response against host constituents.

11.5 Summary

Nature of the immune response

1. The immune response is a specific interaction between the body and a foreign substance to which it has previously been exposed. It involves:

i. **memory** of the previous exposure and

ii. a specificity of interaction based on that memory

The interaction can lead either:

a. to the elimination of the foreign substance (that is, **protection** against it) or

b. to potentiation of the substance (that is, to **sensitivity** to it).

The cells of the body that confer memory, protection, and sensitivity are various types of **lymphocyte.**

2. The substances that initiate an immune response are known as **antigens.** Antigens interact either with soluble substances known as **antibodies** or with specific cells (T lymphocytes). The interaction with antibodies involves a system known as the **humoral** system; the interaction with cells involves a system known as the **cellular** system.

Humoral system

1. Antibodies are proteins: they are the γ-globulin plasma proteins (also known as **immunoglobulins**). An antigen elicits the synthesis of an immunoglobulin, and then reacts with it.

The reaction with immunoglobulin G **(IgG)** leads to elimination of the antigen (protection); reaction with immunoglobulin E **(IgE)** leads to an inflammatory response (sensitivity). Foreign substances such as pathogenic bacteria and bacterial toxins stimulate the synthesis of IgG (that is, this is the mechanism by which **immunity** against **bacterial infection** is achieved); foreign substances such as pollen, bee venom, or certain drugs generally stimulate the synth-esis of IgE (that is, this is the mechanism by which **sensitivity** to substances causing **allergy** is achieved.

2. Immunoglobulins are synthesized by **B lympho-cytes**. In the circulation there are about 10^6 different types of B lymphocyte; an antigen selects that type of cell that carries an immunoglobulin receptor of best 'fit'. Binding results in the secretion of the immuno-globulin corresponding to the receptor.

3. Elimination of antigen is brought about by **phagocytosis** of an antigen–IgG complex by mac-rophages.

4. Inflammation is brought about by stimulation of mast cells by an antigen–IgE complex: stimulated cells secrete **histamine**.

5. Plasma proteins known as **complement** factors potentiate phagocytosis of IgG–antigen complexes and histamine release from stimulated mast cells.

Cellular system

1. Cells infected with **viruses** are antigenic. Such cells stimulate T_c **lymphocytes**, which destroy infected cells; macrophages aid in the process. Foreign cells (which are distinguished from host cells by possession of a different type of **histocompatibility antigen**) are also destroyed by T_c lymphocytes.

2. Delayed hypersensitivity reactions are caused by T_{DH} lymphocytes; T_{DH} cells release soluble factors (known as **lymphokines**), that participate in the pro-cess.

3. The humoral and the cellular immune systems are not distinct, but interact in several ways that are not yet clear.

Disease

An insufficient immune response leads to the establishment of viral and microbial infections. An excess leads to allergies. Certain diseases in which cellular structures become destroyed may be due to the recognition of normal proteins as 'foreign' by the immune system; this is known as autoimmunity.

FURTHER READING

General:
T.R.Bowry (1978). *Immunology simplified.* Oxford University Press.
F.M.Burnet (ed.) (1976). *Immunology* (Readings from *Scientific American*). W.H.Freeman, San Francisco.

I.Roitt (1977), *Essential immunology* (3rd edition). Blackwell Scientific Publications, Oxford.

More specific aspects of lymphocytes:
D.H.Katz (1977). *Lymphocyte differentiation, recognition and regulation.* Academic Press, New York.

Function of complement:
K.B.M.Reid (1976). Activation of the complement system. *Trends biochem. Sci.* **1,** 123.

HLA system
B.A.Cunningham (1977). The structure and function of histocompatibility antigens. *Scient. Am.,* **237,** 96.
W.F.Bodmer (ed.) (1978). The HLA system. *Brit. Med. Bull.* **34**.

The relationship between cancer and immunology:
F.M.Burnet (1976). *Immunology, ageing and cancer: medical aspects of mutation and selection.* W.H.Freeman, San Francisco.
L.J.Old (1977). Cancer immunology. *Scient. Am.,* **236,** 62.

12

Nervous system

12.1 Nature of system

The functions of the body are coordinated by two systems. Metabolism is controlled largely by the endocrine system; movement is controlled largely by the nervous system. Most types of muscular movement, whether voluntary or involuntary, are regulated through the transmission of **nerve impulses** (Fig. 12.1). In addition, the activities of certain glands, including hormone-secreting glands such as the hypothalamus and the adrenal medulla, are controlled by the nervous system. It is through the nervous system that man is made aware of his surroundings, and through the nervous system that he reacts to them.

The environment is monitored by incoming, or afferent, nerves that emanate from sensory receptors; receptors that are sensitive to touch, sight, smell, or sound. Incoming impulses are processed by the brain, as a result of which outgoing, or efferent, nerves are stimulated (Fig. 12.1). The sight and sound of a dangerous animal, for example, causes stimulation of motor neurons that innervate the muscles of the legs; contraction of the muscles allows the individual to run away. At the same time the secretion of adrenalin from the adrenal medulla is stimulated. This in turn stimulates the heart to pump more blood, containing metabolic fuel, to the muscles. Adrenalin also stimulates the breakdown of glycogen in liver and that of triglyceride in fat cells, so that the level of nutrients in the blood-stream is increased.

Efferent nerves originate in the brain, and then pass to the spinal cord; the spinal cord is in a sense a continuation of brain tissue that runs within the vertebral column through the neck and down the back of the spine. Together brain and spinal cord are known as the **central nervous system** (CNS). The rest of the nervous system is referred to as the peripheral nervous system.

The nerves that regulate involuntary activity, in other words the contraction of smooth muscle and to a certain extent that of heart muscle, and secretions from endocrine and other glands, are collectively known as the **autonomic nervous system** (ANS). Nerves supplying voluntary (skeletal) muscles and the ANS together make up the peripheral nervous system. Involuntary activity goes on most of the time: it is not switched on or off, but the rate at which it operates is increased or decreased. The nervous impulses, however, are all-or-nothing events; a nerve cannot pass a small impulse one minute and a large one the next. Hence involuntary organs are controlled by two separate types of nerve working in concert: a stimulatory nerve and an inhibitory nerve. The mechanism by which an impulse is transmitted is the same in each type of nerve; that is, a current generated by the movement of ions. What is different is the way in which the impulses

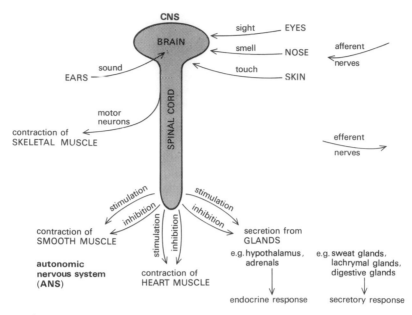

Fig. 12.1 The nervous system

The nervous system may be divided into the central nervous system (CNS) and peripheral nervous system; the latter consists of nerves to voluntary organs (skeletal muscle) and nerves to involuntary organs (smooth muscle, heart muscle, endocrine and other glands). The nerves to involuntary organs are collectively known as the autonomic nervous system (ANS). Each organ of the ANS is controlled by both stimulatory and inhibitory neurons. Note that the contraction of heart muscle is controlled both by the ANS, and independently of it (by its 'pacemaker').

are received by the receptor cells of the organ. Transmission between stimulatory nerve and receptor cell is by one type of chemical substance, between inhibitory nerve and receptor cell by another.

The control of involuntary activity by stimulatory and inhibitory nerve impulses is not unlike the control of metabolic activity by stimulatory and inhibitory hormones. The rate of TSH secretion by the anterior pituitary, for example, is stimulated by TSH-releasing hormone and is inhibited by thyroxine; the rate of Ca^{2+} release from bone is stimulated by parathormone and is inhibited by calcitonin. In fact the similarity is quite close, since the difference between stimulatory and inhibitory nerves lies in the nature of the chemical substance, as in hormonal control, that is transmitted.

Voluntary activity is discontinuous. During sleep, for example, most voluntary muscles are at rest. Following each stimulus, the muscle returns to the relaxed state until another stimulus is received. In other words, only one type of nerve, that is, a stimulatory nerve, is involved in the contraction of skeletal muscle.

12.2 Structure of nerve cells

Nerves, like muscles, are cellular structures. Like skeletal muscle cells, some nerves are extremely long; nerves that extend from the spinal cord to a muscle in the foot, for example, can be as much as a metre long.

The part that conducts the electrical impulse is a specialized structure known as an **axon,** the rest being known as the cell body; the whole cell is referred to as a **neuron** (Fig. 12.2). Some nerve fibres contain neurons that branch out from a single point. Such branching allows an impulse to be 'amplified' and transmitted simultaneously to several thousand receptor cells on a muscle or organ. Equally, incoming nerve fibres are branched so that several sensory cells transmit an impulse simultaneously to a receptor in the central nervous system. A highly branched network of nerve fibres underlies much of the complexity of the nervous system in the brain, which contains some 10^{10} neurons.

Nerve cells are not in continuous contact. Between one neuron and the next, as well as between a neuron and the muscle or gland cell that is being stimulated, there is a space that may be as small as 20 nm; the space is known as a **synapse** or synaptic cleft; in the case of skeletal muscle it is known as a neuromuscular junction. Across the synapse the nervous impulse takes the form of chemical, not electrical, transmission. The substances that are involved in chemical transmission are known as neurotransmitters. Neurotransmitters are stored in granules at the end of an axon. The arrival of an electric stimulus causes the secretion of neurotransmitter; this diffuses across the synapse and in turn stimulates or inhibits the next neuron or the cell of the receptor organ.

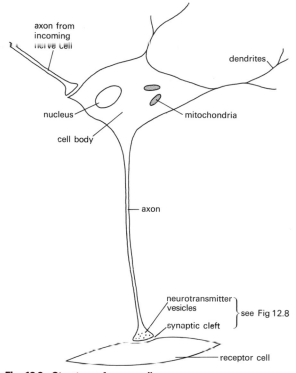

Fig. 12.2 Structure of nerve cells

A nerve cell (neuron) making contact with a receptor cell (e.g. secretory gland cell) is shown. The principle features are the same in nerve cells that make contact with skeletal muscle (motor neuron) or with other nerve cells (pre-ganglionic neuron).

In many types of neuron the axon is surrounded by a layer of lipid-rich material, known as a **myelin sheath.** Just as in a metal wire, electrons move faster if the wire is insulated by some non-conducting material, so in an axon, ions move faster if the axon is insulated by myelin. Myelinated nerves conduct impulses at approximately 10^3–10^5 per second, compared with $<10^3$ per second by non-myelinated nerve. Insulation by myelin takes the form of phospholipid bilayers wound round and round the axon. As in other membranes, the bilayer contains protein and cholesterol as well as phospholipids. It is particularly rich in sphingomyelin and glycolipids; a large part of the glycolipids carry a negative charge in the form of gangliosides and sulphated glycolipids (cerebroside sulphates) (Fig. 12.3). Since one of the major myelin proteins is positively charged (known as myelin basic protein), there is a tight ionic interaction between lipid (negatively charged) and protein (positively charged) in myelin. This probably contributes to the closeness of packing, and hence to the insulating capacity of the myelin sheath.

It is apparent from the composition and structure of myelin (Fig. 12.3) that it is similar to multilayered plasma membrane. In fact myelin *is* plasma membrane, being elaborated by special cells, that lie alongside a developing axon. The cells are known as oligodendrocytes in the case of axons within the central nervous system, and as Schwann cells in the case of axons outside the central nervous system. Exactly how the plasma membrane from each of these cells is spun out and wound around an axon is not clear.

Fully-developed neurons do not divide. The nucleus, which is situated in the cell body, synthesizes RNA but not DNA. The RNA is translated into protein in the cytoplasm. A neuron contains mitochondria, which form the ATP required for the propagation of impulses. Nerve cells have no capacity for anaerobic glycolysis, nor can they metabolize fatty acids or amino acids; they are entirely dependent on oxygen and glucose to synthesize ATP. They can to a certain degree adapt to the oxidation of ketone bodies, in place of glucose.

The axon is not just a hollow tube, but contains a regularly-spaced array of microtubules (as in cilia, Fig. 6.5), which maintain the shape of the axon. The microtubules also provide a series of channels through which proteins, phospholipids, and other molecules pass. Some of the components of neurotransmitter vesicles, for example, are synthesized in the cell body and pass along the axon to the site of vesicle storage and release; neurotransmitter itself is generally synthesized near the site of the storage vesicles. Such axonal transport is a two-way process, and the components of storage vesicles are returned to the cell body region for turnover.

Neurotransmitters are not the only substances transported through axons. The hormones that are secreted by the posterior pituitary, namely oxytocin and ADH, are synthesized in the hypothalamus and are transported by axonal flow through neurons that connect the two organs.

The generation of a mature neuron is thus a process of some complexity; a specific protein, called nerve growth factor, may be involved. Undifferentiated precursor cells lacking an axon begin by developing a protuberance at a point on the cell surface. This then elongates, laying down microtubules, until it meets another neuron or an effector cell such as muscle or secretory gland. A synapse is formed and the nerve is ready to transmit an impulse. Myelination takes place once the axon has formed; it involves the proliferation of Schwann cells or oligodendrocytes, in order to cover the length of the axon.

A damaged nerve has little capacity for repair. Once a brain cell that initiates a series of impulses leading to the contraction of a particular muscle is damaged, that muscle will not function again. In other words temporary damage to a brain cell, by lack of oxygen due to a failure in the blood supply, or by infection with a virus

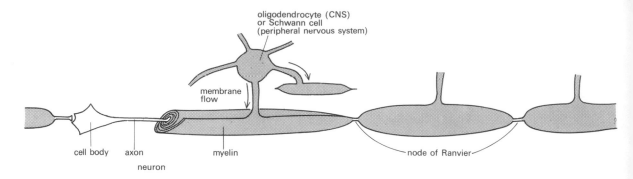

oligodendrocyte (CNS)
or Schwann cell
(peripheral nervous system)

membrane
flow

cell body   axon   myelin   node of Ranvier

neuron

Composition of myelin (% of dry weight)
Protein: 21%
of which
50 % is 'proteolipid'
30 % is basic protein
20 % is higher molecular weight proteins
Lipid: 79%
of which
41 % is cholesterol
16 % is glycolipids (cerebrosides)
14 % is ethanolamine-phospholipids (incl. phosphatidyl ethanolamine)
11 % is phosphatidyl choline
5 % is sphingomyelin
4 % is sulphated glycolipids (cerebroside sulphates)
0.5% is gangliosides
9 % is other lipid

Adapted from M. R. Rumsby (1978). *Biochem. Soc. Trans.* **6,** 448, with permission.

Fig. 12.3 Myelination

such as encephalitis, leads to permanent failure of the respective receptor organ. Only if the cell body of a neuron remains intact, can a damaged nerve be repaired; that is, if an axon is accidentally cut, a new axon can grow out from an undamaged cell body.

12.3 Function of nerve cells

The most important property of nerves is to transmit an impulse. Transmission is extremely rapid; an impulse travelling from the brain takes only milliseconds to reach parts of the body such as the muscles of the foot, that are some two metres away from it. The time taken by a hormone in the blood-stream to travel the same distance is in the order of seconds. The means by which nerve impulses are transmitted is partly by an ion current, partly by chemical diffusion. The current is generated in one cell, and propagated by its axon to the next cell, and so on. In between cells, the impulse is continued by diffusion of neurotransmitter across the synaptic gap between them; the distance is so small (approximately 20 nm), that a speed of milliseconds is maintained.

12.3.1 Generation of current

An electric current is a movement, or rather a displacement, of electrons. In nerve cells, an ion is generated by a change in the **membrane potential.**

All cells of the body, whether nerve cells or not, have a slight excess of negatively charged molecules on the inside relative to the outside (Fig. 12.4). As a result, a potential difference exists across the plasma membrane. The magnitude of the potential difference is between -20 and -100 mV, depending on the cell type; the $-$ sign indicates that the cell is negative on the inside. It may be wondered how such electrical imbalance is maintained.

The underlying cause is the presence in cells of negatively charged proteins (A^-); these cannot diffuse out, and are therefore balanced by an equivalent amount of cations (Na^+ and K^+) within the cell. But

1. assume no Na pump

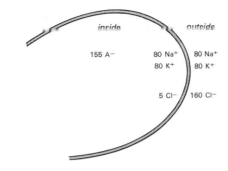

2. Na pump

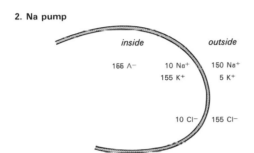

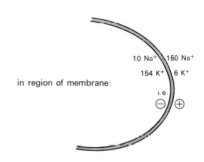

Fig. 12.4   Establishment of membrane potential

The concentrations shown are approximations for illustrative purposes: A^- indicates protein and phosphorylated compounds; the figures for Cl^- include HCO_3^-.

1. *Electrical* neutrality (but there is tendency for Cl^-, and hence Na^+ and K^+, and H_2O, to leak in, to achieve *ionic* balance).
2. Na^+ is pumped out, coupled to entry of K^+ (see Fig. 6.12). Tendency for K^+ to leak out is greater than tendency for Na^+ to leak in (due to difference in permeability of plasma membrane). The result is *neither* electrical neutrality—at least in region of membrane, which leads to membrane potential—*nor* ionic balance.

the cations, which *can* diffuse out, will tend to equilibrate themselves in such a way that their concentration is the same inside as outside the cell. At the same time Cl^- ions, which are the predominant anions that neutralize the Na^+ and K^+ outside cells, and which can also diffuse across the plasma membrane, will likewise tend to equilibrate themselves between the inside and the outside of the cell. The result is a situation in which *either*

a. electrical neutrality *or*

b. ionic equivalence of Na^+, K^+, and Cl^-, between the inside and the outside of cells can be achieved.

To achieve both would require an inflow of water, containing cations and Cl^-, to the point where the presence of the impermeant anions A^- becomes negligible in relation to the permeant ions (Na^+, K^+, and Cl^-); in other words, an inflow of water equivalent to the osmotic pressure of cells. This inflow of water is, however, prevented by the operation of the ATP-driven Na^+ pump (Fig. 6.12). In fact, because of the

existence of the Na^+ pump, *neither* (a) *nor* (b) is achieved. That is, a cell is in a thermodynamically unstable situation, with a tendency for ions to move so as to achieve both electrical neutrality and ionic equivalence.

The reason why the end-result is a negative potential difference across the plasma membrane is the following. Given that the Na^+ pump is actively extruding Na^+, there is an asymmetry of ionic distribution such that cells contain more K^+ on the inside than on the outside, and more Na^+ on the outside than on the inside (Fig. 12.4). There is thus a tendency for K^+ to leak out, and for Na^+ to leak in. However, the plasma membrane is some 100 times more permeable to K^+ than to Na^+, with the result that relatively more K^+ leaks out than Na^+ leaks in. Hence there is a relative deficit of cations within the cell, compared to the outside; that is, an excess of negative charges on the inside. (Cl^- ions, to which the membrane is as permeable as to K^+, do not affect the outcome: on the one hand Cl^- ions tend to move outwards in order to maintain electrical neutrality; on the other hand Cl^-

ions tend to move inwards, in order to achieve ionic equivalence between inside and outside. The two forces just about balance each other.)

If for some reason the relative impermeability of the plasma membrane to Na^+ were to be negated, Na^+ ions would rush into the cell. This is exactly what happens when a nerve or muscle cell is stimulated (Fig. 12.5). Stimulation, or **depolarization,** occurs as a result of specific binding of neurotransmitter to the receptor cell surface. Stimulation of a nerve or muscle cell by an electric current can also lead to depolarization.

The sudden entry of a small number of Na^+ ions, to which the membrane is now more permeable than to either K^+ or Cl^- ions, causes the cell to become positively charged on the inside relative to the outside at this part of the membrane. That is, the potential difference changes from negative to positive. The result is an electrical impulse, which is transmitted along the axon by a flow of ions from one point of depolarization to the next.

Within less than a millisecond of the membrane

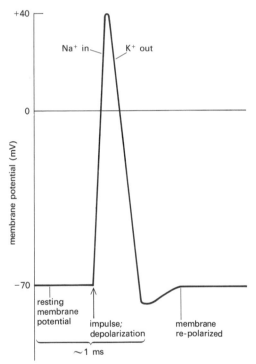

Fig. 12.5 Depolarization of nerve cells

Stimulation of a nerve cell is illustrated. The magnitude of the change in membrane potential, and the speed of the response, vary considerably among different types of nerve. Note that the alteration in overall Na^+ and K^+ levels within a nerve cell during one impulse is negligible; it is the alteration in a small region at the plasma membrane that is responsible for the change in membrane potential.

The events that occur when skeletal muscle, smooth muscle, or gland cells are stimulated (i.e. depolarized) are in general similar to those illustrated in this figure.

increasing its permeability to Na^+, its permeability to K^+ is also increased (its permeability to Cl^- remains the same). Hence a small number of K^+ ions now rush out of the cell, in order to restore electrical neutrality, and to try to attain ionic equivalence between outside and inside. At this point, however, the permeability 'gates' through which Na^+ and K^+ have diffused are 'closed' and the cell is repolarized and ready to receive another impulse. The total time taken between depolarization and repolarization can be as little as a millisecond.

The way in which impulses are transmitted along an axon depends on whether the neuron is myelinated or not (Fig. 12.6). In the case of myelinated neurons, there is a gap, known as the node of Ranvier, every millimetre or so. When an impulse arrives at the gap, the membrane becomes depolarized, and initiates a second impulse; as this reaches the next node of Ranvier, the process is repeated, and this continues until the end of the neuron is reached. At this point the impulse causes a neurotransmitter to be released. In the case of non-myelinated neurons, the process of depolarization and repolarization occurs continuously along the axon; instead of travelling approximately 1 mm, each impulse travels only a fraction of this. Hence the speed of conduction is very much slower.

It should be noted that the amount of Na^+ and K^+ ions that diffuse across the membrane during each impulse is very small in relation to the concentration of Na^+ and K^+ inside and outside the cell. That is, the asymmetric distribution of Na^+ and K^+ is approximately maintained. Only after many impulses have passed would the asymmetry 'run down'. The Na^+ pump, of course, makes good the leakages as they occur, so that the ability to depolarize a neuron is maintained. In short, the asymmetric distribution may be thought of as a kind of battery that is charged by the operation of the Na^+ pump. The battery can be used for energy-requiring processes *other* than nerve conduction; it is used, for example, in the active transport of sugars and amino acids across epithelial cells in the kidney (section 7.2.2.1) and small intestine (section 8.3.2). That is, the **Na^+ pump** converts the **chemical energy** of **ATP** into electrical energy in the form of a nerve impulse, or into osmotic energy in the form of sugar or amino acid transport.

Nerve cells, especially those in the CNS, require a constant supply of oxygen and glucose (or ketone bodies) to maintain their level of ATP. Should the supply of either oxygen or glucose fail, their ability to transmit impulses is impaired. At the same time, the cells begin to die. That is, unlike other cells, nerve cells are particularly sensitive to a failure of nutrient supply. Why is this? The answer is probably threefold. First, nerve cells have no internal stores of energy in the form of fat or carbohydrate. Secondly, once a

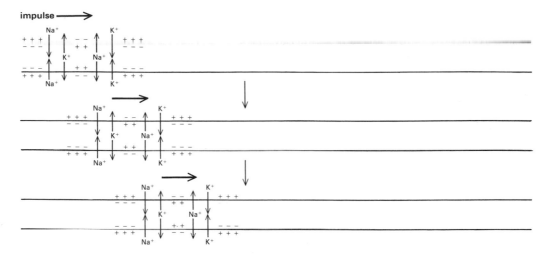

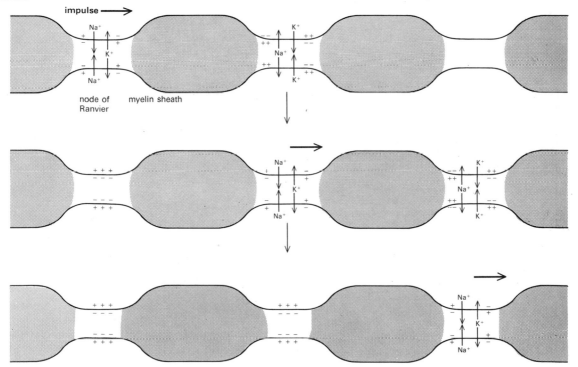

Fig. 12.6 Transmission of nerve impulses

Bold arrows indicate transmission of the nerve impulse. In myelinated neurons, the myelin sheath prevents ion movements across the plasma membrane, except at the nodes of Ranvier. Hence the impulse 'jumps' from one node to the next, with the result that transmission is considerably faster than in non-myelinated neurons.

neuron in the CNS has stopped functioning, it does not seem able to start again. Thirdly, a failure to maintain the activity of the Na^+ pump may have more serious consequences for a cell capable of depolarization, than for one that is not: for a depolarized cell is liable to lose its ionic asymmetry hundreds of times faster than other cells, through repeated opening of the Na^+ and K^+ 'gates'. A nerve cell therefore approaches more rapidly than other cells the situation in which it can no longer prevent osmotic entry of water, with resulting cell death due to lysis; entry of Ca^{2+}, which is toxic in excessive amounts, occurs at the same time as the Na^+ and K^+ 'gates' are open, and this too damages the cell.

12.3.2 Chemical transmission

Between one neuron and the next, as well as between neurons and muscle, or gland cell, is a gap that is normally some 20–50 nm wide. This space, or synaptic cleft, is crossed by the nerve impulse in the form of chemical, not electrical, transmission. **Acetyl choline** and **noradrenalin** are the two most common transmitters in the peripheral nervous system. Nerves that cause acetyl choline to be released are known as cholinergic; nerves that cause noradrenalin to be released, are known as adrenergic (Fig. 12.7).

All nerves leading to voluntary (skeletal) muscle, which are known as motor neurons, are cholinergic; the synapse between nerve and muscle cell, known as the neuromuscular junction, is also cholinergic. In the autonomic nervous system, some neurons are cholinergic, others adrenergic; where a muscle or gland is controlled by both stimulatory and inhibitory

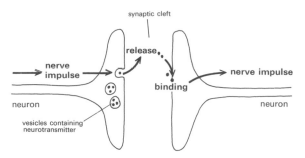

Fig. 12.8 Mechanism of neurotransmission

Excess neurotransmitter is destroyed by metabolism (e.g. Fig. 12.9). The diagram illustrates chemical transmission between one neuron and the next; the same mechanism operates at the neuromuscular junction, and at the surface membrane of cells controlled by the autonomic nervous system.

nerves, one nerve is cholinergic, and the other adrenergic. In the smooth muscle that controls peristalsis of the gastrointestinal tract, for example, cholinergic neurons stimulate contraction, whereas adrenergic neurons inhibit contraction. The ANS is sometimes divided into symphathetic and parasympathetic nerves. Each nerve is made up of two neurons in series, with a cell known as a ganglion linking the two neurons. Sympathetic neurons leading to the ganglion (preganglionic neurons) are cholinergic; the postganglionic neurons are adrenergic. Parasympathetic neurons are cholinergic, whether preganglionic or postganglionic. The nerve supply from the hypothalamus to the adrenal medulla is an exception in that it is sympathetic, postganglionic, yet cholinergic.

The manner in which chemical transmission occurs is as follows (Fig. 12.8). A nerve impulse causes release of the transmitter from the nerve ending. Transmitter, which is stored in vesicles, is released into the synaptic space by exocytosis; a discrete number of molecules, or 'quantum', is released at each transmission. The transmitter is immediately bound by a specific receptor on the surface of the adjacent cell. That cell becomes depolarized and, if it is another neuron that is part of a stimulating nerve, a further impulse is transmitted. If the receptor cell is a skeletal muscle cell, depolarization leads to release of Ca^{2+} from the sarcoplasmic reticulum and to muscle contraction; in the case of smooth or cardiac muscle cells, Ca^{2+} enters from outside the cells. If the receptor cell is a gland cell, depolarization leads to secretion; secretion of adrenalin in the case of adrenal medulla, of tears in the case of lachrymal glands, of sweat in the case of sweat glands, and so forth. Changes in Ca^{2+} concentration may again be part of the trigger.

As soon as the neurotransmitter has bound to its receptor and depolarization of the plasma membrane

Fig. 12.7 Types of neurotransmitters

Compound	Synapse	Type of nerve
Acetyl choline	cholinergic	motor neurone; neuro-muscular junction parasympathetic nerves (ANS) sympathetic nerve from hypothalamus to adrenal medulla (ANS)
Noradrenalin	adrenergic	sympathetic nerves (ANS)
Serotonin	?	CNS
γ-Amino butyric acid (γ-ABA)	?	CNS
Dopamine	?	CNS
Glycine, glutamic acid	?	CNS
Low mol. wt. peptides (enkephalins)	?	CNS

ANS = autonomic nervous system;
CNS = central nervous system.

has taken place, it is inactivated or otherwise removed so that a second impulse can be received.

Inactivation occurs in the case of acetyl choline, by a specific enzyme situated on the cell surface near the receptor. The anzyme is an esterase (**cholinesterase**) that hydrolyses acetyl choline to acetate and choline. Choline is rapidly absorbed back into the neuronal cell by an uptake system that has a very high affinity for choline; acetate diffuses away. Choline uptake appears to be Na^+-dependent. That is, as for sugar and amino acid uptake in epithelial cells (Fig. 6.9), choline uptake is linked to the operation of the Na^+ pump. Inside the neuron, acetyl choline is resynthesized by the pathway shown in Fig. 12.9. The acetyl group is derived not from acetate, but from glucose, via acetyl CoA. This is another reason why nerve cells, or cholinergic nerve cells to be precise, are so dependent on a supply of glucose (or ketone bodies) in order to function. During cholinergic transmission, therefore, ATP, acting via the Na^+ pump, is required for two processes:

i. for maintenance of ionic asymmetry, and

ii. for uptake of choline.

In addition, glucose is required for the synthesis of acetyl CoA.

The mechanism by which adrenergic transmission is terminated is simply by removal of noradrenalin back into the neuron from which it was released. Inactivating enzymes, such as monoamine oxidase and O-methyl transferase (Fig. 9.19) do not participate in neurotransmission. The uptake of noradrenalin is extremely rapid and, like choline uptake, appears to be linked to Na^+ entry. During adrenergic transmission, therefore, ATP is again required for two processes:

i. for maintenance of ionic asymmetry and

ii. for uptake of noradrenalin.

Effector cells of the autonomic nervous system that are innervated by both cholinergic and adrenergic nerves clearly have two types of receptor on their surface; a receptor for acetyl choline and a receptor for noradrenalin. Certain muscle and nerve cells have in addition a receptor for hormones such as adrenalin; this receptor is distinct from the receptor for norad-

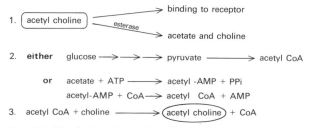

Fig. 12.9 Metabolism of acetyl choline

renalin. In order to distinguish the two catecholamine receptors, they have been termed α- and β-adrenergic receptor. Noradrenalin binds predominantly to the α-receptor, adrenalin binds to both α- and β-receptors. Other cells, such as those of liver or adipose tissue, possess α- and β-receptors also. In such cases both receptors may be stimulated by hormones.

The receptors for neurotransmitters are likely to be proteins. In the case of β-receptors, the protein has been identified as the hormone-binding subunit of the enzyme adenyl cyclase (Fig. 6.14). In the case of α-receptors on non-nerve cells, it is likely that the receptor is involved in mediating a transient entry of Ca^{2+} ions (Fig. 6.14). In nerve cells, the main function of the α-receptor is, as has been seen, the opening of the Na^+ 'gates'. In the case of cholinergic receptors on nerve or muscle cells, the receptor is again part of the trigger by which the membrane becomes depolarized, that is, by which Na^+ entry and K^+ exit are mediated.

Acetyl choline and noradrenalin are not the only neurotransmitters. In the central nervous system, several other substances act as neurotransmitters. Some are listed in Fig. 12.7. All are amino acid derivatives of one sort or another. Glycine and glutamic acid are amino acids. Enkephalins are low molecular weight peptides. Serotonin, dopamine and γ-amino butyric acid are derivatives formed by the reactions shown in Fig. 12.10. The identification of these substances as neurotransmitters is as yet rather circumstantial. The compounds appear to be concentrated in nervous tissue, and have high rates of turnover. Chemical analogues of the putative transmitters affect brain metabolism, and have proved to be effective drugs for the treatment of various types of neurological disorders. But participation of the substances in specific nerve impulses has not been shown.

The nervous system is responsible for functions other than transmission of impulses from skin, eyes, etc, to the central nervous system and from the central nervous system to muscles and glands. Thought, memory, and mood are all processes dependent on nerve transmission in the brain. Exactly what the effector systems are is not clear. It has been suggested that the synthesis of specific proteins is part of the molecular basis of memory, but at present there is little evidence to support or detract from such a hypothesis.

In certain tissues, such as heart muscle, impulses are transmitted from cell to cell not by diffusion of a substance across a synaptic cleft, but by direct flow from the inside of one cell to the inside of another cell. This is achieved through the presence of communicating junctions (section 6.2.5), which allow free diffusion of Na^+ and K^+ between cells. In other words, the cells are electrically coupled, so that an impulse received by one cell is automatically transmitted to other cells.

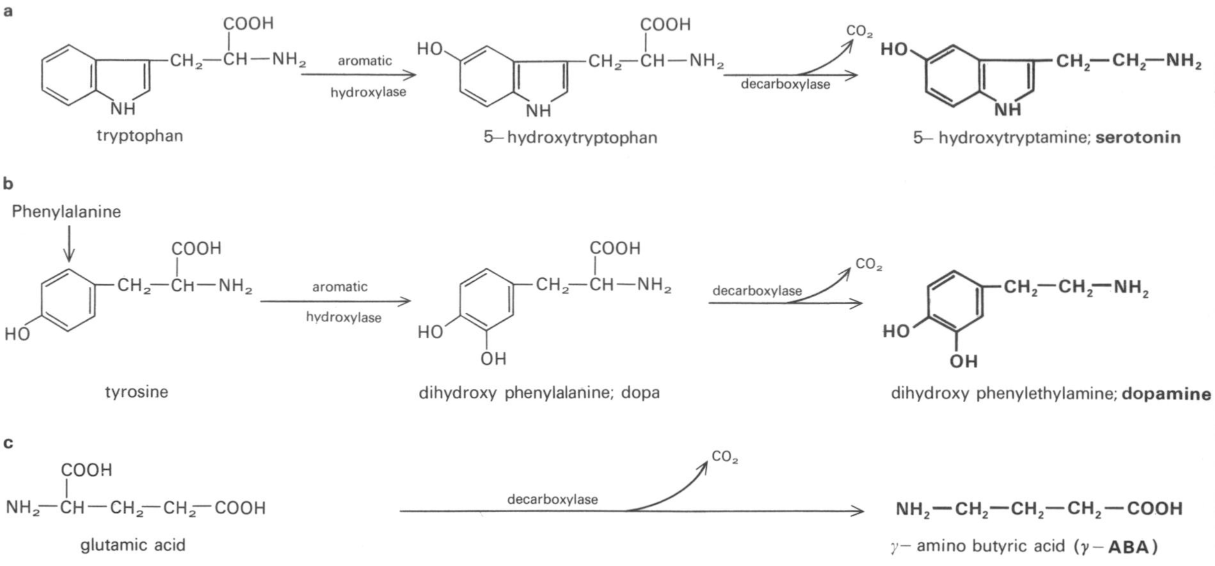
a
COOH
CH2—CH—NH2
aromatic
hydroxylase
HO
NH
tryptophan
5–hydroxytryptophan
CO2
decarboxylase
CH2—CH2—NH2
5–hydroxytryptamine; serotonin
b
Phenylalanine
tyrosine
dihydroxy phenylalanine; dopa
OH
dihydroxy phenylethylamine; dopamine
c
NH2—CH—CH2—CH2—COOH
glutamic acid
NH2—CH2—CH2—CH2—COOH
γ– amino butyric acid (γ–ABA)

Fig. 12.10 Biosynthesis of putative neurotransmitters

Note that serotonin and dopamine are each intermediates in the synthesis of certain hormones (Fig. 9.19).

12.4 Disease

Diseases of the nervous system may be due to a metabolic, infectious, or other cause (Fig. 12.11). The central nervous system (that is, brain and spinal cord) is often affected more than the peripheral nervous system (that is, nerves to voluntary muscle and the autonomic nervous system). Degeneration of myelin is frequently part of the disease.

Metabolic diseases may be due to lack of a particular nutrient or due to failure to metabolize normal nutrients. In few cases is the nervous system a primary target of dietary deficiency diseases. Nevertheless lack of vitamin B_1 (thiamin), B_6 (pyridoxine), B_{12} or other vitamin leads to specific changes in the nervous system. Lack of protein, or of calories in general, leads to mental retardation if the deficiency occurs in early childhood, when the central nervous system is being formed. Later on in life, such deficiency has rather less effect.

Enzyme deficiencies and other metabolic disturbances have a more pronounced effect on the nervous system than do dietary deficiencies. In the case of many of the sphingolipidoses (see Fig. 2.25), failure of the nervous system, leading to mental retardation and other defects, is an important part of the disease.

Important infectious diseases that affect the nervous system are poliomyelitis (now largely controlled by vaccination) and Herpes infections. The causative agents are viruses that attack nerve cells. The mechanism by which such viruses achieve specificity towards a particular cell type is not clear. Bacterial diseases that

Fig. 12.11 Some diseases of the nervous system

Cause	Outcome	Clinical description
Lack of calories, protein or B vitamins	Defective function of peripheral and central nervous system	Polyneuritis; mental disturbances e.g. Korsakow's psychosis
Lysosomal and other enzyme defects	Accumulation of toxic levels of metabolites; degeneration of myelin; defective function of central nervous system	Tay Sachs and other lysosomal diseases (Fig. 2.25); phenylketonuria (all rare).
Viral and bacterial infections	Defective function of peripheral and central nervous system	Poliomyelitis; herpes infections, encephalitis, meningitis.
Auto-immunity (?); viral infection (?)	Degeneration of myelin	Multiple sclerosis
Obstruction of blood-supply to brain	Failure of affected neurons in central nervous system	Cerebral thrombosis (stroke)

affect the nervous system are meningococcal infections, tetanus, and botulism. In the latter two cases the destructive agent is a toxin released by the bacteria. The toxin, which is a protein, binds to the surface membrane of nerve cells and as a result disrupts the permeability characteristics thus impairing conduction.

Among the other diseases that affect the nervous system, the most common is cerebral thrombosis or stroke. That is, cessation of nervous activity in part of the brain resulting from a failure of the blood-supply due to the formation of a blood clot; atherosclerosis is often a precipitating cause.

Another important disease that affects the nervous system is multiple sclerosis. It is possible, though not yet proved, that a very slow-acting virus is a causative factor. The immune system appears also to be involved (Figs 11.15 and 11.16).

12.5 Summary

Nature of system

The nervous system enables man to be aware of his surroundings (by touch, sight, smell, and sound) and to react accordingly (by muscular movement and by the secretion of hormones). The two processes are connected by the transmission of **nerve impulses** from sensory receptors (in skin, eyes, and so forth) to the brain, and from the brain to receptor cells of muscle and endocrine glands. Nerve impulses travel along fibres composed of nerve cells; on reaching a receptor cell, an impulse causes it to become stimulated (or inhibited).

Structure of nerve cells

1. Many nerve cells (which are also known as **neurons**) are extremely long. At one end (known as the cell body) nucleus, mitochondria, and other organelles are present. At the other end, vesicles containing neurotransmitters are present. Most of the cell is specialized into a long tube known as an **axon.**

2. In many types of neuron the axon is covered with a layer of material known as a **myelin sheath.** The myelin sheath (which consists of a specialized plasma membrane wound round and round the axon) serves to insulate the axon and enables impulses to travel along it faster. The major components of myelin are proteins, phospholipids, cholesterol, and glycolipids (especially sialic acid-containing, and sulphate-containing compounds).

3. Between one neuron and the next, and between a neuron and a receptor cell of muscle or gland, is a gap, some 20–50 nm wide, known as a **synaptic cleft.**

Function of nerve cells

1. An impulse that arrives at one end of a neuron causes the plasma membrane to become **depolarized.** That is, this part of the membrane is transiently made many times more **permeable** to Na^+ ions; Na^+ ions therefore start to leak into the cell. As a result, the **membrane potential** (which is largely a consequence of the ionic asymmetry between the inside and the outside of cells) changes (from a negative value to a positive value) at this part of the cell. The membrane now becomes more permeable to K^+ ions also, which start to leak out. Within as little as a millisecond, the permeability to Na^+ and K^+ ions is restored to its original state, and the membrane potential is restored. That amount of Na^+ that has leaked into the cell (and that amount of K^+ that has leaked out) is made good by the Na^+ pump; in other words **ATP** is required.

2. The change in membrane potential at one part of the cell causes the adjoining parts of surface membrane to become depolarized, and by this means an impulse is **propagated** from one end of the neuron to the next. In myelinated neurons, there are gaps (every millimetre or so) where the plasma membrane is free, and able to be depolarized. Impulses are therefore propagated by 'jumping' from gap to gap.

3. At the end of the neuron, the impulse crosses the synaptic gap that separates one neuron from the next neuron, or from a muscle or gland cell. Transmission is achieved by the release of **neurotransmitters;** the type of neurotransmitter varies in different types of nerve cells: in some nerves it is **acetyl choline;** in others it is **noradrenalin.** The neurotransmitter (released by exocytosis) diffuses across the gap, and on arrival at the next cell causes it to become depolarized. In the case of a neuron, this sets up a further impulse; in the case of a receptor cell, such as of muscle or endocrine gland, depolarization results in muscle contraction or secretion of a specific hormone. An increase in cytoplasmic Ca^{2+}, caused by depolarization, is probably the trigger in each case.

4. In order to maintain the speed of nerve transmission (which can be as rapid as 10^5 impulses per second), neurotransmitter has to be rapidly removed from the synaptic gap in order for repolarization to take place prior to the arrival of a second impulse. In the case of acetyl choline, removal is achieved through hydrolysis by an acetyl choline **esterase,** situated on the surface membrane of the receptor cell. In the case of noradrenalin, removal is achieved through active uptake back into the neuronal cell. In each case **ATP** is expended during the process of re-storage of the neurotransmitter.

Disease

Failure of the oxygen supply (for example through a blood clot), to nerve cells in the brain results in a stroke; as in failure of the oxygen supply to heart muscle, this is a common cause of death. Other diseases involve the degeneration of the myelin sheath (as in multiple sclerosis and some inherited diseases in which glycolipid metabolism is impaired).

FURTHER READING

Nerve transmission in general

R.H.Adrian (1974). *The nerve impulse* (Oxford Biology Reader no.67). Oxford University Press.

J.R.Cooper, F.E.Bloom, and R.H.Roth (1978). *The biochemical basis of neuropharmacology* (3rd edition). Oxford University Press, New York.

E.G.Gray (1973). *The synapse* (Oxford Biology Reader no.35). Oxford University Press.

S.W.Kuffler and J.G.Nicholls (1976). *From neuron to brain: a cellular approach to the function of the nervous system.* W.H.Freeman, San Francisco.

More specific aspects:

H.A.Lester (1977). The response to acetyl choline. *Scient. Am.* **236,** no.2. 106.

T.C.Norman (1976). Neurosecretion by exocytosis. *Int. Rev. Cytol.* **46,** 2.

D.J.Triggle and C.R.Triggle (1976). *Chemical pharmacology of the synapse.* Academic Press, London.

V.P.Whittaker (1976). The biochemistry of the cholinergic neuron. *Trends biochem. Sci.*, **1,** 172

Diseases of the nervous system:

A.J.McComas (1977). *Neuromuscular function and disorders.* Butterworths, London.

Chapter on Diseases of the nervous system, in R.H.S.Thompson and I.D.P.Wotton (eds.) (1970). *Biochemical disorders in human disease* (3rd edition). J. and A.Churchill, London.

Index

Page numbers in *italics* refer to Figures; page numbers in **heavy type** refer to the structure (formula) of a particular compound. The summaries at the end of each chapter have not been indexed.